NOTES

ON

MILITARY EXPLOSIVES

BY

ERASMUS M. WEAVER

MAJOR GENERAL, U. S. ARMY; CHIEF OF COAST ARTILLERY

FOURTH EDITION, REVISED AND ENLARGED

TOTAL ISSUE FOUR THOUSAND

NEW YORK

JOHN WILEY & SONS, INC.

LONDON: CHAPMAN & HALL, LIMITED

1917

PRESS OF
BRAUNWORTH & CO.
BOOK MANUFACTURERS
BROOKLYN, N. Y.

PREFACE TO FOURTH EDITION.

EXHAUSTION of the third edition of these "Notes on Military Explosives" has given an opportunity to bring them up-to-date and to include such changes bearing upon the manufacture, use, storage, and transportation of military explosives as have developed during the last four years. It has particularly given the opportunity to introduce certain changes that have developed in connection with the European war. The more important of these latter changes have been the substitution of wood pulp for cotton in the manufacture of the nitrocellulose explosives, and the fixation of the nitrogen of the air by the three separate processes which are now employed. Both of these important changes have been due to the ingenuity, cleverness, and skill of the German chemists. Generally speaking, there have been no new explosives introduced, and it would seem that in the matter of explosives the field is limited, apparently somewhat definitely, to the nitrocellulose series, the nitroglycerin series, the nitrobenzene series, the alkaline-metallic nitrate mixtures, and to a combination of two or more of these with the others. The great propellent explosive for guns continues to be nitrocellulose, alone or in combination with nitroglycerin. The explosive for charging shells appears to have been quite definitely reduced to picric acid or some derivative thereof; that for submarine mines and torpedoes to trinitrotoluol or guncotton. As to the old nitrate mixtures, they appear to be limited to hand grenades, rockets, and pyrotechnics.

There has been inserted in the appendix of this edition a discussion of "The Rôle of Chemistry in the War" by Allerton S. Cushman, Ph. D., Director of the Institute of Industrial

Research, Washington, D. C., which sets forth better than anything that has come to the attention of the author, the basic chemical action of nitrogen, carbon, hydrogen, and oxygen in all explosives.

The regulations of the Interstate Commerce Commission in regard to the transportation of explosives having been revised during the last few years, the new regulations are substituted in the appendix for the old.

It is desired to make acknowledgments to the following individuals for assistance in the collection of new material and in the modifications introduced into this edition: Lieut. Col. Wirt Robinson, Professor of Chemistry, Mineralogy, and Geology, at the United States Military Academy, who has very kindly gone over the notes of Chapter I, Principles of Chemistry, and has made a number of recommendations in regard to changes therein which have been adopted; to the Chief of Ordnance, U. S. Army, for recent data in regard to powders, shell fillers, and methods of testing and storing explosives; to the President of the Army War College, Washington, D. C., for information of a general character in regard to recent data pertaining to military explosives as furnished by the files of the information division of the War College; to Professor F. W. Clarke of the Geological Survey, Washington, D. C., for information in regard to recent changes in atomic weights; to Major George A. Nugent, Coast Artillery Corps, for valuable suggestions in regard to changes in these Notes; to Mr. B. H. Meyer, Chairman of the Interstate Commerce Commission, Washington, D. C., for the revised regulations of the Commission governing the transportation of explosives; and to my secretary, Mr. S. W. Sowerbutts, for valuable clerical assistance and for suggestions in regard to the arrangement of these Notes.

E. M. Weaver.

Washington, D. C.
November 20, 1916.

CONTENTS.

I.

PRINCIPLES OF CHEMISTRY

II.

SUBSTANCES USED IN THE MANUFACTURE OF EXPLOSIVES.

III.

GENERAL REMARKS ON EXPLOSIVES.

IV.

PROGRESSIVE EXPLOSIVES.

V.

DETONATING EXPLOSIVES.

VI.

VII.

SERVICE TESTS OF EXPLOSIVES.

VIII.

STORAGE OF EXPLOSIVES.

IX.

HANDLING HIGH EXPLOSIVES.

X.

DEMOLITIONS.

APPENDIX.

NOTES ON MILITARY EXPLOSIVES.

I.

PRINCIPLES OF CHEMISTRY.

1. Before entering upon a study of explosives it is desirable that some knowledge be had of the fundamental chemical principles involved in the composition of explosive substances and in the changes which take place in connection with explosive phenomena. To this end a brief review will be given of the simple chemical laws, the system of notation, the meaning of chemical reactions, the relations of volumes and weights in these reactions, and problems arising thereunder.

Forms of Matter.

2. As a foundation, it is well to have some conception of the forms of matter as generally conceived at the present time. With this in view it is convenient to consider matter as occurring in the three following forms:

(*a*) In the *mass;* including all aggregations from the smallest quantity perceptible to our senses to the great masses of the heavenly bodies.

(*b*) In the *molecule;* which is defined to be that portion of a substance which has reached the limit of subdivision by physical means; that smallest portion of a given mass of a substance which, in a progressive process of subdivision, would *retain all and only the properties of the substanee.* If by any means a further subdivision be effected, some or all of the properties of the substance would be changed.

(*c*) In the *atom;* which is the smallest portion of any given kind of simple matter that has been differentiated in scientific investigations by reasoning processes.[1]

3. The *atom* is the ultimate unit of matter so far as known;[1]

[1] As a result of investigations consequent on the discovery of radium and the properties of other radio-active substances, a new theory of the

the *molecule* is an aggregation of atoms; the *mass*, or *body*, is an aggregation of molecules.

4. A body may be *homogeneous*, like a piece of copper or salt, or *heterogeneous*, like a piece of granite. Homogeneous bodies contain only *one kind* of matter; heterogeneous, more than one kind; granite, for example, is made up of three kinds of matter called, respectively, feldspar, quartz, and mica.

5. Homogeneous matter implies only similarity of the molecules; it is made up of similar molecules. With similar molecules there would of course, from the definition of a molecule, be similar physical properties throughout the mass.

6. It is known that molecules are of two kinds also: those made up of *atoms of the same kind* and those made up of *atoms of different kinds*. The former are called *elementary molecules;* the latter, *compound molecules*.

---

constitution of matter has been enunciated. According to this theory, the atom of any elementary substance is made up of particles charged with negative electricity, suspended throughout a larger mass charged with positive electricity. The number of negative particles and the resulting attractions and repulsions between the charged negative and positive masses determine the constitution of the atom. The negative particles are called *corpuscles*, and the theory, from these, is called *the corpuscular theory of matter*. Since the atom is ordinarily neutral, the quantity of negative charge must equal the quantity of positive charge. The mass of a corpuscle is constant, and is computed to be about $\frac{1}{1700}$ of the mass of the hydrogen atom. Assuming the corpuscle to be a sphere, its radius is computed to be about 10^{-13} cm., and its ratio to the radius of the hydrogen atom about 10^{-5}. The masses positively charged appear to vary; the smallest, however, is at least equal to the hydrogen atom.

The foregoing theory is that as enunciated by J. J. Thomson. A recent modification of this theory by Sir Ernest Rutherford contemplates that the positive electricity is not diffused, as in Thomson's theory, but concentrated in a central nucleus which is surrounded by rings of electrons; 50 of these rings mean 50 charges. Experiments show that this nucleus must be exceedingly small, 10^{-12} cm. in diameter, the orbit diameter of the rings being 10^{-8} cm. Thomson's theory contemplates an *extended* positive nucleus, within which the electrons revolve in Saturnian orbits; Rutherford's theory, of an extremely small *concentrated* nucleus with electrons in planetary or Saturnian orbits. Two facts seem to be generally accepted: (1) All atoms contain electrons as part of their constitution; of these electrons the atoms may lose a certain number without altering their chemical identity, while the loss of other sets change them into different elements; this is certainly true for radio-active elements, and the law probably extends to all the elements. (2) There exist also positively-charged nuclei associated with the atomic mass, containing multiples of the fundamental electric charge, and the chemical nature of the element seems to be determined by this multiple

7. Homogeneous bodies made up of the same elementary molecules are called *elements;* if made up of compound molecules, they are called *compounds.*

8. Any body will be either (1) homogeneous and an *element* or a *compound,* or (2) *heterogeneous,* made up of different elements or compounds, or a mixture of these two classes; this form of matter is also called a *mixture.*

9. More than eighty elements have been isolated; that is, so far as known at present there are at least this number of different *atoms.* Future investigations may discover new elements, or disclose that some now thought to be elements are compounds.

Properties of Atoms.

10. The atom of each element has its own proper weight, which is different from the weight of any other atom. The lightest known atom is that of the element Hydrogen; the weights of all other atoms are expressed in terms of the weight of the hydrogen atom as a unit.

11. The elements are grouped into two classes, namely:

(1) The *metals;* those possessing properties like copper, iron, gold, etc.

(2) The *non-metals;* those possessing properties like carbon (charcoal, graphite, diamond), sulphur, phosphorus, etc.

12. The following are the names of the most important elements, and opposite each name is placed the weight of its atom to the nearest unit in terms of the hydrogen atom.

METALS.

Name.	At. Wt.	Symbol.	Valency.
1. Potassium	39	K′ (Kalium)	I.
2. Sodium	23	Na′ (Natrium)	I.
3. Barium	136	Ba″	II.
4. Strontium	87	Sr″	II.
5. Calcium	40	Ca″	II.
6. Magnesium	24	Mg″	II.
7. Aluminum	27	Al‴	III.
8. Zinc	65	Zn″	II.

METALS—*Continued.*

Name.	At. Wt.	Symbol.	Valency.
9. Nickel	58	Ni''/'''	II or III.
10. Cobalt	59	Co''/'''	II or III.
11. Iron	56	Fe''/''' (Ferrum)	II or III.
12. Manganese	55	Mn''/iv	II or IV.
13. Chromium	52	Cr'''/vi	III or VI.
14. Copper	63	Cu'/'' (Cuprum)	I or II.
15. Lead	205	Pb'' (Plumbum)	II.
16. Tin	118	Sn''/iv (Stannum)	II or IV.
17. Tungsten	183	Wvi (Wolframium)	VI.
18. Antimony	119	Sb'''/v (Stibium)	III or V.
19. Mercury	199	Hg'/'' (Hydrargyrum)	I or II.
20. Silver	107	Ag' (Argentum)	I.
21. Gold	196	Au''' (Aurum)	III.
22. Platinum	194	Pt''/iv	II or IV.

NON-METALS.

Name.	At. Wt.	Symbol.	Valency.
1. Oxygen	16	O'' [1]	II.
2. Hydrogen	1	H'	I.
3. Nitrogen	14	N'''/v	III or V.
4. Carbon	12	Civ	IV.
5. Silicon	28	Si iv	IV.
6. Sulphur	32	S''	II.
7. Phosphorus	31	P'''/v	III or V.
8. Chlorine	35	Cl'	I.
9. Iodine	126	I'	I.
10. Bromine	79	Br'	I.
11. Fluorine	19	F'	I.

13. These twenty-two metals and eleven non-metals, either separately or in combination, make up more than ninety per cent of all known matter. The weights of these atoms are the constants in all chemical computations in which they enter.

14. Besides weight, atoms possess another important property. They have *mutual attractions for the same kind and for*

[1] As to the standard for atomic weights, some chemists prefer to take the weight of the oxygen atom as the standard, calling it 16, instead of that of the hydrogen atom, unity. The reason for this is that oxygen forms a greater number of compounds, and they are susceptible of more exact analysis than many of the hydrogen compounds. An uncertainty exists as to the ratio between the atomic weights of H and O. Late determinations make O = 15.8 when H = 1, or H = 1.008 when O = 16. The International Atomic Weights for the year 1917 on the latter basis are given in the Appendix, page 372.

certain different kinds of atoms. The *intensity* of these attractions vary for different atoms, but, like the weights, are always constant for the same atom. This attraction existing among atoms is called *affinity*, or *chemical affinity*. Just as *gravity* or *weight* is a property of matter in mass, by means of which bodies fall to the earth, so affinity is a property of atoms, by means of which they come together and combine, when released from one set of conditions in a molecule, to form a new set in a new molecule. Atoms do not as a rule exist separately in nature; if free, they will associate themselves either with atoms of the same kind or with atoms of a different kind, forming thereby the *elementary* and *compound* molecules described above.

15. Atoms have still another important property. In the molecules formed by the action of the so-called force of affinity, as described in the last paragraph, it is found that one atom requires one, two, three, and so on, atoms of other elements to combine with it to form molecules. This property of atoms which determines *the relative number of atoms*, in any case, that enter into chemical combination in forming molecules is called *valency*.

16. Elements are classified according to the valency of their respective atoms. The valency of the hydrogen atom is taken as the *unit of valency*.

17. There are certain atoms that do not combine with the hydrogen atom. The valency of the atoms of elements whose atoms do not combine directly with the hydrogen atom is determined through their combination with the atom of some element that does combine with the hydrogen atom. Thus: the lead atom and the zinc atom do not combine with the hydrogen atom, but all three of these atoms combine separately with the oxygen atom, and from this fact the relative valencies of the lead and zinc atoms may be obtained with respect to hydrogen.

18. An element whose atom has the same combining power (valency) as the hydrogen atom, that is, combines atom for atom with the hydrogen atom or its equivalent, is said to be

univalent, or is called a *monad.* An element whose atom has twice the combining power of the hydrogen atom, that is, will combine with two atoms of hydrogen, or two atoms of any *univalent* element, is said to be *bivalent,* or is called a *dyad.* An element whose atom has three times the valency of the hydrogen atom is said to be *trivalent,* or is called a *triad,* and so on. The degree of valency is represented by small ticks or Roman numerals placed to the right and above the atomic symbol, thus: H', O'', N''', C^{iv} (see table, pages 3 and 4, for valencies).

Notation.

19. For convenience, atoms are represented in chemistry by *symbols.* These symbols are the initial letters of the ordinary or Latin names of the elements, or the initial and one other letter selected therefrom. These symbols are also often used as abbreviations of the name of the element. These two uses should be kept distinctly in the mind. In all chemical equations and computations the symbols represent definitely the weights of atoms. The symbols of the more important elements will be found in the table on pages 3 and 4.

20. A single atom is represented by the simple symbol. Thus: one atom of hydrogen, H; one atom of calcium, Ca; one atom of lead, Pb.

21. Two or more atoms may be represented either by placing the number as a coefficient in front of the symbol, or writing it as a subscript to the right and below. Thus: two atoms of oxygen, 2O or O_2; three atoms of iron, 3Fe or Fe_3.

22. An elementary molecule composed of two atoms would be indicated as explained in the last paragraph. Thus, the molecule of nitrogen contains two atoms; it is represented by N_2. The molecule of phosphorus contains four atoms; its molecule would be expressed by P_4.

23. A compound molecule is represented by writing the symbol of each element which enters it side by side, and giving to each symbol a numeral subscript to indicate the number of atoms of each element. Thus: the molecule of sulphuric acid

is known to contain two atoms of hydrogen, one atom of sulphur, and four atoms of oxygen; it would be represented in symbolic notation by H_2SO_4. In the same way, the molecule of alcohol is known to contain two atoms of carbon, six atoms of hydrogen, and one atom of oxygen; its molecular symbol would be C_2H_6O. The group of symbols used to represent a compound molecule is called the *formula* of the compound, or the *molecular formula* of the compound.

24. In case two or more molecules of the same compound are considered, the proper coefficient is placed before the symbol, or a parenthesis may be placed about the symbol and the number of molecules indicated by a numeral subscript. Thus: two molecules of sulphuric acid, $2H_2SO_4$ or $(H_2SO_4)_2$; three molecules of alcohol, $3C_2H_6O$ or $(C_2H_6O)_3$.

Reactions.

25. These symbols are made use of in chemical writings in indicating the changes which take place when chemically interactive substances are brought together under conditions which excite or permit interaction among their constituents. This is done by representing the substances which are brought together by their proper symbols, writing the sign plus (+) between the symbols of the separate substances used, writing the equality sign (=) after the last substance used, then writing, in the same way, the symbols of the substances resulting from the chemical combinations which have taken place. That is, the *form of an equation is made use of to abbreviate the description that would otherwise be necessary.* For example, the fact that 58 parts by weight of common salt (symbol NaCl) mixed with 63 parts of nitric acid (symbol HNO_3) produces 85 parts of sodium nitrate (symbol $NaNO_3$) and 36 parts of hydrochloric acid (symbol HCl) would be represented thus:

$$NaCl + HNO_3 = NaNO_3 + HCl.$$

Such an equation is only a means to abbreviate the description of chemical changes by using symbols. It is called a *reac-*

tion. The substances on the left of the equality sign are called *reagents*; those on the right, *products*.

26. It should be kept clearly in mind that such equations are quite different from algebraic equations. No mathematical operations can be performed with them. They simply express the fact that the substances on the left of the equality sign will produce those on the right. The total numbers of each kind of atom and the total weights must, of course, be the same on each side; in this sense, only, are *reactions* equations.

27. There are three kinds of reactions, namely, *analytical, synthetical, metathetical*. An *analytical reaction* involves a disintegration of a compound, separating the constituent elements, or reducing it to simpler chemical forms. For example, limestone is a compound of carbon, oxygen, and calcium, and if a piece of limestone be heated, some of the carbon and oxygen will pass off, in combination, as a gas, leaving the calcium and the rest of the oxygen in combination. This reaction may be represented as follows:

$$\underset{\text{Limestone}}{CaCO_3} + \text{heat} = \underset{\text{Lime}}{CaO} + \underset{\text{Carbonic-acid gas.}}{CO_2}$$

A *synthetical reaction* involves a combination of elements or compounds and the formation of substances of a more complex nature than the original ones. Thus, if sulphur be heated to a high temperature in an atmosphere of oxygen, the oxygen and sulphur will combine, forming a sulphur-oxygen compound. The reaction would be represented as follows:

$$S + O_2 = SO_2.$$

If this compound be mixed with water, a new compound is formed, the reaction being represented as follows:

$$SO_2 + H_2O = H_2SO_3.$$

A *metathetical reaction* involves the interchange of atoms between two substances, or the displacement of one element in a compound by a single separate element or a group of elements. Thus, if a solution of common salt (sodium chloride) be treated with a solution of silver nitrate, the sodium of the salt and the silver of the nitrate will exchange places, giving silver chloride and sodium nitrate, the reaction being represented as follows:

$$NaCl + AgNO_3 = AgCl + NaNO_3.$$

Again, if metallic zinc be immersed in hydrochloric acid, the zinc will displace the hydrogen of the acid, the reaction being represented as follows:

$$2HCl + Zn = ZnCl_2 + H_2.$$

Nomenclature.

28. There are certain rules followed in the naming of the elements and compounds which may be briefly stated as follows:

29. The more recently discovered *metals* have names ending in *um*, and some of the more recently discovered *non-metals* have names ending in *ine*. Examples: metals—sodium, ferrum; non-metals—chlorine, iodine.

30. *Compounds composed of two elements* are called *binary* compounds. Such compounds are written with the symbol of the non-metal or the more non-metallic element last, and the name of the compound is given by the name of the first element followed by the name of the second element with the ending *ide*. Thus: common salt is a compound of the metal sodium and the non-metal chlorine; its symbol would be written thus, NaCl, and its name is given by the name of the metal followed by the name of the non-metal, replacing the ending *ine* by *ide*, making the full name of the binary *sodium chloride*. In the same way, FeO is iron oxide; NO, nitrogen oxide; CO, carbon oxide.

31. The combination of oxygen with another element follows this nomenclature rule, forming a large class of binary compounds called "oxides." Oxygen combines with a great many elements, some metallic and others non-metallic;[1] the resulting binary compounds constitute two distinct classes of oxides. These two classes have distinct properties, and are called, respectively, the *metallic or basic oxides* and the *non-metallic or acid oxides.*

32. The terms *base* and *basic, acid* and *acidic* have important meanings in chemistry. They are suggestive of the manner in which the force of affinity will act in any particular case. Bases and acids are the opposites in chemical action. A substance that possesses basic properties suggests chemical union with a substance possessing acidic properties. The tendency of bases and acids to combine depends on their strengths as bases and acids; the strongest or most pronounced bases have the greatest tendency to unite chemically with the strongest acids. As the two classes—bases and acids—approach each other in the scale of chemical affinity, the tendency to unite is less marked. Difference of chemical affinity is, as it were, a difference of chemical potential. As difference of electrical potential suggests *capacity* for electrical work, so the relative basic or acidic properties of substances suggest capacity for chemical combination.

33. Speaking generally, the result of the combination of basic and acidic substances is a third class of substances called *salts.* Many salts possess neither basic nor acidic properties: they are the chemical neutrals; such represent zero difference of chemical potential under the particular conditions.

34. There are simple tests to determine whether certain particular substances are basic, acidic, or neutral. A substance that is chemically active as an acid will turn blue litmus red; one that is chemically active as a base will turn reddened litmus blue. A salt that is perfectly neutral will have no effect on either red or blue litmus. There are other color tests for acids and bases, and, of course, the whole range of chemical reactions to determine the basic, acidic, or neutral properties of sub-

[1] See Experiments Nos. 1 and 3.

stances and the degree thereof, but the litmus test is sufficient for the limits of these notes.

35. The principles given in paragraphs 32 and 33 give rise to a general classification of substances into *bases, acids,* and *salts.*

36. There are other rules governing the naming of compounds which may be introduced here.

37. Both *prefixes* and *suffixes* are resorted to to specify particular compounds. For example, nitrogen combines with oxygen in several proportions, forming separate oxides; these may be written as follows:

1. N_2O .... Nitrogen *mon*oxide.
2. N_2O_2............ Nitrogen *di*oxide.
3. N_2O_3............ Nitrogen *tri*oxide.
4. N_2O_4............ Nitrogen *tetr*oxide.
5. N_2O_5............ Nitrogen *pent*oxide.

They are designated by using the prefixes *mon-, di-, tri-, tetra-,* and *pent-* before the word oxide, as indicated above.

38. Binary compounds in which there are three atoms of the second element to two atoms of the first element may be designated by the prefix *sesqui-* placed before the second with its proper ending. Thus, N_2O_3 is nitrogen sesquioxide; Fe_2O_3 is iron sesquioxide; Sb_2S_3 is antimony sesquisulphide.

39. The suffixes *-ous* and *-ic* are used after the first element of a binary compound to indicate which of *two* compounds is meant, in cases where but two compounds are formed between the two elements considered, or in cases where there are several and two are more important. Thus: sulphur forms two principal oxides, namely, SO_2 and SO_3; the first, or *lower*, degree of oxidation takes the suffix *-ous,* being called sulphur*ous* oxide (or sulphur dioxide); the second, or *higher*, oxide takes the suffix *-ic* and is called sulphuric oxide (or sulphur trioxide). Also, iron forms three oxides, FeO, Fe_2O_3, and Fe_3O_4; the first is called ferr*ous* oxide, and the second ferr*ic* oxide.

40. The prefix *hypo-* is sometimes used before a compound to indicate a still lower degree of oxidation than the *-ous*. Thus, there is a *hypo*sulphurous acid which contains less oxygen than sulphurous acid.

41. The prefix *hyper-* is similarly used before compounds to indicate a higher oxidation; and the prefix *per-* to indicate the highest degree of oxidation. Thus Fe_3O_4 above is the *per*oxide of iron, or iron *per*oxide.

42. While these uses of prefixes and suffixes are explained for oxides only, they may be used also in the case of other compounds; in all cases they indicate the degree of combination of the non-metallic element. Thus, mercury has two chlorides, HgCl and $HgCl_2$. The former is mercury *mono*-chloride, or mercur*ous* chloride; the latter is mercury *di*chloride, or mercur*ic* chloride, or mercury *per*chloride.

43. Instead of using the metal or more metallic element as an *adjective* and the non-metal or more non-metallic element as a noun, it is just as correct to use *the prepositional phrase equivalent.* For example, instead of nitrogen dioxide, *the dioxide of nitrogen;* instead of mercury perchloride, *the perchloride of mercury*, etc.

44. The prefix *proto-* is used to indicate the lowest combination with the non-metallic element; thus, HgCl above is sometimes called the protochloride of mercury, PbO, the protoxide of lead, etc.

45. Many of the *acid* oxides, like SO_2, SO_3, CO_2, N_2O_5, etc., unite with water, H_2O, forming a class of compounds known as *oxyacids*.[1] These possess in a marked degree acid properties, combining readily with bases to form salts.

46. Oxides which thus unite with water to form oxyacids are sometimes called *acid anhydrides*, or simply *anhydrides*.

47. The *oxyacids* are designated by the same suffixes as the acid oxides which form them; thus, sulphur*ous* oxide (SO_2) forms sulphur*ous* acid, and sulphur*ic* oxide (SO_3) forms sulphur*ic* acid, etc. This may be represented by reactions thus:

---

[1] See Experiment No. 5.

$$\underset{\text{Sulphurous oxide}}{SO_2} + \underset{\text{Water}}{H_2O} = \underset{\text{Sulphurous acid}}{H_2SO_3}$$

$$\underset{\text{Sulphuric oxide}}{SO_3} + \underset{\text{Water}}{H_2O} = \underset{\text{Sulphuric acid}}{H_2SO_4}$$

48. The salts formed from acids having the *-ous* suffix are designated by the suffix *-ite.*[1] Thus, salts formed from sulphur*ous* acid are called *sulphites;* from nitr*ous* acid, nitr*ites*, etc.

49. The salts formed from acids having the *-ic* suffix are designated by the suffix *-ate.*[2] Thus, salts formed from sulphur*ic* acid are called *sulphates;* from nitr*ic* acid, nitr*ates;* from carbon*ic* acid, carbon*ates*, etc.

50. There is another class of acids which do not contain oxygen. These are called *hydracids.*[3] They contain only hydrogen and some non-metal. Such acids are HCl, called hydrochloric acid, and H_2S, called sulphydric acid.

51. The salts formed from hydracids take names according to the *binary rule;*[4] salts from HCl are called *chlorides;* from sulphydric acid, *sulphides.*

52. Both oxyacids and hydracids contain hydrogen, and the fundamental characteristic and most important chemical property of these acids is that they *will often exchange all or a portion of the hydrogen they contain for a metal,* whether the metal be alone or in combination with other elements,[4] forming thereby salts.

53. The term *basicity* is used with respect to acids to indicate the number of hydrogen atoms which are replaceable by a metal or equivalent in chemical union. Thus, H_2SO_4 is a *bibasic* acid, HCl is *monobasic*, etc., since in the former two atoms of hydrogen are replaceable by a metal or equivalent, and in the latter there is but one atom to be so replaced.

54. Some of the common acids are indicated by the following names and formulas of their molecules:

---

[1] See Experiment No. 10.
[2] See Experiment No. 11.
[3] See Experiment No. 6.
[4] See Experiment No. 7.

MONOBASIC ACIDS.		BIBASIC ACIDS.	
Hydrochloric. . . .	HCl	Sulphydric. . . .	H_2S
Nitrous.	HNO_2	Sulphurous. . .	H_2SO_3
Nitric.	HNO_3	Sulphuric. . . .	H_2SO_4
		Carbonic.	H_2CO_3
		Hydric.	H_2O (see par. 58).

55. Acids may be graded, according to their respective *avidities*, with respect to nitric acid as a standard. The term *avidity* is used to indicate the proportion of a base that any given acid will combine with, when chemically equivalent quantities [1] of the given acid and nitric acid are mixed separately, with a solution of a given base. Any base may be used. The avidities of the three standard acids at ordinary temperatures have been established as follows: $HNO_3=1$; $H_2SO_4=0.5$; $HCl=1$. That is, in solutions of equal concentration HCl and HNO_3 are stronger acids than H_2SO_4. But if heat be applied, the greater volatility of the first two will enable H_2SO_4 to displace them from salts.

56. A bibasic acid may form three kinds of salts, depending on whether all of the hydrogen or a portion only is replaced, and whether one or two metals are used. These salts are named as follows:

Acid salt, when only half the hydrogen is replaced.[2]

Normal salt, when all of the hydrogen is replaced and by one [3] metal.

Double salt, when all of the hydrogen is replaced and by two metals.

EXAMPLES.

H_2SO_4 Sulphuric acid	+	Na Sodium	=	$NaHSO_4 + H_2$ Acid sodium sulphate
H_2SO_4 Sulphuric acid	+	Na_2 Twice as much sodium	=	$Na_2SO_4 + H_2$ Normal sodium sulphate
H_2SO_4 Sulphuric acid	+	Na + K Sodium and potassium	=	$NaKSO_4 + H_2$ Double sodium-potassium sulphate

[1] See Par. 75. [2] See (*a*), Experiment No.10. [3] See (*b*), same experiment.

57. Compounds containing three different elements are called *ternary* compounds; e.g., H_2SO_4; those containing four different elements are called *quaternary* compounds; e.g., $NaKSO_4$; etc.

58. The principal basic substances are the metallic oxides[1] and another group of substances called *hydroxides*. Oxygen combines with hydrogen in two proportions: first, one atom of oxygen to two atoms of hydrogen, forming *water;* and, secondly, one atom of oxygen to one atom of hydrogen, forming *hydroxyl*. Water exists in nature as a stable liquid; hydroxyl does not exist separately in nature, only in combination with some metal or other chemical equivalent. In the table of acids on page 14 it is to be noted that water is classed as an acid. It comes under this classification only in that it has the property of exchanging its hydrogen for certain metals. (It is neutral to blue litmus and has no other characteristic acid property.) The most important of these metals in a chemical sense are *potassium, sodium, lithium, cæsium, rubidium;* especially the first two. These act on water directly to decompose it, displacing one of the two hydrogen atoms,[2] thus:

$$\underset{\text{Water}}{2H_2O} + \underset{\text{Metallic potassium}}{K_2} = \underset{\text{Potassium hydroxide}}{2KHO} + \underset{\text{Free hydrogen}}{H_2}$$

The oxides of these metals form hydroxides, as follows:

$$K_2O + H_2O = 2KHO,$$

without giving off free hydrogen.

59. Metallic oxides which combine with water to form hydroxides are sometimes called basic anhydrides.

60. The rule for writing and naming oxides applies to

[1] The oxides of the metals as a rule neutralize acids, forming salts, and behave in this way as bases. There are some few metallic oxides like SnO_2 and Sb_2O_5, which are "anhydrides," forming acids with water. No non-metallic oxide is known to have basic properties. There is another class of oxides, both metallic and non-metallic, which are neutral, such as water (H_2O), and the black oxide of manganese, MnO_2. But the general rule is that *metallic oxides are basic and non-metallic oxides are acid.*

[2] See (*a*), Experiment No. 2.

hydroxides. HO is written after the metal and the ending *ide-* is used; thus, KHO is *potassium hydroxide,* or *hydroxide of potassium.*

61. The hydroxides of the metals named in paragraph 58 constitute a group of the strongest bases and are called *alkalies.* One other hydroxide is included in the alkalies, namely, ammonium hydroxide, $NH_4(HO)$.

62. A second group of hydroxides, formed by the *direct* action of metals or their oxides on water,[1] are those known as the *alkaline earths.* These are the hydroxides of calcium, $Ca(HO)_2$; barium, $Ba(HO)_2$; strontium, $Sr(HO)_2$; and magnesium, $Mg(HO)_2$. These rank next to the alkalies in strength as bases.

63. The hydroxides of other metals cannot be formed directly by the action of the metals or their oxides on water.[2] They are formed by combining one of the alkalies or alkaline earths in solution with a solution of some soluble salt of the metal. Thus, zinc hydroxide may be formed by mixing a solution of zinc chloride with a solution of potassium hydroxide, the reaction being represented thus:

$$\underset{\substack{\text{Zinc-}\\\text{chloride}\\\text{solution}}}{ZnCl_2} + \underset{\substack{\text{Potassium-}\\\text{hydroxide}\\\text{solution}}}{2KHO} = \underset{\substack{\text{Zinc-}\\\text{hydroxide}\\\text{solid}}}{Zn(HO)_2} + \underset{\substack{\text{Potassium-}\\\text{chloride}\\\text{solution}}}{2KCl}$$

64. In general and for the purposes of these notes, it may therefore be said that substances may be classified chemically as follows:

Acids.
- Oxides of the non-metals (acid oxides).
- Oxyacids (union of acid oxides with water).
- Hydracids (union of hydrogen with certain non-metals, but not oxygen).

Bases.
- Oxides of the metals (basic oxides).
- Hydroxides (union of basic oxide or metal with water).

Salts. Neutral substances resulting from the combination of acids and bases.

---

[1] See (*b*), Experiment No. 2. [2] See (*c*), Experiment No. 2.

Radicals.

65. It has been stated that oxygen may be considered as existing in combination with hydrogen in chemical substances in the proportion of one atom of oxygen to one atom of hydrogen, HO, and that the name hydroxyl has been given to this particular combination. It should be understood here that there is no substance in nature existing separately, having the molecular formula HO. The oxides of hydrogen which do so exist are H_2O_2 and H_2O. The assumption of its existence is made because, in the chemical changes which take place in the formation and decomposition of the class of hydroxides the proportions of hydrogen and oxygen represented by HO are found invariably associated together. *Groups of atoms which are found thus to persist together throughout chemical reactions are called compound radicals, or often simply radicals.* (The *atoms* of the elements are the "*elementary radicals.*") Often such groups are written either inclosed in parentheses or pointed off by periods thus: K.HO or K(HO); $Zn(HO)_2$; $Ca(HO)_2$. There are many *possible* groupings of atoms, but only those which *are found to exist in chemical analysis and synthesis* are *legitimate* radicals.

66. Compound radicals are considered to have valencies the same as atoms of elements. Hydroxyl, for example, is univalent and will combine with only one univalent atom or another univalent compound radical. Other important compound radicals are:

Amidogen.	NH_2',	valency 1.
Methyl.	CH_3',	valency 1.
Carbonyl.	CO'',	valency 2.
Nitroxyl or nitryl.	NO_2',	valency 1.
Cyanogen.	CN',	valency 1.

67. Compound radicals have basic or acid properties or are neutral, the same as elementary radicals. Radicals composed of the two elements carbon and hydrogen only, are usually basic; if oxygen is also present, the radical is usually acid.

68. Basic radicals, whether compound or elementary, are electropositive; acid, electronegative.[1]

Graphic Formulas.

69. The valency of atoms and compound molecules, and the manner in which the units of valency in any molecule are satisfied or grouped, are often represented graphically by joining together the symbols with small lines, each line representing a unit of valency. Thus, for

Hydrochloric acid,	$H'Cl'$,	we may write	H—Cl
Water,	$H_2'O''$,	" " "	H—O—H
Ammonia,	$H_3'N'''$,	" " "	H—N—H with H below N
Marsh-gas,	$H_4'C^{iv}$	" " "	H—C—H with H above and H below C

$$\begin{array}{c} \mathrm{H-N-H} \\ | \\ \mathrm{H} \end{array} \qquad \begin{array}{c} \mathrm{H} \\ | \\ \mathrm{H-C-H} \\ | \\ \mathrm{H} \end{array}$$

The manner in which valency is satisfied by such graphic formulas may be understood better, perhaps, by imagining each atom, as represented by its symbol, to have bonds or hooks extending from it, and each bond or hook having capacity for engaging with a free bond or hook of another atom. Suppose, for example, that the H's, in the above formulas, are connected with the other bonds or hooks as indicated by the lines between the letters.

The hooks linking, or the bonds attaching them together, in a measure represent the idea involved in "satisfying" units of valency. Such formulas are called *graphic* or *structural formulas*. They merely indicate how valency may be supposed to be satisfied in combinations. They do not represent the relative positions of atoms in molecules.

---

[1] This fact has a bearing on the corpuscular theory of matter (see note bottom of page 2).

70. When all the units of valency are satisfied, as in the groups in paragraph 69, the molecule is said to be *saturated.*

71. Elements whose atoms have an even number of units of valency are called *artiads;* those whose atoms have an odd number of units of valency are called *perissads.* In any *saturated* molecule the *sum of the perissad atoms* is always *even.* This is the *law of even numbers.*

72. An unsaturated molecule is one having one or more units of valency unsatisfied. The compound radicals in paragraph 66 are unsaturated. The free units of valency and the consequent combining power of these radicals respectively may be determined by writing out their graphic formulas, thus:

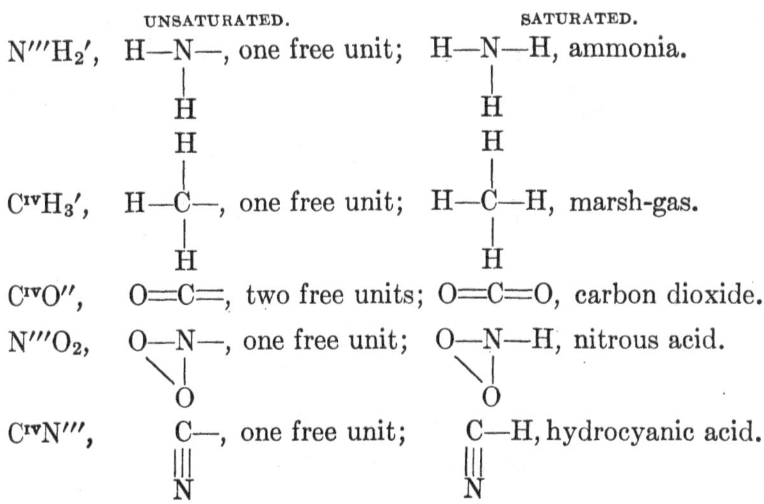

73. If valency be a definite property of atoms, it is necessary to account for what appear to be variations in valency, or variable valency. Thus, it is known that chlorine has but one unit of valency, yet tin and mercury unite in two proportions with chlorine, as follows:

1. $SnCl_2$
2. $SnCl_4$
3. $HgCl$
4. $HgCl_2$

In 1, Sn has a valency of II; in 2, of IV; in 3, Hg has a valency of I; in 2, of II.

The question arises, How can such variations as these be reconciled with a constant atomic property? The use of graphic formulas may assist in explaining such seeming contradictions.

If the graphic formulas of the compounds referred to be written as follows, all units of valency are satisfied, and in each case there is the proper proportion by weight and constant valency for each atom. For $SnCl_2$ we may write Sn_2Cl_4 or $(SnCl_2)_2$, preserving the proportions by weight; that is, consider *two molecules* instead of *one molecule* to be involved in the condition of saturation. The graphic formula for *this* condition would be, assuming Sn to have a valency of IV, the highest:

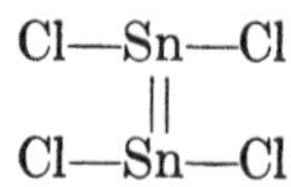

and for $SnCl_4$ the graphic formula would be

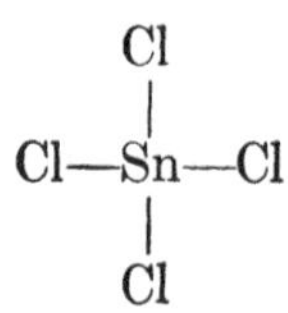

For HgCl we may write Hg_2Cl_2 or $(HgCl)_2$, and the graphic formula of this is, assuming Hg to have valency of II, the highest:

Cl—Hg—Hg—Cl.

Again, for $HgCl_2$, valency still II:

Cl—Hg—Cl.

Other cases of seeming variable valency may be similarly explained by considering the proper grouping of molecules.

The series of the nitrogen oxides may be represented by the following graphic formulas, taking valency of nitrogen III:

$N_2O =$ 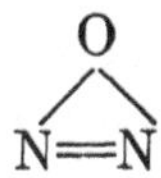

$NO = N_2O_2 =$ O=N—N=O

$N_2O_3 =$ 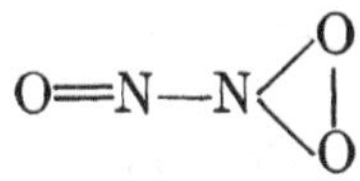

$NO_2 = N_2O_4 =$ 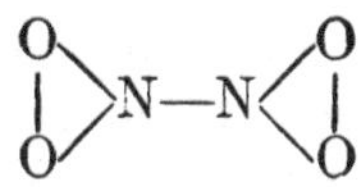

$N_2O_5 =$ 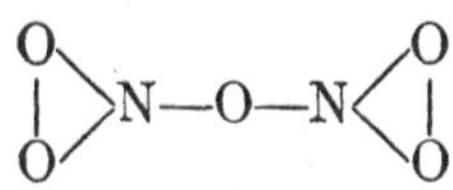

74. If variable valencies may be thus explained, the original definition of valency may be adhered to, namely, it is the greatest number of univalent atoms an atom will combine with.[1]

75. The equivalent weight of any element (or compound) is that weight of it which combines with, is substituted for, or otherwise is chemically equivalent to, one part by weight of hydrogen. Thus:

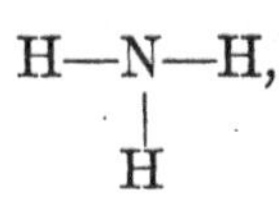

In ammonia (NH_3) the equivalent weight of nitrogen is $\frac{1}{3}$ of the atomic weight of nitrogen, since it combines with 3 atoms of hydrogen; that is, its equivalent weight is $\frac{14}{3} = 4.67$. In H_2O the equivalent weight of oxygen is $\frac{16}{2} = 8$.

The equivalent weight of NH_4 is 18, since 18 parts displace 1 part of H in HCl, giving NH_4Cl.

76. Accepting these definitions, the valency of any element is equal to its atomic weight divided by its equivalent weight.

Organic and Inorganic Chemistry.

77. Substances which result from the operation of life functions, either animal or vegetable, are called organic sub-

[1] The model atoms displayed by Professor J. J. Thomson in his lectures on the corpuscular theory of matter appear to support this conclusion.

stances, and that portion of chemistry which treats of them is called organic chemistry. Substances obtained as minerals from the earth and which are not directly the result of life are called inorganic substances, and that portion of chemistry which treats of them is called inorganic chemistry.

Objects of Chemistry.

78. The objects of chemistry may be enumerated as follows:

(1) To study the properties of a substance so as to be able to identify it with certainty under whatever conditions it may be met with.

(2) To ascertain a method of producing it at pleasure.

(3) To determine its precise composition by weight and volume.

(4) To investigate its action with other substances and the phenomena associated therewith.

Physical and Chemical Phenomena.

79. In studying the properties of substances it is important to distinguish between physical properties, changes, and effects and chemical properties, changes, and effects. All mass effects, outside the limits of molecules and between molecules, which do not affect the integrity of any of the molecules of the mass, pertain to physical phenomena. All effects within the limits of molecules and between the atoms of different molecules, which accomplish disintegration of molecules and the rearrangement of atoms as constituents of new molecules, pertain to chemical phenomena.

Thus the physical properties of a substance include the state of aggregation of its molecules, as gas, liquid, or solid; its color, odor, taste, hardness, specific gravity, form of crystal, fusing-point, and boiling-point.

The chemical properties of a substance include its classification as an acid, a base, or a salt; the action of acids, bases, or salts with it; its composition.

Mixtures, Solutions, Alloys, Amalgams.

80. In addition to the elementary substances and the homogeneous compounds there are other aggregations of matter which may be classified as mechanical mixtures, solutions, alloys, and amalgams.

81. A mechanical mixture consists of two or more substances mixed together in any proportions. It differs essentially from a chemical compound in that the proportions of the constituents of the latter are always the same by weight. Each constituent of a mechanical mixture always retains its own distinguishing physical properties, whereas in a true compound the characteristic physical properties of the separate constituents disappear. Granite is a mechanical mixture of quartz, feldspar, and mica; these ingredients may vary throughout all possible proportions, and although the physical properties of the separate constituents remain the same, those of the conglomerated mass vary to correspond to the varying proportions of the constituents; it is nevertheless always granite. Marble, on the contrary, is a chemical compound, and the proportion by weight of calcium, carbon and oxygen is always the same; the physical properties of the mass are always the same, but the physical properties of the constituents have completely disappeared. Among explosives, black and brown powders are mechanical mixtures of potassium nitrate, carbon, and sulphur; guncotton is a chemical compound, composed of carbon, hydrogen, oxygen, and nitrogen.

82. A solid, liquid, or gas may be in solution in a given liquid; the latter is called the solvent. In passing into solution the solid liquefies and mixes with the solvent; the liquid mixes directly, and, when a homogeneous solution obtains, the two liquids are said to be mixable or miscible; gases are absorbed, so to speak, into the body of the solvent, the amount of gas passing into solution being directly proportional to its pressure on the surface of the solvent and inversely proportional to the temperature of the solvent. Usually the quantity of a solid that will dissolve increases with the temperature of the solvent.

Simple solution appears often to be a *quasi* chemical as well as physical phenomenon, though there is usually a reduction of temperature due to the physical change of the solid to liquid state. In chemical solution there is chemical combination.

83. A solvent will usually take up only a limited quantity of a soluble substance; when this quantity has been taken up further addition only causes an accumulation in the liquid of the solid in the solid state. At this stage the solvent is said to be saturated and the solution is called a saturated solution. Fractional solutions may be made in the percentage quantities required from a saturated solution. A saturated solution is sometimes improperly called a normal solution. A normal solution is one in which each litre contains the number of grams of the substance equal to its molecular weight. A standard solution is such that each litre contains a known and definite amount of the substance. There may be an infinite number of standard solutions of a substance.

84. Proximity of molecules favors chemical action. The form of solution is particularly favorable, both for the reason that the molecules are closer together than in the gaseous state, and the action of affinity is not interfered with by the force of cohesion which acts between the molecules of substances in the solid form.

85. Alloys partake of the nature of solidified solutions of two or more metals mixed together in the molten state. The constitents may vary in any proportion.

86. An amalgam is a union of a metal with mercury. Iron is the only metal in common use which does not form amalgams readily with mercury. Amalgams approach more nearly to compounds than alloys or solutions.

87. The single molecule is invisible. In order that matter become visible the molecules must be brought to within certain limits of nearness to each other. In the state of gas the molecules are not sufficiently close to each other to produce visibility. The passage from visibility to invisibility is well illustrated in the disappearance of condensed steam escaping from an engine. The proximity of molecules in the liquid and solid states causes visibility.

88. The passage from a liquid or solid state to gaseous is called evaporation, or vaporization. Water evaporates whether in the liquid or solid form (ice or snow). Camphor and a few other solids vaporize directly, like ice; notably $(NH_4)Cl$ and S.

89. The passage from the solid state to liquid by the application of heat is fusion, and the temperature at which the change of state takes place is the fusing-point. If the temperature be raised from the fusing-point until vaporization begins in the interior of the liquid as well as on the surface, the latter temperature is the boiling-point. As a rule, fusible substances have definite, characteristic fusing- and boiling-points.

90. The change of state from vapor to liquid or vapor to solid is condensation. The cycle of change from solid or liquid to vapor back to liquid is distillation; from solid to vapor back to solid, sublimation.

91. When a solid absorbs moisture directly from the air at ordinary temperatures and combines therewith to form a liquid, the phenomenon is called deliquescence.

92. Change of state from solid to liquid, solid to vapor, or liquid to vapor causes a disappearance of heat; that is, there is a lowering of temperature. The reverse series of changes cause a corresponding and equal development of heat—elevation of temperature.

93. As a rule chemical actions resulting in the building up of compound molecules from elementary molecules, or which increase the complexity of the molecules (synthetical reactions), involve evolution of heat. Reactions resulting in a separation of the constituents into elements or simpler molecules involve, as a rule, disappearance of heat. In any particular case precisely the number of heat-units made evident in synthesis are made latent or disappear in analysis.

94. There are certain exceptions to the rule given in the last paragraph. There are some molecules, like nitrous oxide,

N_2O, and potassium chlorate, $KClO_3$, and fulminate of mercury, $HgO_2C_2N_2$, which absorb heat in formation and give off heat in disintegration. This property has an important bearing in explosives. Such molecules are said to be endothermic. Molecules which give off heat in formation and absorb heat in disintegration, according to the usual rule, are said to be exothermic.

95. The number of heat-units involved in the synthesis of a molecule is to some extent a measure of the stability of the compound. It will require an equal quantity of heat or some form of equivalent energy to disrupt the bonds forged in the heat of chemical union. Water, for example, is one of the more stable molecules, and the heat given off by H_2, combining with O to form water (H_2O), (that is, the burning of hydrogen in oxygen) amounts to 68,400 units; [that is, 2 grams of hydrogen combining with its equivalent weight (16 grams) of oxygen will give off enough heat to raise 68,400 grams of water 1° C.

96. The effect of high temperature on complex molecules is to weaken the molecular bonds and to favor disruption and a rearrangement of the atoms in new molecules depending on the kind of atoms within the scope of chemical union and their relative affinities for each other under the existing conditions. Heat also weakens the cohesive bonds between molecules, as stated above in connection with changes of physical states of matter.

97. The molecular bonds may be so weakened by the application of heat that the constituents part company. The phenomenon which includes the separation of the constituents of a compound under the influence of heat and their recombination as the temperature falls, by operation of the original chemical affinities which have not at any time been diverted into other relations, is called dissociation. The molecules of elements are sometimes dissociated.

98. When the constituents of a molecule are separated and do not reunite after the disturbing cause has ceased to operate,

having taken up new relations, the phenomenon is termed decomposition.

Fundamental Laws.

99. There are three laws of special importance in chemical science; these are:

1. The Law of Fixed Proportions.
2. The Law of Multiples.
3. The Law of Avogadro.

100. The Law of Fixed Proportions is, that a chemical compound always contains the same elements in the same proportion by weight. For example, pure water contains oxygen and hydrogen and only these two elements, and they are always associated in the proportion of 1.111 parts by weight of hydrogen to 8.889 parts by weight of oxygen in every 10 parts by weight of pure water.

101. The Law of Multiples is, that when two elements unite in more than one proportion the weights of one which combine with the fixed weight of the other bear to each other a ratio that may be expressed by simple whole numbers. Thus nitrogen combines with oxygen to form five separate compounds, and the weight of oxygen entering this series increases by multiples of the smallest weight when a fixed weight of nitrogen is taken in each compound. If we say that the weight of nitrogen shall be 28 pounds in each compound, then the weight of oxygen in the first of the series would be 16 pounds, and it would increase by 16 pounds for each of the subsequent compounds of the series, as follows:

1.	Nitrogen,	28 lbs.;	oxygen,	16 lbs.			
2.	"	"	"	"	32	"	$=2\times16$.
3.	"	"	"	"	48	"	$=3\times16$.
4.	"	"	"	"	64	"	$=4\times16$.
5.	"	"	"	"	80	"	$=5\times16$.

102. The Law of Avogadro may be stated as follows:

All gases under the same conditions of pressure and temperature have the same number of molecules in equal volumes. That is, a cubic foot of hydrogen will have the same number of molecules as a cubic foot of oxygen, or a cubic foot of the vapor of water, or of the vapor of alcohol, or of any other gas; provided all of these are at the same temperature and subjected to the same pressure.

The law may also be stated as follows: The same number of molecules of all gases occupy equal volumes under the same pressures and temperatures. This law being true of any number of molecules is true of one. If, therefore, we consider the law as applying to the volumes occupied by single molecules, it is evident that the volumes of all single molecules are equal. That is, the space occupied by a single molecule of hydrogen is equal to that occupied by a single molecule of oxygen, or a molecule of water, or a molecule of alcohol. The volumes of all single molecules therefore are equal whether they be elementary or compound.

103. It has been ascertained by experiment that the molecules of most of the elements contain two atoms. Some of the exceptions to this are the following:

Cadmium, Mercury, Zinc } have but one atom in a molecule.

Phosporus, Arsenic } have four atoms in a molecule.

For purposes of discussion the conditions existing among diatomic elements only will first be considered.

104. The hydrogen molecule may be taken as the type of diatomic molecules. The space occupied by the molecule, that is the molecular volume, may reasonably be assumed to be equally divided between or occupied by the two hydrogen atoms. The space occupied by one hydrogen atom, that is half the volume of the hydrogen molecule, may be taken as the unit of

volumes; that is, the ultimate standard volume for comparing specific gravities is half the volume of the hydrogen molecule, or the space occupied by the hydrogen atom. The expression, "space occupied by the hydrogen atom," is used for the reason that the atom is supposed not to occupy solidly the limits of the half-molecule; that is, while it occupies the space of the half-molecule, it does not fill it. Calling such space the atomic space, to distinguish it from the true volume of the atom, *the standard volume may be considered the atomic space of the hydrogen atom.*

105. Since the volumes of all molecules are equal, it may be assumed that the atomic spaces of all diatomic elements are equal. That is, the space occupied by any atom of a diatomic element occupies a space equal to that occupied by the hydrogen atom, and the weights of atoms of diatomic elements are the weights of equal volumes. Keeping in mind the fact that the atomic weight of hydrogen is unity and that the atomic weights of other elements are expressed in terms of this unit, it is evident that the atomic weights of diatomic elements express the relative weights of equal volumes, and if hydrogen be taken as the standard of specific gravity for gases, *the atomic weights of diatomic elements are the specific gravities of these elements in gaseous state referred to hydrogen as a standard.* For example, the specific gravity of oxygen referred to hydrogen is 16, of nitrogen 14, etc., the same as their atomic weights.

106. For elements whose molecule contain but one atom, that is monatomic elements, the atomic weight represents the matter occupying two "standard volumes" (atomic space of hydrogen atom). The weight of the matter corresponding to one standard volume would therefore be one-half the atomic weight. That is, the specific gravities of monatomic elements in the gaseous state are one-half their atomic weights; e.g., the specific gravity of the vapor of mercury is $\frac{198.5 \text{ (at. wt.)}}{2} =$ 99.25.

107. For elements whose atoms occupy one-half the standard volume, or have four atoms to the molecule, that is tetratomic elements, the atomic weight is the weight of matter in a half-volume; therefore, to get the weight of a whole volume, the atomic weight must be multiplied by two. That is, the specific gravities of tetratomic elements are obtained by multiplying atomic weight by two. Thus the atomic weight of phosphorus is 30.7; its specific gravity in gaseous state is $30.7 \times 2 = 61.4$.

108. A compound gas, like marsh-gas (CH_4) or acetylene (C_2H_2), or a compound vapor like water (H_2O) or alcohol (C_2H_6O), has as its smallest volume the molecular volume, because by definition the molecule is the smallest quantity that possesses all and only the properties of the substance. Hence the specific gravities of all compound gases are based on the weight of matter in a molecular volume, which is twice the standard volume. Therefore *the specific gravity of all compound gases is obtained by dividing the weight of the molecule by two.* The specific gravity of marsh-gas (CH_4) is $\frac{12+4}{2}=8$; of water-vapor (H_2O) is $\frac{2+16}{2}=9$; of alcohol-vapor (C_2H_6O) is $\frac{24+6+16}{2}=23$, etc.

109. A very important principle is based on the fact that the volumes of all molecules are equal. It is this: *Whatever number of elementary or compound gases combine chemically to form a single compound gas, the latter will occupy but two volumes.* Let the reaction for the formation of water be taken as follows:

$$H_2 + O = H_2O.$$

From paragraph 104 each symbol of an atom of a diatomic element represents a standard volume, provided the symbols stand alone, as in the first member of this equation. That is, in the first member of this equation there are two standard volumes of hydrogen represented, and one standard volume of oxygen, or three standard volumes altogether. When chemical union takes place forming the molecule, water, but one molecule is formed, and it cannot occupy more than two standard volumes.

Again, one volume of nitrogen combines with three volumes of hydrogen to form two volumes of ammonia, thus:

$$\underset{\text{1 vol.}}{N} + \underset{\text{3 vols.}}{H_3} = \underset{\text{2 vols.}}{NH_3}$$

This fact, which is based upon the truth of Avogadro's law and is confirmed by experiment, is sometimes referred to as the *principle or law of gaseous condensation.*

110. The examples in paragraph 109 contemplate strictly theoretical standard volumes, that is the spaces occupied by single atoms; but of course such spaces cannot be dealt with in practical work. However, it is axiomatic that what is true of these theoretical volumes will be equally true of any multiple of the volumes, and it follows that the practical standard volume may be assumed as one cubic foot, or one thousand cubic feet, or one litre, or multiple or fraction thereof, and the first reaction of the last paragraph might just as truly have been stated thus:

$$\underset{\text{2 cu. ft.}}{H_2} + \underset{\text{1 cu. ft.}}{O} = \underset{\text{2 cu. ft.}}{H_2O}$$

and the second reaction, thus:

$$\underset{\text{1 cu. ft.}}{N} + \underset{\text{3 cu. ft.}}{H_3} = \underset{\text{2 cu. ft.}}{NH_3}$$

Determination of Atomic Weights.

111. In paragraph 3 it is stated that the atom is the ultimate unit of matter so far as known. It is convenient here to explain how these smallest known quantities of matter have been ascertained. For this purpose the elements may be divided into, first, those which may be volatilized and dealt with in the form of gas or vapor, and, secondly, those which cannot conveniently be so experimented with.

112. The determination of the atomic weights of gaseous elements is based on the principles of the Law of Avogadro and chemical analysis.

Let it be assumed that the atomic weight of hydrogen is desired.

All possible gaseous compounds in which hydrogen enters as a constituent are collected.

(1) According to Avogadro's Law and the deductions therefrom the molecular weights are the weights of equal volumes (all molecular volumes being equal). But the standard theoretical volume is the half-molecular volume. That is, the molecular weights are the weights of double the standard volume, or, in other words, twice the specific gravities of gases, hydrogen being taken as the standard for specific gravity. If, therefore, equal volumes of hydrogen and all its compound gases be weighed under the same conditions of temperature and pressure, and the resulting weights of the compound gases, expressed in terms of the weight of the volume of hydrogen as unity, be multiplied by two, the products will be the molecular weights [1] in terms of the weight [1] of the hydrogen atom.

For example, it is known that water contains hydrogen; if a cubic foot of water-vapor be weighed, it will be found to weigh 9 times more than an equal volume of hydrogen under the same pressure and at the same temperature. Multiplying 9 by 2, the product 18 is the weight of the water-molecule; that is, the water-molecule weighs 18 times more than the weight of hydrogen which occupies the atomic space.

---

[1] The word weights has been used throughout, but it should be kept in mind that quantity of matter, mass, is the exact idea that should be carried along.

In any case mass $=\dfrac{\text{weight in pounds}}{\text{acceleration due to gravity at the place}}$, or $m=\dfrac{w}{g}$. To be correct, we should speak of atomic masses and not atomic weights. The masses are constant, the weights vary with the force of gravity at different latitudes. Atomic weights are expressions for the relative weights of atoms, hydrogen being unity. The weights of all atoms vary with the latitude, but as they all vary according to the same law, their relative weights are as constant as the masses themselves. Therefore no numerical error is introduced by using atomic weights instead of atomic masses.

(2) By chemical analysis the constituents in each one of the compounds may be separated, and the proportion by weight of hydrogen which enters each sample can be found. For example, suppose that the sample of water was 10 pounds. By chemical analysis it can be accurately determined that 1.111 pounds of this was hydrogen gas and 8.889 pounds was oxygen gas. Or, $\frac{11.11+}{100}$ by weight of water consists of hydrogen.

(3) It was ascertained in (1), above, that the molecular weight of water is 18, in terms of the weight of the hydrogen atom. But it now appears that $\frac{11.11+}{100}$ of any mass of water is hydrogen, whether it be a ton or a molecule. Hence $\frac{11.11+}{100}$ of 18 will be the proportional part of hydrogen in the water-molecule, expressed in terms of the weight of the hydrogen atom, or $.1111 \times 18 = 1.999+$, that is 2, and the hydrogen in the water-molecule is represented by H_2.

(4) Any of the compounds of hydrogen may be dealt with as explained for water. Take hydrochloric acid, for example. Its vapor weighs 18.25 times more than equal volumes of hydrogen, hence, from (1), its molecular weight is 36.5. It may be ascertained by chemical analysis that in every part by weight of hydrochloric acid there are $\frac{2.74}{100}$ parts by weight of hydrogen. This is as true of a single molecule as of any larger quantity. Hence of the 36.5 units of the molecular weight $36.5 \times .0274 = .999+$ of them are units of hydrogen, that is 1 atom of hydrogen, and the quantity of hydrogen in the molecule of hydrochloric acid is therefore represented by H.

(5) All other compounds of hydrogen may be treated in the same way, and the smallest quantity of hydrogen in terms of the weight of the half-hydrogen molecule may be ascertained.

The data resulting from such a series of experiments may be tabulated as follows:

Hydrogen Compounds.	Specific Gravities.	Molecular Weights in Microcriths.	Proportion of Hydrogen.	Weight of Hydrogen in Microcriths.[1]	Symbols.
Water-vapor	9.0	18.0	$.1111=\frac{2}{18}$	2	H_2O
Hydrochloric acid	18.25	36.5	$.0274=\frac{1}{36.5}$	1	HCl
Hydrobromic acid........	40.5	81.0	$.0123=\frac{1}{81}$	1	HBr
Sulphydric acid..........	17.0	34.0	$.0588=\frac{2}{34}$	2	SH_2
Ammonia...............	8.5	17.0	$.1765=\frac{3}{17}$	3	NH_3
Phosphorus trihydride...	17.0	34.0	$.0882=\frac{3}{34}$	3	PH_3
Marsh-gas..............	8.0	16.0	$.25 =\frac{4}{16}$	4	CH_4
Olefiant gas.............	14.0	28.0	$.142 =\frac{4}{28}$	4	C_2H_4
etc.	etc.	etc.	etc.	etc.	etc.

If, in any case, a value less than unity were obtained for this smallest quantity, say ½, that would be taken as the standard atomic weight instead of the one now assumed; if this were made equal to unity, it would necessitate doubling all existing atomic weights. But no weight of hydrogen less than the weight of the half-hydrogen molecule has ever been separated by any procedure or reasoning. The hydrogen atom is, therefore, to be understood to be the smallest quantity of hydrogen that is now known to exist.

(6) All of the compounds of any other *gaseous* element may be analyzed chemically and experimented with physically in the same manner, and the smallest weight of that element which is found in any compound is taken as its atomic weight.

113. The atomic weights of some of the *solid* elements have been determined by a comparative study of the specific heats[2]

[1] The weight of the half-hydrogen molecule is often called a *microcrith*.

[2] The specific heat of a body at any temperature is the ratio of the quantity of heat required to raise the temperature of the body one degree to the quantity of heat required to raise an equal weight of water at its temperature of maximum density (4° C., 39.2° F.) through one degree. The unit of heat is the quantity of heat required to raise the temperature of one unit weight of water one degree. Depending on the weight units involved and the tem-

of the elements in the solid state and a comparison of these specific heats with known atomic weights.

Two investigators, Petit and Dulong, developed the fact that the specific heats of the solid elements are nearly inversely proportional to their atomic weights. That is, the quantity of heat required to raise weights proportional to atomic weights through one degree is practically constant and approximately equal to 6.4 units of heat. This number is called the *atomic heat*. If, therefore, the specific heat of a solid element be determined, and the atomic heat, 6.4, be divided by the specific heat, the quotient will be approximately the atomic weight.

Used in conjunction with chemical analysis, the principle of atomic heat will give sufficiently reliable results. For example, by analyzing silver chloride chemically it is found that 108 parts by weight of silver and 35.5 parts of chlorine are obtained. If there be two atoms of silver in this compound its atomic weight is 54; if three atoms, 36; if four, 27; if one, 108. The specific heat of silver at ordinary temperature is .057; the quotient, 112, obtained by dividing 6.4 by .057, suggests that the number 108 should be taken as the true atomic weight, instead of 54, 36, or 27. Chemical analysis is a more exact process than the determination of specific heat, therefore the number 108 is taken in preference to 112.

114. The number of atoms in an elementary molecule is obtained in any case by first ascertaining what the molecular weight is, then the atomic weight, and then dividing the molecular weight by the atomic weight.

Conditions Influencing Affinity.

115. In paragraph 14 it is stated that one property of atoms is that those of one kind have an attraction for certain other kinds. This attractive force is, as already stated, called

perature scale used, it may be either the number of units of heat required to raise one pound of water at 39.2° F. through 1° F., or one pound of water at 4° C. through 1° C., or one kilogram of water through 1° C. (large calorie), or one gram of water at 4° C. through 1° C. (small calorie).

affinity or chemical affinity. It operates between atoms only. Chemical changes which result in the formation of new substances, by new groupings of the atoms involved, are due to the operation of this force. The intensity of its action varies between different atoms and is modified by different conditions. The quantity of heat evolved in the formation of new substances is, in any given case, to some extent a measure of this intensity, as well as of the stability of the resulting molecules.

116. There are certain causes and conditions which influence the action of chemical affinity. Among these the following may be enumerated:

Temperature.—Substances that do not combine at one temperature will combine at another; and conversely, through the action of temperature alone, decomposition may be effected. Increase of temperature may cause either a synthetical or an analytical reaction; for example, the synthetical reaction where heat is used in forming metallic oxides, and the analytical reaction where lime is formed from marble by heat.

Solution.—In order to have the force of chemical affinity act, it is necessary that the molecules be very close together. Chemical affinity acts at very short distances only. The form of solution is particularly favorable to the action of chemical affinity. Therefore it is used to get chemical combination where other methods have failed. The objection to a solid form is that the force of cohesion opposes combination by impeding or preventing the mutual penetration and close proximity of the particles of the different substances. In gases cohesion does not interfere with chemical action, but owing to the distance between the particles preventing the necessary close proximity, bodies evince but little disposition to combine when in the gaseous state and under normal pressure. If any reaction will take place at all, it will take place in the case of solution.

Insolubility—The principle of insolubility may be stated thus: when two soluble substances, which contain between them the constituents of an insoluble or sparingly soluble substance, are brought together in the form of solutions, the insoluble or less soluble substance is formed and appears in the combined liquids as a suspended solid, called a *precipitate,* which eventually will settle to the bottom. For example, if a solution of silver nitrate ($AgNO_3$) be mixed with a solution of common salt ($NaCl$), a metathetical reaction will take place, the metals silver and sodium exchanging places in the molecules, forming silver chloride ($AgCl$) and sodium nitrate ($NaNO_3$), the former appearing suspended in the resulting liquid as a white curdy precipitate. The reaction would be represented thus:

$$AgNO_3 + NaCl = AgCl + NaNO_3.$$

Volatility.—The principle of volatility may be stated as follows: if two substances contain between them the constituents of a volatile substance, and these two substances be mixed and heated together, the volatile substance will be formed and separate as gas. For example, if pulverized ammonium chloride (NH_4Cl) be mixed with pulverized sodium carbonate (Na_2CO_3) and the mixture heated, the volatile substance ammonium carbonate ($(NH_4)_2CO_3$) will be formed and pass off as a gas, leaving sodium chloride.

Physical Surroundings.—The atmosphere surrounding a substance has an influence on the chemical reactions which may take place. For example: If iron oxide be heated in an atmosphere of hydrogen, the oxygen combines with the hydrogen, passing off as water vapor and leaving ultimately metallic iron. Conversely, if water vapor be passed over heated iron filings, iron oxide will be produced and hydrogen gas liberated.

Nascent State.—By nascent state is meant the state of the element or substance just in the act of being separated in chemical decomposition. The nascent state is particularly favorable to chemical combination. Reactions which will not otherwise take place may take place at the instant that atoms are freed from the bonds that have held them in a molecule.

Pressure.—The retarding influence of pressure is seen in such cases as the action of acids on metals, or the electrolysis of water in sealed tubes. In these cases the elimination of a gas is an essential condition of the change, and this being prevented, the action is retarded. On the other hand, there are numerous reactions which are greatly promoted by increased pressure—those, namely, which depend on the solution of gases in liquids, or on the prolonged contact of substances which under ordinary pressure would be volatilized by heat.

The relation of chemistry to explosives has recently been admirably enunciated in a paper entitled "The Rôle of Chemistry in the War" (Senate Document 340, 64th Congress, 1st Session) by Allerton S. Cushman, Ph.D., Director of the Institute of Industrial Research, Washington, D. C., which, with the permission of Dr. Cushman, is republished in Appendix III of these "Notes," pages 349–371.

Stoichiometry.

117. Stoichiometry is that part of chemistry which deals with the computations of the weights of substances used in chemical reactions and resulting therefrom, and in the volumes of gases connected therewith. The foregoing principles may be applied, now, in the solution of chemical problems involving weights and gaseous volumes.

118. It has been seen that symbols represent atoms; that the atoms have definite weights for each element, and that the weight of the molecule of any substance is the sum of the weights of the atoms which compose the molecule.

It may now be stated that the symbols may be used not only to represent atomic weights of the elements, but any weights proportional to their atomic weights. In stoichiometry they are so used. That is, to the abstract numbers in the second column of the table on pages 3 and 4 the name of any unit of weight may be applied, such as grams, ounces, pounds, tons.

A reaction that is true for the atomic weights proper is equally true *if the same proportions by weight be observed, using any unit of weight.*

For example: one atom of oxygen unites with two atoms of hydrogen to make water. Since the weight of the oxygen atom is 16 and the hydrogen atom 1, it follows that *any weights whatever of oxygen and hydrogen in the proportions of 16 to 2 will produce 18 parts by weight of water.* That is, 16 lbs. of oxygen will unite with 2 lbs. of hydrogen to make 18 lbs. of water, and we may write $\underset{16}{O} + \underset{2}{H_2} = \underset{18}{H_2O}$. Any unit of weight may be applied to the numbers written below the symbols.

In the same way any reaction may be utilized to solve problems involving weights.

119. Reactions may also be used to solve problems relating to volumes of gases, and these problems are often of value in dealing with explosives.

The symbols of the atoms of gaseous elements may be considered to represent the atomic spaces as well as atomic weights, it being kept in mind that the ultimate standard volume for the comparison of gases is the space occupied by the half-molecule, and that all single molecules, whether simple or compound, have equal volumes. These principles were enunciated in paragraphs 102 and 109, and it was seen in the latter paragraph that *one* volume of oxygen united with *two* volumes of hydrogen to make *two* volumes of water-vapor, or that, giving concrete values to the volumes, one cubic foot of oxygen will combine with two cubic feet of hydrogen to make two cubic feet of water-vapor, considering all gases at the same temperature and pressure. Expressed in connection with the reaction, this may be written

$$\underset{\text{1 cu. ft.}}{O} + \underset{\text{2 cu. ft.}}{H_2} = \underset{\text{2 cu. ft.}}{H_2O}.$$

In the same way, any reaction involving gases may be made use of to write out the volume relations existing among the reagents in the first member of the equation and the products in the second member. If any solids appear in the reaction they are not, of course, to be considered in these volume relations.

120. In solving problems in stoichiometry, it will be useful to keep certain units and numbers in mind; among these may be enumerated the following:

1 cubic foot of hydrogen at 60° F. and 30 inches barometer weighs about 37 grains; at 0° C., 40 grains.

1 pound of hydrogen under same temperature and pressure occupies about 189 cubic feet.

1 gram = 15.43 grains.

1 litre = 61.02 cu. inches = 1.76 pints.

1 gram of hydrogen at 0° C. and 760 mm. barometer occupies 11.16 litres.

Volumes of gases under the same pressure vary with temperature, increasing as the temperature increases, or decreasing as the temperature decreases; as follows: A volume of gas at 60° F. will increase or decrease its volume $\frac{1}{519.4}$ of its volume for each degree of temperature Fahrenheit above or below 60° F., or will increase or decrease its volume at 0° C. $\frac{1}{273}$ of that volume for each degree centigrade of increase of temperature or decrease of temperature above or below 0° C.

Pressure of Gases.—If the volume of a gas remains constant and the temperature changes, the pressure of the gas will increase or decrease according to these same ratios—by $\frac{1}{519.4}$ of the pressure at 60° F. or $\frac{1}{273}$ of the pressure at 0° C. for each degree Fahrenheit or centigrade, respectively.

The ratio giving the rate of change in terms of volume at any other temperature than 60° F. or 0° C. may be obtained from the denominators of the fractions given for 60° F. and 0° C., by adding the number of degrees of higher temperature or subtracting the number of degrees of lower temperature. For example, the ratio for volume at 0° F. would be $\frac{1}{519.4-60}=\frac{1}{459.4}$, and for 20° C. would be $\frac{1}{273+20}=\frac{1}{293}$.

The coefficient of expansion at any temperature is one divided by the corresponding absolute temperature.

PROBLEMS.

1. To find the relative weights of the constituents in any quantity of a compound, as, for instance, H_2O. It is seen that in this formula the constituents of the compound are in the proportion by weight, of 16 to 2. It makes no difference whether we deal with a single molecule or a pound of water, this same relation obtains. In the first case the unit is the microcrith, in the second, the unit is the pound. If required, therefore, to find the number of pounds of hydrogen to make a ton of water, we have this proportion:

$$2:18::x:2000 \text{ pounds.}$$

2. To find the percentage composition of a substance, given the molecular formula. Let us take, for example, cellulose:

$$C_6H_{10}O_5.$$

The following form will be found convenient in solving such problems:

	Atomic weights.	No. of atoms.	Total weights.	Per cent.
C_6	12	6	72	$\frac{72}{162}=44.4$
H_{10}	1	10	10	$\frac{10}{162}=6.2$
O_5	16	5	$\frac{80}{162}$	$\frac{80}{162}=49.4$

3. To find the empirical formula of a substance, given the percentage composition and atomic weights. The empirical formula is the simplest expression for the numerical relations of the atoms as determined by analysis, and this is directly

connected with the percentage composition. It is found by first determining by analysis the composition of a substance and then dividing each part by weight by the atomic weight of the corresponding element. For example, take cellulose as in the last problem:

	Part by weight.		Atomic weights.	
C=	72	÷	12	= 6
H=	10	÷	1	=10
O=	80	÷	16	= 5

4. To find the molecular formula having the empirical formula and the molecular weight. By chemical analysis determine the relative parts by weight. Divide the weight of each element thus obtained by the atomic weight of that element for the proportional number of atoms in the compound. The empirical formula will be the smallest number of whole atoms consistent with this proportion. The molecular formula will be that combination of atoms whose weight in the aggregate is equal to the molecular weight. For example: A compound of carbon and hydrogen is analyzed. The total weight of the compound is 6½ grams. It is known that its molecular weight is 78. Analysis gives 6 grams of carbon and ½ gram of hydrogen.

$$\therefore\ C_6H_{\frac{1}{2}} = C_{\frac{6}{12}}H_{\frac{\frac{1}{2}}{1}} = C_{\frac{1}{2}}H_{\frac{1}{2}}.$$

Taking the lowest number of whole atoms we have for the empirical formula CH, the combined weight of the atoms of which is 12+1=13.

It is evident that the molecular formula will therefore be 78÷13=6 times greater than that of the empirical formula. Consequently, the molecular formula is C_6H_6.

5. Since the atomic weights of substances represent not only the actual weights of atoms, but also the weight of quantities proportional thereto, if we fix on the weight of any one

element, all the others are fixed by that act. For example, (*a*) take the reaction $Cu+O=CuO$. Assume 5 grains of copper.

Then $63.2:16::5:x \quad \therefore x=1.26$ grains of O. The atomic weights of Cu and O being 63.2 and 16 respectively, x gives the weight of O in grains.

$63.2+16=79.2.$

Then $63.2:79.2::5:x \quad \therefore x=6.26$ grains of CuO.

(*b*) Take the reaction $CaCO_3+\text{heat}=CaO+CO_2$. Assume 30 pounds of $CaCO_3$. The weights of the resulting products would be found as follows:

$$
\begin{aligned}
CaO &= 40+16=56 = \text{mol. wt.} \\
CO_2 &= 12+32=\underline{44} = \quad " \\
CaCO_3 &= \qquad\quad 100 = \quad "
\end{aligned}
$$

$100:56::30:x$, giving 16.8 pounds CaO

$100:44::30:x$, " 13.2 " CO_2.

6. As each molecule occupies two volumes, we can from inspection of a chemical equation readily determine the number of molecules, and from these the volumes of the gaseous reagents or products.

Take CH_4. It is a combustible gas (marsh-gas). Both C and H unite with O in burning. C will burn to CO_2, and for this we must have 2 atoms of O. H_4 will burn to $2H_2O$, and for this we must also have 2 atoms of O. In order, therefore, to burn CH_4 we must supply it with 4 atoms of O. We may therefore write:

$$\underset{\text{2 vols.}}{CH_4}+\underset{\text{4 vols.}}{2O_2}=\underset{\text{2 vols.}}{CO_2}+\underset{\text{4 vols.}}{2H_2O}.$$

These volumes may refer to any unit of volume. For example, assume 20 cubic feet of CH_4. The problem would then be, How many cubic feet of O are required to burn 20 cubic feet of CH_4? We have, $2:4::20:x \quad \therefore x=40$ cubic feet.

Again, take the reaction, $\underset{\text{1 vol.}}{N}+\underset{\text{3 vols.}}{H_3}=\underset{\text{2 vols.}}{NH_3}$. Note that the sums of the volumes in the two members of the equation do not have to balance; the sums of the weights on both sides of the equality sign must, however, always balance.

7. In order to pass from weights to volumes, we have the following relation:

$$\frac{\text{Wt. of gas}}{\underset{\text{(usually 1 cubic foot)}}{\text{Wt. of unit vol.}}}=\text{volume.}$$

Therefore weight of gas = volume in cubic feet × weight of 1 cubic foot of gas.

8. To find the weight of a cubic foot and the specific gravity of a *mixture* of gases; for example, atmospheric air.

Assume the weight of 1 cubic foot of H (barometer 30″, thermometer 32° F.) = 40 grains.

Any given weight or volume of air consists approximately of

$$\underbrace{\underset{\text{2 vols.}}{O_2} + \underset{\text{8 vols.}}{4N_2}}_{\text{10 vols.}}$$

$$\begin{aligned} O_2 &= 16\times 2 = 32 \\ 4N_2 &= 14\times 8 = 112 \\ & \overline{\quad 144} \end{aligned}$$

$\therefore$ Wt. 1 vol. air $=\frac{1}{10}$ of 144 = 14.4 = specific gravity
" 1 cubic foot H = 40 gr.
$\therefore$ " 1 " " air = 40 × 14.4 = 576 gr.

9. To find the number of cubic feet of air that will be required to burn 100 pounds of wood. Assume wood to have the molecular formula $C_6H_{10}O_5$ and the reaction of combustion to be as follows:

$$C_6H_{10}O_5+6(O_2+4N_2)=\underbrace{6CO_2+5H_2O+48N}_{}$$

Mol. wts.: $162 \quad +6(32+112) = \quad 1026$

It therefore takes 1026 pounds of air to burn 162 pounds of wood.

How much will it take to burn 100 pounds?

$$162:864::100:x. \quad \therefore\ x=533.33 \text{ pounds.}$$

To reduce to cubic feet: 7000 gr. = 1 pound.

1 cu. ft. air = 576 gr. 576)3733300 grs.

6481 cu. ft.

11. Assume that the following represents the reaction involved in the burning of illuminating-gas. If there be in this group of mixed gases 2 cubic feet of hydrogen, what are the other volumes?

$5SH_2$ +	$2C_2H_4$ +	CH_4 +	$2H_2$ +	$16O_2$ +	O =	$13H_2O$ +	$5CC_2$ +	$5SO_2$
10 vols.	4 vols.	2 vols.	4 vols.	32 vols.	1 vol.	26 vols.	10 vols.	10 vols.
5 cu. ft.	2 cu. ft.	1 cu. ft.	2 cu. ft.	16 cu. ft.	0.5 cu. ft.	13 cu. ft.	5 cu. ft.	5 cu. ft.

Since there are 2 cubic feet of hydrogen and 4 standard volumes of hydrogen ($2H_2$), one standard volume in this case is $\frac{2 \text{ cubic feet}}{4}=0.5$ cubic foot. Multiply each number of "vols." by 0.5 cubic foot and we have the volume of each gas in cubic feet, as shown below each molecular formula.

12. *a*. Find the percentage of iron in limonite or brown hæmatite, $2Fe_2O_3.3H_2O$.

$$\begin{aligned} 2Fe_2 &= 56\times4 = 224 \\ 2O_3 &= 16\times6 = \ \ 96 \\ 3H_2 &= \ \ 1\times6 = \ \ \ \ 6 \\ 3O &= 16\times3 = \ \ 48 \\ \text{Mol. wt.} &= 374 \end{aligned}$$

From which the per cent of Fe is found to be 59.9.

b. Same for hæmatite or specular iron ore, Fe_2O_3.

$$\begin{aligned} Fe_2 &= 56\times2 = 112 \\ O_3 &= 16\times3 = \ \ 48 \\ \text{Mol. wt.} &= 160 \end{aligned}$$

From which the per cent of Fe is found to be 70.

c. Same for magnetite or magnetic oxide, Fe_3O_4.

$$\begin{array}{l} Fe_3 = 56 \times 3 = 168 \\ O_4 \;\; = 16 \times 4 = \;\; 64 \\ \hline \qquad\qquad\qquad\; 232 \end{array}$$

From which the per cent of Fe is found to be 72.4.

d. Same for spathic ore, clay ironstone, or blackband, $FeCO_3$.

$$\begin{array}{l} Fe = 56 \times 1 = \;\; 56 \\ C \;\; = 12 \times 1 = \;\; 12 \\ O_3 = 16 \times 3 = \;\; 48 \\ \hline \qquad\qquad\qquad\; 116 \end{array}$$

From which the per cent of Fe is found to be 48.3.

e. Same for iron pyrites, FeS_2.

$$\begin{array}{l} Fe = 56 \times 1 = \;\; 56 \\ S_2 \;\; = 32 \times 2 = \;\; 64 \\ \hline \qquad\qquad\qquad\; 120 \end{array}$$

From which the per cent of Fe is found to be 46.7.

13. The celebrated Russian chemist Mendeléeff has suggested a method of comparing explosives by finding the number of volumes in the explosive reaction corresponding to 1000 parts by weight; this volume is indicated by the symbol V_{1000}. The effect of the temperature of the explosion is not considered.

a. Determine Mendeléeff's relation of V_{1000} in the following reaction for black gunpowder:

$$\underbrace{\underset{202}{2KNO_3} + \underset{32}{S} + \underset{36}{3C}}_{270} = \overset{\text{solid}}{K_2S} + \underbrace{\underset{6\text{ vols.}}{3CO_2} + \underset{2\text{ vols.}}{N_2}}_{8\text{ vols.}}\,^{1}$$

$$270:1000::8:V_{1000}; \quad V_{1000} = 29.6 \text{ volumes.}$$

[1] The numbers under the reagents (first members of the equations) are *molecular weights*, those under the products (second members of the equations) are *volumes*.

b. Same for brown gunpowder:

$$\underbrace{\underset{606}{6KNO_3} + \underset{160}{2C_5H_4O}}_{766} = \overset{\text{solid}}{3K_2CO_3} + \underbrace{\underset{14\text{ vols.}}{7CO} + \underset{8\text{ vols.}}{4H_2O} + \underset{6\text{ vols.}}{3N_2}}_{28\text{ vols.}}$$

$$766:1000::28:V_{1000}; \quad V_{1000}=36.5 \text{ volumes.}$$

c. Same for nitroglycerine:

$$\underset{908}{4C_3H_5O_3(NO_2)_3} = \underbrace{\underset{24\text{ vols.}}{12CO_2} + \underset{20\text{ vols}}{10H_2O} + \underset{12\text{ vols.}}{6N_2} + \underset{2\text{ vols.}}{O_2}}_{58\text{ vols.}}$$

$$908:1000::58:V_{1000}; \quad V_{1000}=63.9 \text{ volumes.}$$

d. Same for guncotton:

$$\underset{594}{C_6H_7O_5(NO_2)_3} = \underbrace{\underset{6\text{ vols.}}{3CO_2} + \underset{18\text{ vols.}}{9CO} + \underset{14\text{ vols.}}{7H_2O} + \underset{6\text{ vols.}}{3N_2}}_{44\text{ vols.}}$$

$$594:1000::44:V_{1000}; \quad V_{1000}=74.1 \text{ volumes.}$$

e. Same for smokeless powder; nitrocellulose having 12.75 per cent of nitrogen, N, which is about the percentage in smokeless powder for cannon in the United States:

$$\underset{1098}{2C_{12}H_{15}O_{10}(NO_2)_5} = \underbrace{\underset{2\text{ vols.}}{CO_2} + \underset{46\text{ vols.}}{23CO} + \underset{30\text{ vols.}}{15H_2O} + \underset{10\text{ vols.}}{5N_2}}_{88\text{ vols.}}$$

$$1098:1000::88:V_{100}; \quad V_{1000}=80.1 \text{ volumes.}$$

f. Same for C burned with sufficient supply of O:

$$\underbrace{\underset{12}{C} + \underset{32}{O_2}}_{44} = \underset{2\text{ vols.}}{CO_2}$$

$$44:1000::2:V_{1000}; \quad V_{1000}=45.5 \text{ volumes.}$$

g. Same for C burned with insufficient supply of oxygen:

$$\underbrace{\underset{12}{C} + \underset{16}{O}}_{28} = \underset{2\text{ vols.}}{CO}$$

$$28:1000::2:V_{1000}; \quad V_{1000}=71.4 \text{ volumes.}$$

h. A comparison of the values of V_{1000} in *f* and *g* is important because on this basis depends the whole argument of Mendeléeff as to the desirability of so arranging the percentage of C, H, O, and N in nitrocellulose as to have all the C burn to CO. When this is done, we get the reaction for Mendeléeff's pyrocellulose, which is nitrocellulose containing 12.44% of nitrogen:

$$\underset{1350}{C_{30}H_{38}O_{25}(NO_2)_{12}} = \underbrace{\underset{60 \text{ vols.}}{30CO} + \underset{38 \text{ vols.}}{19H_2O} + \underset{12 \text{ vols.}}{6N_2}}_{110 \text{ vols.}}$$

$$1350:1000::110:V_{1000}; \quad V_{1000}=81.5 \text{ volumes.}$$

i. Same for cordite (used in our service in Armstrong guns); a mixture of nitroglycerine and nitrocellulose:

$$\underbrace{\underset{454}{2C_3H_5O_3(NO_2)_3} + \underset{1764}{7C_6H_8O_5(NO_2)_2}}_{2218} = \underbrace{\underset{96 \text{ vols.}}{48CO} + \underset{66 \text{ vols.}}{33H_2O} + \underset{20 \text{ vols.}}{10N_2}}_{182 \text{ vols.}}$$

$$2218:1000::182:V_{1000}; \quad V_{1000}=82 \text{ volumes.}$$

j. Same for picric acid, used in shell (lyddite, melinite, etc.):

$$\underset{916}{4C_6H_2(NO_2)_3HO} = \underbrace{\underset{12 \text{ vols.}}{6H_2O} + \underset{44 \text{ vols.}}{22CO} + \underset{10 \text{ vols.}}{5N_2} + \underset{4 \text{ vols.}}{2CN}}_{70 \text{ vols.}}$$

$$916:1000::70:V_{1000}; \quad V_{1000}=76.4 \text{ volumes.}$$

General Properties of Important Substances.

The following general properties of important substances may be committed to memory with advantage, preparatory to laboratory work:

The *nitrates* are all soluble in water.

The *dichlorides*, except that of lead, are soluble in water.

The *monochlorides*, except those of silver and mercury, are soluble in water.

The *sulphates* are soluble in water, except that of calcium, which is but slightly soluble, and that of barium, strontium, and lead, which are insoluble.

The *sulphates* are insoluble in alcohol.

The *carbonates* are insoluble in water, except those of the alkalies; they are all soluble in water containing CO_2 in solution, i.e., carbonated water.

The *carbonates*, except those of the alkalies, are decomposed by heat (CO_2 passing off).

The *carbonates* are decomposed by sulphuric, hydrochloric, and nitric acid, with evolution of CO_2 and effervescence.

The *chlorates* are soluble in water.

The *acetates* are soluble in water.

The *oxides* are insoluble in water, except those of the alkalies and barium, which are soluble; those of the alkaline-earth metals, except barium, are slightly soluble.

The *sulphides* are insoluble in water, except those of the alkalies and alkaline earths.

The *hydroxides* are insoluble in water, except those of the alkalies and alkaline-earth metals, the latter being but slightly soluble.

The *phosphates* are insoluble in water, except those of the alkalies.

II.

SUBSTANCES USED IN THE MANUFACTURE OF EXPLOSIVES.

BEFORE treating directly of explosives proper, it will be advantageous to consider apart the substances used in their manufacture.

Regarded from the point of their composition, explosives may be divided into two classes, namely:

1. Explosive mixtures.
2. Explosive compounds.

The former consist of an intimate mixture of distinct substances, properly prepared and conglomerated mechanically in varying proportions to meet the requirements of different demands. Each particle of such explosive mixtures must have at least a particle of some oxygen-supplier, such as a nitrate or chlorate, and some combustible, such as carbon or sulphur. The old black and brown powders and the new explosive called ammonal are typical examples of such mechanical mixtures. The characteristic quality of such explosives is that the nature of the explosion may be graded by varying the proportions of the ingredients.

The latter class consist of substances whose molecules contain within themselves the oxygen and carbon and hydrogen necessary for combustion. Any substance whose molecule contains oxygen, carbon, and hydrogen in the proportions to give CO, or CO_2 and H_2O, may become an explosive. One which is so constituted and at the same time has weak molecular bonds due to the presence of the radical NO_2 or other weak-binding radical is an explosive compound. The characteristic quality of this class of explosives is that the elements constituting the explosive are always present in the molecule in the same quantities, according to the law of fixed proportions, and the nature of the explosion cannot be graded by varying the quantities of the constituent elements, as in the case of mechanical mixtures.

The substances used in the manufacture of these two classes of explosives may be considered conveniently in the following order:

1. The nitrates and chlorates, used as oxygen-suppliers in explosive mixtures.
2. The combustibles, charcoal and sulphur, used in explosive mixtures.
3. The hydrocarbons and other compounds of organic origin used in the manufacture of high explosives, and of the more recently developed nitro-powders. This includes hydrocarbons proper, alcohol, ether, acetone, phenol, glycerine, cellulose, and certain nitro-derivatives of some of these, including the nitrobenzenes, nitronaphthalene, nitrophenol.

Reference is made in this connection to the part played chemically by nitrogen, hydrogen, and carbon in explosives, as given on pages 349–368, Appendix III.

Potassium Nitrate, (KNO_3). Nitre. Saltpetre.

This salt is found in nature as an incrustation on the surface of certain soils in hot countries. It results in such instances from the decomposition of nitrogenous organic matter in the presence of moist alkaline earths. The decomposition of both animal and vegetable matter produces ammonia; the oxidation of ammonia in nature appears to be furthered by the growth of certain low forms of vegetable life; this combines with the atmospheric oxygen, yielding nitric acid, and this, in turn, acts on other potassium salts to produce the nitrate; the solution evaporating at the surface leaves the solid as an incrustation.

It may be produced artificially by the nitre-bed process. Vegetable and animal matter are piled together in large heaps, with limestone, old mortar, wood ashes, and any alkaline material, on an impervious floor protected from the weather.

One side is made nearly vertical and this side is exposed to the prevailing wind; the opposite side is cut into terraces. Urine from stables and other sources is poured over the terraces, which have a slight pitch toward the body of the heap, with a small gutter cut at the inner junction of the step with the body of the heap. The temperature is kept at 60° to 70° F. The liquid seeps through the mass, and the chemical action described above takes place in the body of the heap; the soluble nitrates percolating through the heap finally reach the vertical side exposed to the wind, and evaporation occurring there leaves on this surface an incrustation of nitrate mixed with other salts.

The nitre coming from these sources is known as *crude*. An analysis of crude saltpetre from India gave the following: nitre, 97.40; potassium chloride, 0.84; sodium chloride, 0.20; insoluble, 0.21; water, 1.35. The crude nitre from the beds contains in addition, as a rule, chlorides of calcium, magnesium, and ammonium. Crude nitre must be refined before using in explosives. The chlorides are separated from the nitre by dissolving the mass in hot water. The nitre crystallizes first in cooling and is skimmed off. The chloride remaining in solution is converted into nitre by mixing with a solution of sodium nitrate.

It is important that nitre used for explosives should contain no chlorides because of their hygroscopic properties. A sample should therefore always be subjected to the standard test for chloride. The sample solution should be tested also with barium chloride or nitrate for any sulphate, and with ammonium oxalate for lime.

Nitre is distinguishable by the form of its crystals (long striated or grooved six-sided prisms), and by its deflagration when heated on charcoal. It fuses at 635° F. (335° C.) to a colorless liquid which solidifies on cooling to a translucent crystalline mass. Heated to red heat it effervesces from the escape of oxygen and becomes reduced to the nitrite (KNO_2). If heated beyond this, the nitrite is decomposed, leaving a mixture of K_2O and K_2O_2.

Its value in explosives is due to the fact that it acts as a supplier of oxygen to the combustible element present. Five-sixths of its oxygen is available for combination with any combustible, the nitrogen coming from its decomposition being given off in the free state. The reaction for the decomposition of nitre by charcoal may be represented as follows:

$$2KNO_3 + C_3 = K_2CO_3 + CO_2 + CO + N_2.$$

Owing to the concentrated form in which the oxygen is presented to the carbon by nitre, carbon burning to CO_2 or CO gives a much higher temperature than in ordinary combustion where the O is supplied by the air.

The specific gravity of nitre is 2.07 compared with water. Since one cubic inch of water weighs 252.5 grains, one cubic inch of nitre weighs $(252.5 \times 2.07 =)$ 523 grains.

Write $2NO_3K = N_2O_5OK_2$.

The five atoms of oxygen of the N_2O_5 only are available for combustion; that is, in 202 grains (weight proportional to the weights of two molecules) of nitre there are 80 grains of oxygen free to unite with a combustible. Since one cubic inch of nitre weighs 523 grains, it will contain $(523:202::x:80)$ 207 grains of oxygen available for combustion, and since 16 grains of oxygen gas has a volume of 46.7 cubic inches at 60° F. and 30″ barometer, the 207 grains of oxygen in one cubic inch of solid nitre will be equivalent to 607 cubic inches of oxygen gas at 60° F. and 30″ barometer. And since there is but one volume of oxygen in five volumes of air $(4N+O)$, we arrive at the result that one cubic inch of nitre contains as much oxygen as is found in 3000 cubic inches of air at 60° F. and 30″ barometer.

It is this fact that causes the high temperature of explosions of black gunpowder. The nitre presents the oxygen to the charcoal and sulphur in concentrated and pure form, and the reaction between the minute particles of the mixture takes place at each point in a very short time. All of the nitrates that are used in explosives are, like nitre, oxygen-carriers.

Almost all of the nitre now used in the manufacture of gunpowder is obtained by the conversion of sodium nitrate into potassium nitrate by means of potassium chloride. When sodium nitrate and potassium chloride are mixed and the solution boiled down, sodium chloride is deposited and potassium nitrate remains in the boiling liquid, the reaction being $NaNO_3 + KCl = KNO_3 + NaCl$.

The potassium chloride required for this conversion is obtained from the refuse of the sugar-beet root, and from certain salt deposits, notably the salt-mines of Stassfurt, Saxony; also from sea-salt, seaweed. The mineral carnallite is a double chloride of potassium and magnesium.

Sodium Nitrate, ($NaNO_3$). Peruvian or Chile Saltpetre. Cubical Saltpetre.

This salt is found in large beds beneath the surface of the soil in the provinces of Atacama and Tarrapaca, Chile. It occurs at depths of from one to five yards, and in strata from two to twelve feet thick. The mined earth contains from fifteen to sixty-five per cent of sodium nitrate and a large quantity of other salts, such as sulphates, chromates, chlorates, iodates, borates, etc.

The crude nitrate is extracted from the earth by a process of boiling and crystallization. As thus obtained it contains about one or one and a half per cent of impurities, chiefly sodium chloride and sodium sulphate.

The sodium-nitrate crystal is different from that of potassium nitrate; the former being a rhombohedron, the latter a six-sided prism.

Sodium nitrate is more hygroscopic than potassium nitrate, and on this account cannot be used with advantage in the manufacture of explosives, unless the explosive be kept absolutely protected from the air. As mentioned under potassium nitrate, its chief value in connection with explosives is as a source from which potassium nitrate may be obtained by chemical reaction. It is also used in the manufacture of nitric acid.[1]

[1] It is estimated that at the present rate of consumption, the Chilian saltpetre beds will be exhausted in about thirty years. This fact has caused a revival of the old process of producing nitrates and nitric acid by the oxida-

Ammonium Nitrate (NH_4NO_3).

This salt has properties resembling in a general way those of the two nitrates just considered. It was formerly looked upon with favor as an oxygen-carrier in explosives on account of the fact that the basic part, NH_4, on explosion, gave free gases instead of solids. Its excessive hygroscopic properties, however, have eliminated it from use, except in a few special explosives which are so prepared as to be protected against the action of the air. It is used in certain blasting-powders and dynamites with a view to reducing the temperature of the explosion, the weak affinity of the element nitrogen for other products of explosion causing a comparatively low temperature of explosion; dissociation of the products also favors a lower temperature.

The "fire-damp" gas (marsh-gas, CH_4) is explosive when mixed with two volumes of oxygen or ten volumes of air. This mixture ignites at about 2200° C. The temperature of explosion of most explosives is above this; therefore, when used in mines, they may serve to ignite the fire-damp.

---

tion of the nitrogen of the air. Over one hundred years ago Cavendish observed that the electric spark would oxidize the nitrogen of the air, which is composed of about 79 parts of nitrogen by volume and 21 parts of oxygen. It is only recently, however, and in the face of the prospective disappearance of natural sources, that this fact has been considered of use in a commercial way. A plant has been installed at Notodden, Norway, where nitrate of calcium is being manufactured, applying the Cavendish principle. Air is forced at a carefully regulated rate through a disk of electric arcs. The high temperature of the arcs causes the oxidation, but unless removed speedily from this temperature the nitrogen oxide is decomposed by the same heat. By forcing the air through the disk its movement is so regulated *that at a certain velocity the oxide is not reduced*, and is conducted on to a tower down which milk of lime is made to trickle, and this latter absorbs the nitrogen oxide. The electricity used at Notodden is generated by water-turbines. About 75,000 litres of air are passed through the plant per minute. Each unit produces about 325 tons of calcium nitrate per year, the chemical equivalent of 250 tons of nitric acid, 100 per cent, or 337 tons of nitrate of sodium. The total capacity of the Notodden plant, three units at present, is equivalent to about 1000 tons of Chilian saltpetre per year.

Another process of fixing the nitrogen of the air by oxidation is known as the cyanamid process. (See page 357, Appendix III.)

Good types of these so-called "safety" explosives are the Favier Explosives. P. A. Favier of Paris suggests the following safety mixture:

Favier No. 1—	Ammonium nitrate.......	88 per cent.
	Dinitronaphthalene.......	12 per cent.

The nitrate is dried in a steam-heated tube, pounded in a heated mortar, and, while still heated, sprinkled with melted dinitronaphthalene, pressed into cylinders, dipped in melted paraffin, and wrapped in paraffined paper.

When exposed to gentle heat, ammonium nitrate melts at 150° C., boils at 210° C., and disappears in the form of steam and nitrous oxide:

$$NH_4NO_3 + \text{heat} = N_2O + 2H_2O.$$

It deflagrates if heated suddenly to a high temperature, as by throwing it on a red-hot surface. If very carefully heated it may be sublimed.

Interest has lately been revived in this substance by the fact that it is an ingredient in the explosive, "ammonal." This explosive consists essentially of ammonium nitrate and pulverulent metallic aluminum, the latter being prepared by a special process. Some potassium nitrate and charcoal are also present in varying proportions in different grades of ammonal.

The chief claim for ammonal is that the aluminum protects the ammonium nitrate from moisture and thus eliminates the objection heretofore held against its excessive hygroscopic properties.

Ammonal has given excellent results as a charge for shell and for disruptive purposes. As a mechanical fixture it possesses the insensitiveness of this class of explosives. It must, however, prove itself to be a thoroughly stable mixture when stored for long periods of time under the conditions to be found in ordinary service and storage magazines. It is too new and untried an explosive, as yet, to merit a place among standard military explosives.

Closely allied to ammonal is ginite, which is one of the explosives used by the French in their hand grenades. It is composed chiefly of nitrate of ammonium, the other component substances being trinitrotoluol, (T.N.T.), powdered aluminum, and dicyanadiamide. It is claimed for it that it can be exploded by a simple fulminate, is unaffected by heat or moisture, is not detonated by shock, and has an explosive strength considerably greater than that of picric acid.

Barium Nitrate.

Of all the metallic nitrates used in explosives, barium nitrate is least hygroscopic. It is, on this account, used in some cases instead of KNO_3. It is much heavier than the other alkaline and alkaline-earth nitrates.

It is found in nature as the mineral witherite. Artificially it is produced by dissolving the carbonate in dilute nitric acid.

It is decomposed by heat, leaving the oxide of barium and giving off oxygen with some form of nitrogen oxide, depending on the degree of temperature used.

It is an ingredient in some of the modern military and sporting smokeless powders.

Its rate of combustion is slower, its temperature of ignition higher, and the quantity of free oxygen available is less than for potassium nitrate.

The per cent of oxygen in the several nitrates just considered is given in the following table:

Sodium nitrate	56.47%	$NaNO_3$.
Ammonium nitrate	60.00%	NH_4NO_3.
Potassium nitrate	27.49%	KNO_3.
Barium nitrate	36.78%	$Ba(NO_3)_2$.

Of the 60% of oxygen in ammonium nitrate only 20% is available as free oxygen, 80% being required for combination with hydrogen to form water, as shown by the molecular formula when written thus: $N_2O(H_2O)_2$.

Although barium nitrate gives the lowest percentage of O by weight, by *volume* it gives about the same as nitrate of sodium on account of its high specific gravity.

The Chlorates.

The chlorates are oxygen-carriers like the nitrates. They act more readily as oxidizers and at lower temperatures. Indeed they part with their oxygen so readily that the heat of even ordinary friction will cause the union of their oxygen with a combustible.[1] This is favored in the case of potassium chlorate by the fact that its molecule gives off heat in breaking up. At high temperatures the chlorates act violently on all combustible substances. Potassium chlorate is the oxidizing ingredient in signal and pyrotechnic compositions, being usually mixed with sulphur and some metallic compound to give the color desired to the flame. The following combinations may be given:

Red Fire—(1) 40 grains strontium nitrate thoroughly dried over a lamp are mixed with 10 grains of potassium chlorate and reduced to the finest powder. *In another mortar* 13 grains of sulphur are mixed with 4 grains of black sulphide of antimony. The two powders are then placed upon a sheet of paper and very intimately mixed with a bone-knife, avoiding great pressure.

(2) Another prescription: Charcoal 1 part, shellac 2 parts, sulphur 8 parts, potassium chlorate 12 parts, strontium nitrate 40 parts.

Blue Fire.—Potassium chlorate 15 parts, potassium nitrate 10 parts, oxide of copper 30 parts; mix in mortar; transfer mixture to paper and mix with a bone-knife with sulphur 15 parts.

Green Fire.—Barium chlorate 10 parts, barium nitrate 10 parts; mix in mortar; transfer to paper; mix these with sulphur 12 parts.

A composition of friction-primers for cannon consists of twelve parts of potassium chlorate, twelve parts of sulphide of

---

[1] A mixture of pulverized potassium chlorate and sulphide of antimony explodes if struck with a hammer. A grain or two of potassium chlorate rubbed in a mortar with a little sulphur will explode.

antimony, and one part of sulphur worked into a paste with a solution of an ounce of shellac in a pint of grain alcohol.

The explosive used in fire-crackers is a mixture of potassium chlorate and lead ferrocyanide.

All mixtures of chlorates with combustible substances are liable to spontaneous combustion.

Sulphur, S.

Sulphur is found in the uncombined state in nature in certain volcanic districts. It is found in the combined state especially in the sulphide ores of many metals, and in some mineral waters as hydrogen sulphide. Among the ores may be mentioned iron pyrites (FeS_2), copper pyrites ($CuFeS_2$), galena (PbS), blende (ZnS), crude antimony (Sb_2S_3), cinnabar (HgS). Also with oxygen and the metals as sulphates, such as gypsum ($CaSO_4.2H_2O$), heavy spar ($BaSO_4$), Epsom salts ($MgSO_4.7H_2O$), Glauber's salts ($Na_2SO_4.10H_2O$).

Sulphur is obtained from native veins in volcanic districts. It is obtained also by reduction from the sulphides (either the ores or the tank-waste residue of alkali works).

The process of getting sulphur from the alkali works is known as the Chance-Claus process. Calcium sulphide was formerly a useless by-product in the making of sodium carbonate.[1] Now it is a paying by-product. The calcium sulphide waste is mixed with water, stirred into a paste, and run into large cast-iron vessels (carbonizers); through this mass CO_2 is forced. The effect of heat, moisture, and CO_2 is to form $CaCO_3$ and liberate SH_2. The SH_2 is passed into a gas-holder, where it is mixed with air and burned:

$$SH_2 + (4N + O) = S + OH_2 + N_4.$$

[1] The production of CaS in the alkali works is as follows:

$$2NaCl + SO_4H_2 = SO_4Na_2 + 2HCl.$$

$$\left\{\begin{array}{l} SO_4Na_2 + C_4 = Na_2S + 4CO. \\ CO_3Ca + C = CaO + 2CO. \end{array}\right.$$

$$Na_2S + CaO + CO_2 = Na_2CO_3 + CaS.$$

The sulphur obtained by the Chance-Claus process is of great purity and requires no refining.

Native sulphur obtained from the veins is purified by direct distillation and subsequent refining to free it of earthy impurities. The same process is followed in obtaining sulphur from iron and copper pyrites.

The refining process is conducted in large retorts connected with a subliming chamber and distilling tank; it consists of melting down the crude sulphur and distilling it from the molten state.

In refining crude sulphur, whether from native sulphur or the pyrites ores, a charge of seven hundred pounds, or over, of the crude sulphur is put in a large cast-iron retort. A fire is started under the retort. The sulphur will begin to melt at 239° F. This will be evidenced by the appearance of a light yellow vapor above the mass. The vapor of sulphur rises and passes into a subliming-chamber, where it is condensed and falls as "flowers of sulphur." When the temperature of the mass is about 560° F., red fumes will be observed in the retort. Distillation then takes place instead of sublimation. The vapor of sulphur now passes over into a condensing-tank which is cooled by circulating cold water, and it is condensed as a thick yellow liquid. The sulphur which first passes over is known as *sublimed sulphur* or *flowers of sulphur;* it is not used for making gunpowder, as it sometimes contains a small percentage of foreign substances; it is returned to the retort for reworking. That which is distilled over at the higher temperature is known as *distilled* or *roll sulphur;* it is this that is used in the manufacture of gunpowder.

As an ingredient of gunpowder, sulphur is valuable on account of the low temperature (500° F.) at which it ignites, thus facilitating the ignition of the mixture; its combination with the oxygen of the nitre gives also a higher temperature than would obtain if charcoal alone were used; this higher temperature has the effect of increasing the rate of combustion and pressure of the gases evolved.

Heat has an extraordinary effect on the physical condition of sulphur. If a quantity of sulphur be placed in a glass flask and heated, the following changes will be observed:

At about 120° C. it is a pale yellow, limpid liquid. As the temperature rises from 120° C. the color grows darker and the liquid more viscous until, at 180° C., it is nearly black and opaque, and so viscous that the flask may be inverted without spilling the sulphur. At this point the temperature remains constant, although the application of heat continues, showing changes taking place within the molecular structure of the sulphur. On continuing the heat, the sulphur becomes liquid again at 260° C., though not so mobile as at first. At 444° C. it boils and is converted into a brownish-red, very heavy vapor, and an explosion often takes place between the red vapor and the air.

If the flask be now removed from the flame and decanted into water, the sulphur will descend through the water in the form of a brown, soft, elastic, rubber-like string. If a portion be allowed to remain in the flask and to cool therein, it will pass successively through the same states as described above in the inverse order, becoming black and viscous at 180° C., and a pale-yellow thin liquid at 120° C.; if it now again be poured into cold water, it will descend through it in small button-like drops of ordinary sulphur. As the portion still left in the flask cools it will deposit small tufts of crystals, and finally solidify into a yellow crystalline mass.

The brown, rubber-like sulphur after a few hours will become yellow and brittle; the change is accelerated by gentle heat and is attended with an evolution of the heat made latent at the 180° C. stage.

The *roll sulphur*, or *distilled sulphur*, used in the manufacture of powder is always easily soluble in carbon disulphide; the flowers of sulphur only partially so.

Charcoal. Carbon. C.

Carbon is the combustible element of most explosive mixtures, and it is present in combination with hydrogen in most explosive compounds. Its function in all cases is to combine with oxygen, producing either CO or CO_2, the heat resulting from this chemical reaction causing increased volume of the gases produced.

In black gunpowder and other mechanical mixtures the carbon is supplied in the form of pulverized charcoal. The charcoal used in the manufacture of powder is obtained by the destructive distillation of certain woods and woody fibres, such as willow, alder, dogwood,[1] and rye straw; the lighter woods being used because they give a charcoal more easily combustible than the heavier ones.

The charring is done in a metallic cylinder placed in a retort over a furnace-fire. The effect of the heat is to drive off the volatile parts of the wood; these pass off for the most part in the form of wood naphtha (CH_4O), pyroligneous acid ($C_2H_4O_2$), carbon dioxide, carbon monoxide, and water, leaving a residue containing from 70 to 85 per cent of carbon, associated with small quantities of hydrogen (5% to 3%), oxygen (23% to 10%), and ash (about 2%) consisting of the carbonates of K, Ca, Mg, calcium phosphate, potassium sulphate and silicate, sodium chloride, oxides of Fe and Mg.

The wood consists of sticks about $\frac{1}{2}$ to $\frac{3}{4}$ of an inch in diameter, cut into short lengths. It is cut when in full sap, in the spring of the year, is stripped of its bark, and dried for a considerable time either in the open air or in hot-air drying-chamber. Charcoal that is charred in cylinders is called *cylinder* charcoal, to distinguish it from the common *pit* charcoal.

After charring, the charcoal is kept for about two weeks exposed to the air; it is then ready for grinding for powder-making. If ground at once after charring, there is danger of spontaneous combustion from combination with oxygen of air.

The charring process takes from $2\frac{1}{2}$ to $3\frac{1}{2}$ hours; its completion is known by the blue flame of CO burning to CO_2 at the mouth of the pipe which conducts the volatile products of distillation from the retort to the flame of the furnace under the retort. The charred wood weighs about 30% of its original weight.

[1] European dogwood.

If charred at temperatures above 400° C., the product is not sufficiently friable. At very high temperatures, 1000° to 1500° C., the charcoal is very hard, dense, and rings with a metallic sound.

The temperature of ignition varies directly with the temperature of charring. Charcoal that has been charred at 260° to 280° C. will ignite at from 340° to 360° C.; that made at 290° to 350° C., at from 360° to 370° C.; that at 432° C., at about 400° C.; that at 1000° to 1500° C., at 600° to 800° C.

If mixed with sulphur, it ignites at lower temperatures; that made at temperatures under 400° C. mixed with powdered sulphur will ignite at 250° C. If the charcoal has been made at higher temperatures, the sulphur burns, leaving the charcoal unchanged.

The capacity of charcoal to decompose the nitrates varies in the same way. Charcoal made at temperatures between 270° and 400° C. will combine with saltpetre at 400° C.; if made at temperatures of 1000° to 1500° C., it combines only when heated to redness.

Freshly made charcoal has remarkable powers to absorb certain gases into its pores. One cubic inch of charcoal will absorb 100 cubic inches of ammonia oxygen gas, 50 cubic inches of sulphuretted-hydrogen gas, 10 cubic inches of oxygen, and 7 cubic inches of water-vapor. This is purely a mechanical effect, but the intimate association of such gases in the mass of charcoal in time develops chemical action and leads to spontaneous combustion. Freshly prepared charcoal, pulverized and stored in that form, will ignite spontaneously if the mass is over two feet deep. The ignition begins at the bottom or near the bottom. Samples thus treated have ignited in 36 hours.

The property of freshly made charcoal to absorb gases is made use of in deodorizing sewers, cesspools, etc.

The charcoal formerly used in the manufacture of brown powder was made from rye straw. The straw was carefully selected, only the large, firm, perfect stalks being taken. The charring was done by superheated steam at a relatively low temperature. The charcoal contained about 48 per cent of carbon, 5.5 per cent of hydrogen, 45 per cent of oxygen, 1.5 per cent of ash.

Compounds of Organic Origin.[1]

Most of the recently developed explosives, whether used for propulsion or disruptive effects, are derived from organic substances. Substances of organic origin are also used in their manufacture. It therefore becomes necessary to present some of the more simple relations existing among these substances and to define certain general terms.

The organic substances enumerated below may be regarded as the most important ones in connection with explosives.

1. *The Hydrocarbons.*—Compounds of C and H only, in various modes of grouping, starting with the saturated hydrocarbon, C_nH_{2n+2}, the isologous series down to C_nH_{2n-6}, with their derivatives constitute the fatty group, because many of them exist in fats; the C_nH_{2n-6} group and its derivatives constitute the aromatic group, because many are obtained from balsams, essential oils, gum resins, etc. The physical state of a hydrocarbon may generally be known from the number of C atoms present in its molecular formula. If there be 4 or less, the substance is gaseous; if more than 4 and less than 12, it is liquid; if more than 12, solid. Most hydrocarbons are obtained by the fractional distillation of organic substances and are volatile; they have characteristic odors, are insoluble in water, soluble in alcohol, ether, and carbon disulphide.

---

[1] The following organic radicals should be noted:

$(HO)'$	occurring in	alcohols and phenols,	called	*hydroxyl.*
$(CO)''$	"	" ketones,	"	*carbonyl.*
$(CO.HO)'$	"	" acids,	"	*carboxyl.*
$(CH_3)'$	"	" wood-alcohol derivatives,	"	*methyl.*
$(C_2H_5)'$	"	" grain " "	"	*ethyl.*
$(C_6H_5)'$	"	" benzene derivatives,	"	*phenyl.*
$(CH_3CO)'$	"	" acetic "	"	*acetyl.*
$(NO_2)'$	"	" nitro-compounds,	"	*nitryl.*

The ending "yl" indicates an unsaturated radical; the unsatisfied valency units are indicated by the marks to the right and above the parentheses inclosing the radicals.

The most important of the hydrocarbon series in explosives are:

(*a*) *The Paraffins.*—General formula C_nH_{2n+2}, in which n represents any whole number. They are derived from the fractional distillation of mineral oil.

(*b*) *The Olefins.*—General formula C_nH_{2n}, in which n represents any whole number not less than 2. They are found in the products of distillation of coal, wood, etc.

(*c*) *The Acetylenes.*—General formula C_nH_{2n-2}, in which n represents any whole number not less than 2. The first member of this series, acetylene, C_2H_2, is formed by the direct union of carbon and hydrogen under the influence of high temperature. The molecule is endothermic, 61,100 units of heat being absorbed in its formation. It is the only hydrocarbon that has been formed by direct union of its elements. The acetylenes are found in the products of distillation of all substances rich in carbon and hydrogen.

(*d*) *The Benzenes.*—General formula C_nH_{2n-6}, in which n represents any whole number not less than 6. The hydrocarbons of this series are extracted from the coal-tar obtained by the distillation of coal in manufacturing illuminating-gas.

2. *The Alcohols.*—From their chemical behavior they may be considered as hydroxides of the paraffin hydrocarbons, and represented by the general formula $C_nH_{2n+2-x}(HO)_x$, in which n represents any whole number, and x any whole number not greater than n. Example:

$$C_2H_5.HO, \quad \begin{array}{c} \quad\ H\ \ H \\ \quad\ |\ \ \ | \\ H-C-C-(HO), \\ \quad\ |\ \ \ | \\ \quad\ H\ \ H \end{array} \quad \text{ethyl alcohol.}$$

3. *The Ethers.*—They may be regarded as derived from the alcohols by the replacement of one or more atoms of hydrogen

of the hydroxyl radicals of alcohols by a univalent paraffin hydrocarbon radical. Example:

$C_2H_5.O.C_2H_5$, 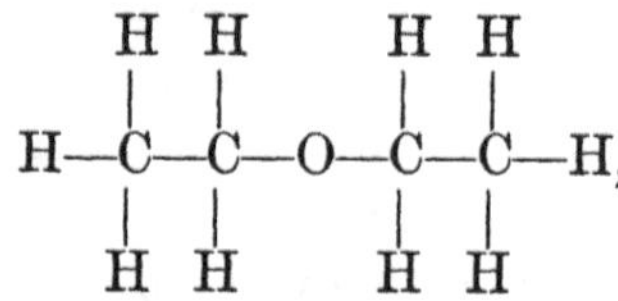 ethyl ether.

They may also be regarded as the oxides of the paraffin hydrocarbons. Under this conception, the molecular formula for ethyl ether would be written $(C_2H_5)_2O$.

4. *The Ketones.*—They may be regarded as combinations of hydrocarbon radicals of the paraffin series with carbonyl (CO). Example:

$(CH_3)_2CO$, 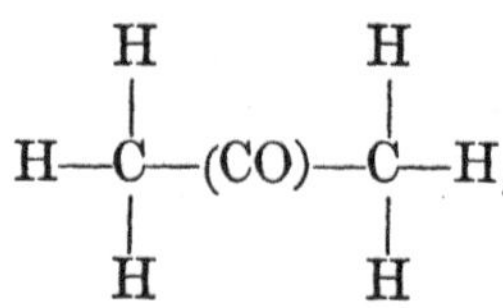 acetone.

5. *The Phenyls.*—They are derived from benzene (C_6H_6) by substituting hydroxyl (HO) for one or more atoms of hydrogen. Example:

$C_6H_5.HO$, 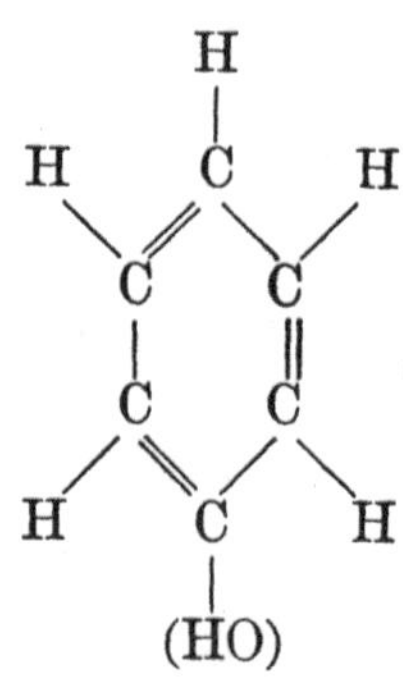 , phenol.

6. *The Quinones.*—They may be considered derived from benzene by substituting two oxygen atoms for two hydrogen atoms. Example:

$C_6H_4O_2$, 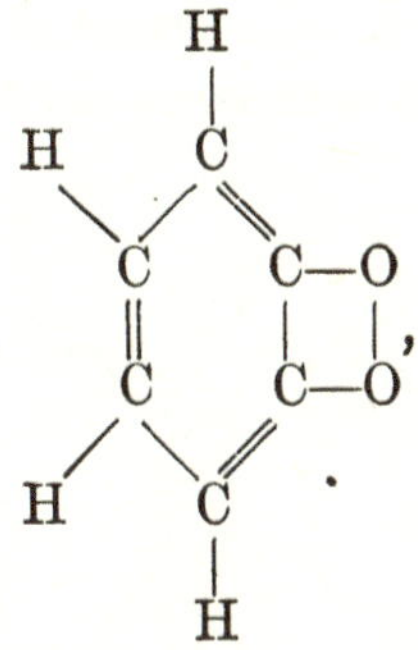 quinone.

7. *The Carbohydrates.*—These are combinations of six atoms of carbon, or some multiple of six, with some multiple of the water group (H_2O). Example: $C_6(H_2O)_5$, cellulose.

The Benzene Series.

Benzene itself (C_6H_6) is not used as an explosive, but lately certain of its derivatives have come into prominence as disruptive explosives, particularly as charges for shell.

The chief source of benzene is coal-tar. In the distillation of coal-tar, that portion of the distillate which passes over between 79° and 82° C. consists chiefly of benzene; it is purified by cooling below 0° C., at which temperature it solidifies and the lighter hydrocarbons then may be squeezed out by pressure. It boils at 80° C. It is insoluble in water, soluble in ether, acetone, chloroform, and alcohol. It is a solvent for fats and india-rubber, resin, sulphur and essential oils. It is inflammable, burning with a smoky flame. It is very volatile and its vapor is heavier than air. This vapor mixed with a certain proportion of air is explosive. These facts make it necessary to be careful about exposing benzene to evaporation in laboratory or elsewhere. The lower stratum of air in a room may be heavily charged with benzene vapor and the odor of it not be detected by a person standing. It has a strong characteristic odor.

Nitric acid acts upon it, converting it into nitrobenzene. The structural formula of benzene may be written as follows:

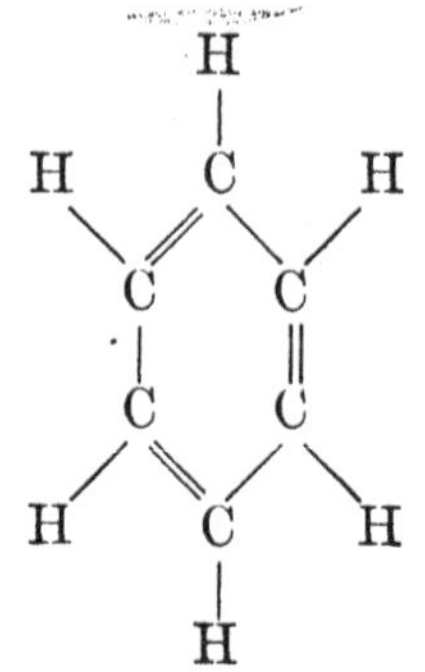

or by way of abbreviation, it is frequently indicated by the lines of a hexagon, thus:

The action of nitric acid is to substitute one or more nitryl groups (NO_2) for one or more atoms of hydrogen, giving rise to the following molecular relations:

$C_6H_5.NO_2$, $C_6H_4.(NO_2)_2$, $C_6H_3.(NO_2)_3$; or, structurally,

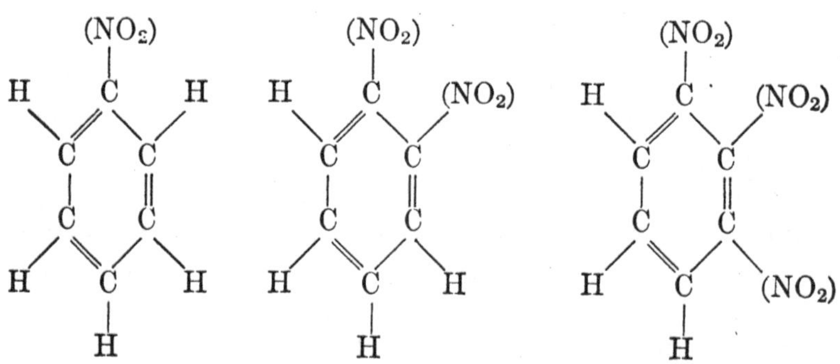

Mononitrobenzene. Nitrobenzene. Mirbane Oil, $C_6H_5(NO_2)$.

This substance is produced by adding one part of benzene to three parts of a mixture of nitric acid (sp. gr. 1.40) and sulphuric acid (sp. gr. 1.84), this mixture being made up of 40 parts of the former to 60 parts of the latter.

The benzene is added gradually, avoiding too violent chemical action. The heat due to this action must not be allowed to rise too high, the reaction being conducted in running water.

Mononitrobenzene may be made also by dropping benzene into the strongest nitric acid, or into a mixture of equal volumes of ordinary nitric acid and sulphuric acid. A violent chemical action results, giving rise to red fumes and the liquid becomes red. On pouring the liquid into several times its volume of water, a heavy oily liquid separates, which is mononitrobenzene. The reaction is

$$C_6H_6 + NO_2.HO = C_6H_5.NO_2 + H_2O.$$

The red fumes result from a secondary reaction not represented.

The sulphuric acid, if used, is present merely to maintain the nitric acid at efficient strength by combining with the water formed; it undergoes no resultant chemical change.

When the chemical action ceases the mixture is allowed to cool. The nitrobenzene will be found floating on the top of the waste acids. The latter are separated from the former by a siphon. The liquid remaining is "purified" of free acid by washing with water containing a small quantity of sodium carbonate. In order to avoid the formation of dinitrobenzene, an excess of benzene must be used in the process. A certain quantity of unnitrated benzene, therefore, remains mixed with the nitrated product. These are separated from each other by a process of vaporization, benzene volatilizing at 80° C., and mononitrobenzene not until 205° C.

Mononitrobenzene has the characteristic odor of bitter almonds. It is sold commercially as mirbane oil, which consists of the substance dissolved in alcohol. In this form it is used in perfumery and as a flavoring in confectionery. It is poisonous in large doses both as a vapor and a liquid. It is only slightly soluble in water. It dissolves readily in alcohol, benzene, and concentrated nitric acid.

Cold mononitrobenzene dissolves nitrocellulose, reducing it to a pasty or jelly-like mass. Indurite, a smokeless powder invented by Professor C. E. Munroe, consists of guncotton freed of the lower nitrocellulose by treatment with methyl alcohol and mixed with mononitrobenzene (9 to 18 parts of nitrobenzene to 10 of guncotton). Suitable oxidizing salts may be added. The mixture is then treated with hot water or steam, which has the effect of hardening it to the consistency of bone or ivory, hence its name.

Mononitrobenzene is not explosive alone, but, under the application of heat, decomposes with evolution of nitrous fumes.

If heated to a high temperature in the presence of oxygen, as when a small amount is placed on a red-hot iron plate, it will detonate.

Ignited in the open air, it burns with a reddish smoky flame, owing to the fact that the oxygen of the air does not, under these conditions, combine with the freed carbon. If mixed with explosive substances, such as guncotton, nitroglycerine, etc., the mixture may be detonated by a suitable fulminate of mercury primer.

Mixed with nitroglycerine it serves to lower the freezing-point of that explosive.

Mixed with potassium chlorate it forms the explosive known as rackarock.

Dinitrobenzene, $C_6H_4(NO_2)_2$.

There are three dinitrobenzene isomers having the same molecular formula but having different physical characteristics, viz.: meta- melts at 89° C., ortho- at 118° C., para- at 172° C.

The dinitrobenzene molecule may be represented as follows, illustrating the principle of isomerides, having the same number of atoms in a molecule and the same elements, but possessing different physical properties, due to the different structural arrangement of the atoms within the molecule.

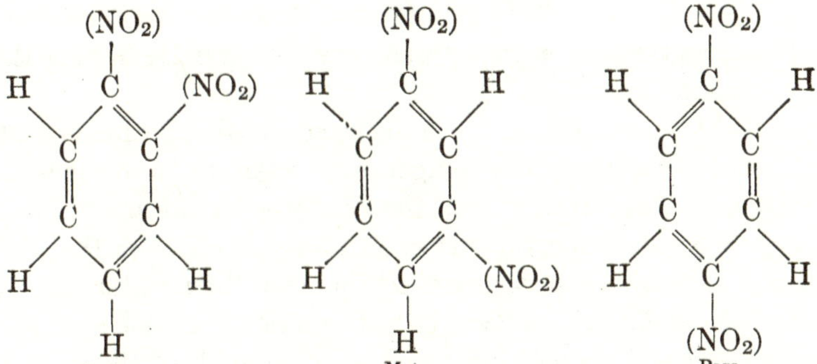

Ortho- Meta- Para-

The theory is, that when adjacent atoms are displaced one substance is produced; when alternate, another; and when opposite atoms, still another. That is, the benzene ring of six carbon atoms may give rise to the three isomerides.

Dinitrobenzene is made as explained for mononitrobenzene. except that the acid mixture is maintained at boiling temperature. On cooling, a yellowish crystalline solid separates from the liquid in long brilliant prisms. This solid is a mixture of the three dinitroisomerides with the *meta-* predominating.

It is soluble in warm water and alcohol, and like the mono-compound is poisonous.

Heated in open air it melts, and if the temperature be raised it ignites and burns with a smoky flame.

When mixed with oxidizing substances it forms an explosive. In this way it is an ingredient of many modern explosives (see Cundill's Dictionary of Explosives).

Trinitrobenzene, $C_6H_3(NO_2)_3$.

This explosive is prepared by treating meta-dinitrobenzene with concentrated nitric acid and fuming Nordhausen sulphuric acid.

While the substance possesses possibilities of use as an ingredient of explosives, little use has been made of it up to the present time.

Naphthalene, $C_{10}H_8$.

This substance is a transparent crystalline solid having the characteristic odor of coal-gas.

Its chief source is coal-tar. In the fractional distillation of coal-tar it passes over when the temperature rises just above 200° C.

When coal-tar is distilled the benzene hydrocarbons first pass over, constituting what is known as *light oil.* As the temperature rises, a heavier yellow oil, heavier than water, passes over. This is known as *dead oil;* it is much more in quantity than the light oil, amounting to about one-fourth of the bulk of the tar; it contains those constituents which have a high specific gravity and high boiling-point. As the temperature of the distillation gets above 200° C., a solid is formed in the distillate as it cools; this is crystalline naphthalene. It is separated from the liquid by pressure. It is freed from the heavier products by sublimation. If heated gently at about 200° C., it sublimes over and may be collected in the form of small transparent white crystals.

It is inflammable, burning in air with a smoky flame.

It is insoluble in water; soluble in alcohol, ether, and benzene.

In its chemical relations it is closely allied to benzene.

The substitution products derived from naphthalene have many isomerides, depending on which atoms of hydrogen are displaced.

Its relation to benzene is illustrated by its structural formula, which is written as follows:

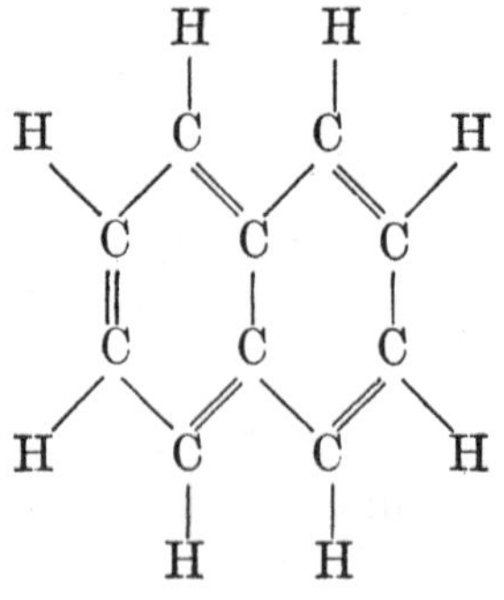

Its nitro-substitution products are more numerous, as a matter of course, than those of benzene, since there are a greater number of hydrogen atoms available for replacement by the nitryl radical (NO_2).

While this substance is not explosive alone, some of its derivatives are susceptible of forming explosives, and the many possibilities presented by a study of its derivatives marks it as one of the most promising organic substances in connection with further developments of explosives.

Mononitronaphthalene, $C_{10}H_7(NO_2)$.

Pulverized naphthalene is added to a mixture of four parts of nitric acid (sp. gr. 1.40) and five parts of sulphuric acid (sp. gr. 1.84). The naphthalene is added little by little and constantly stirred. The temperature of the mixture is kept so that it does not fall below 71° C., in order that the nitronaphthalene formed will not solidify. When the nitration is completed the charge is run off into lead-lined tanks, wherein the mononitronaphthalene crystallizes out. It is separated by pressure from the waste acids, washed in hot water, then granualted in cold water and washed until all trace of free acid is removed.

It melts at 61° C., and crystallizes from the fused state in needle-like yellow crystals. It is only slightly volatile when warmed or heated by steam.

It is insoluble in water; soluble in alcohol, ether, benzene, carbon disulphide.

If heated above 300° C., it decomposes.

It is not explosive alone, but in connection with oxygen-carriers may become explosive, as, for example, in the Favier explosives of France, in which it is associated with ammonium nitrate.

Dinitronaphthalene, $C_{10}H_6(NO_2)_2$.

This is made from the mononitronaphthalene by heating it with cold concentrated nitric acid, or from the unnitrated naphthalene by nitrating at boiling-heat until entirely dissolved, using the strongest acid, or a mixture of a weaker nitric acid (1 part) with sulphuric acid (2 parts).

It is a bright-yellow crystalline solid, the crystals forming in long slender needles.

It melts at 185° C. It is insoluble in water, slightly soluble in ether and in alcohol, less so in carbon disulphide and cold nitric acid. It is readily soluble in hot xylene, benzene, acetic acid, and turpentine.

If crystallized from its solution in acetic acid, it appears to take the form of an isomeride having a melting-point of 216° C.

It is chiefly used in the "safety" explosives in association with ammonium nitrate.

Trinitro- and **tetranitro-naphthalene** may be formed by repeated nitration of dinitronaphthalene at higher temperatures. While possibly available as ingredients of explosives, associated with oxygen-carriers, little use has as yet been made of them.

Phenol, $C_6H_5(HO)$. Carbolic Acid.

Also called *phenic acid, hydroxybenzene, benzene hydroxide* and *monohydrate of benzene.*

It results from the oxidation of benzene.

Its chief source is coal-tar. It passes over in the fractional distillation of coal-tar between 150° and 200° C. It forms a part of the "heavy oil" in this process. After the distillation of heavy oil is allowed to cool and the naphthalene has crystallized out and been separated, the remaining liquid is treated with caustic soda and stirred. On standing, two layers of liquids are observed. The upper layer consists of the higher hydrocarbons of the benzene series, the lower of a solution of sodium phenylate. This is acted upon by sulphuric acid and purified by further fractional distillation. The *phenol* distils over at between 180° and 190° C.; from the distillate it crystallizes out on cooling in needle-like crystals.

It fuses at 42° C.; boils at 182° C.; is soluble in 15 parts of cold water; readily soluble in ether and alcohol. 198 parts by weight combine with 18 parts by weight of water, when heated together, forming the aquate $(C_6H_5HO)_2Aq$, which forms on cooling six-sided prisms; the aquate fuses at 16° C. and is readily soluble in water. The commercial phenol is usually the aquate and soon becomes liquid when the bottle is placed in warm water. Once fused it has a tendency to remain in that state, but solidifies suddenly if the cork is removed.

It blisters the skin and is very poisonous.

It is used as an antiseptic and to arrest fermentation and putrefaction.

Its structural formula is:[1]

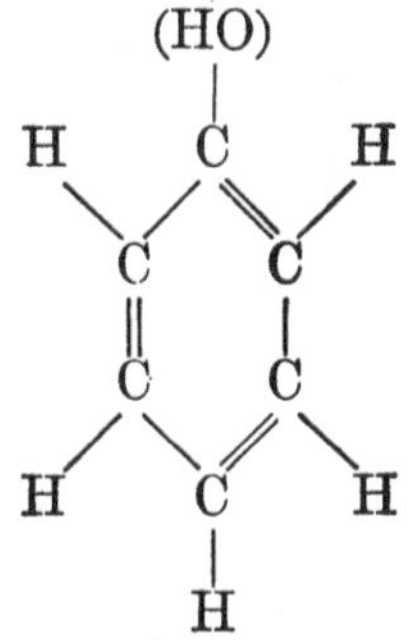

Phenols combine more readily with alkalies than alcohols do, and this property gave rise originally to the designation "acid" used with it. It may be deoxidized by passing its vapor over heated zinc-dust, $C_6H_5(HO) + Zn = C_6H_6 + ZnO$.

Certain compounds of phenol are used as color tests for acids and alkalies.

[1] Benzene forms other hydroxides, including dihydroxides $C_6H_4(HO)_2$ and the trihydroxide $C_6H_3(HO)_3$, pyrogallol.

The aqueous solution of phenol gives the following color indications:

With ferric chloride: purple-blue.

With ammonium hydroxide and calcium chloride: blue.

With mercury dissolved in nitric acid: yellow precipitate. The yellow precipitate dissolves with *dark-red color* in *nitric acid.*

The most important of its color-test compounds is *Phenolphthalein.* This is used in the manufacture of all nitro-explosives to test for the presence of the salts of sodium or potassium, the presence of the carbonates or hydroxides of these metals being indicated by a red color. If a carbonate is tested, it should be in *boiling* solution, driving off free CO_2, as *free* CO_2 will neutralize the test, phenolphthalein giving no color in excess of CO_2.

Picric Acid, ($C_6H_2.HO(NO_2)_3$). Trinitrophenol.

Its structural formula is:

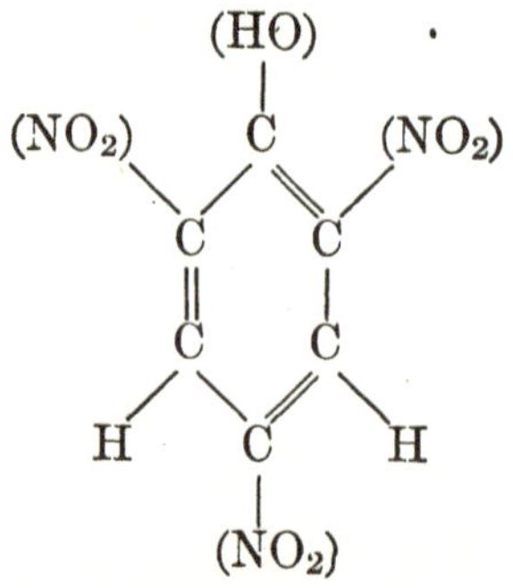

When phenol is treated with nitric acid it may form three nitrates, namely: mononitrophenol ($C_6H_4.HO.NO_2$), the dinitrophenol ($C_6H_3.HO(NO_2)_2$), and the trinitrophenol. The last only has, as yet, found application in explosives. It recently has found use not only as an explosive itself, but more particularly as an ingredient of special explosive mixtures. It and its salts (the picrates) find application in detonating or disruptive explosives only. Most of the new so-called "shell-filler"

explosives are either picric acid, mixtures with it or derivatives thereof. Among these may be mentioned *Ecrasite*, Austrian; *Lyddite*, English; *Melinite*, French; *Shimose*, Japanese; Abel's picric powder and Brugère's powder (nitre and picrate of ammonium); one form of Rackarock (nitrobenzene and picric acid).

MANUFACTURE OF PICRIC ACID.

Equal quantities by weight of concentrated H_2SO_4 and phenol are mixed in an iron vessel, stirred and heated by steam to from 212° to 250° F. From time to time tests are made to see if the phenol-sulphonic acid formed is soluble in cold water. When this is so the mixture is allowed to cool and twice the quantity of water is added.

The nitration then takes place in earthen vessels standing in running water which can be heated by steam-pipes. Three parts by weight of nitric acid is placed in these receivers and one part of the sulphonic solution is added. The latter is allowed to run in gradually, as at first the reaction is violent. Afterwards it becomes sluggish and then steam is turned on and the temperature of the solution raised to restore the chemical action.

The picric acid formed separates at first as a sirupy liquid and becomes crystalline on cooling. It is separated from the mother-liquor in a centrifugal machine, and is washed in the same machine with pure warm water. The crystals are further purified by redissolving in warm water, recrystallizing, and finally drying at 95° F.[1]

---

[1] It is reported that Professor Arthur G. Green, of the Department of Technical Chemistry of the University of Leeds, has invented a method of manufacturing picric acid directly from benzene instead of by the indirect method of first producing carbolic acid from benzene, and then picric acid from carbolic acid. The yield by the direct process is said to be much greater.

The reactions of the process are:

$$C_6H_5HO + H_2SO_4 = \underset{\text{Phenol-sulphonic acid}}{C_6H_4(SO_3H)HO} + H_2O;$$

$$C_6H_4(SO_3H)HO + 3HNO_3 = C_6H_2(NO_2)_3OH + H_2SO_4 + 2H_2O.$$

Picric acid has an extraordinarily bitter taste.
It always gives an acid reaction.

It is sparingly soluble in cold water; it dissolves in hot water, giving a bright-yellow color to a large volume of water.

It dissolves readily in alcohol. Its solution stains the skin and other organic matter yellow and is used in dyeing for this purpose. It is one of the few acids which form sparingly soluble potassium salts. A cold aqueous solution of picric acid is an excellent test for any soluble potassium salt, giving, when added, a yellow, adherent, crystalline precipitate of potassium picrate. This salt in the solid state and dry is very sensitive, exploding with violence if heated or struck.

Considerable diversity of opinion has existed as to whether picric acid is explosive if subjected to simple heating. There is no doubt that it is less explosive than nitroglycerine and guncotton. If a small mass is heated in a capsule or flask, it melts and gives off vapors which ignite and burn without causing an explosion. A very small quantity may be sublimed if carefully heated in a glass tube. It is a mistake, however, to think that picric acid is incapable of explosion by simple heating. If it is heated to a high temperature, it decomposes with disengagement of heat, developing a process of oxidation. When a decomposition liberates heat, its rapidity increases with the pressure or confinement for a given temperature, or with the temperature for a given pressure; in the latter case, the decomposition increases very rapidly. This principle suggests that picric acid would explode if either the temperature or pressure of its environment should increase, and still more rapidly if both temperature and pressure increased together: this is the condition existing when it is heated in a closed space.

Picric acid may, in accordance with these principles, be made to detonate if heated very suddenly to a high temperature in an open vessel at the ordinary pressure, especially if the vessel be heated itself beforehand, so that there is little loss of heat by conduction.

If a glass tube 25 to 30 mm. long be heated to redness and one or two small particles of picric acid be thrown into it, they will explode before they can vaporize. If the mass be consideraby increased, the walls of the tube may be sufficiently cooled by the mass of the picric acid to modify or destroy entirely the explosive effect.

Similar experiments may be conducted with mononitrobenzene, dinitrobenzene, mono-, di-, and tri-nitronaphthalene.

The nature of the decomposition, whether explosive or non-explosive, and the degree thereof depend on the temperature of the enclosure, the temperature and mass of the explosive used.

If, however, a large mass of an explosive, like any of those just named, were to ignite in a closed space, its decomposition would generate more and more heat, the temperature would rise higher and higher, and the phenomenon might cause a detonation at some particular point, and the explosive wave there started might be transmitted throughout a very large mass.

In 1887 a disastrous explosion of picric acid took place in the chemical works of Messrs. Roberts, Dale & Company, at Manchester, England. An investigation at that time, and experiments since made, have revealed the fact that if picric acid is in contact with some metals or the oxides or nitrates of some metals, such as lead, iron, strontium, potassium, it is quite likely that very sensitive explosive salts may be formed. Litharge, the oxide of lead, particularly, has a tendency to form very sensitive compounds if in contact with picric acid, and may cause the detonation of a large mass of it.

Many accidents have resulted in handling shells charged with lyddite which are presumed to have been due to the formation of such sensitive compounds.

For these reasons red or white lead should not be used to seal the screw-threads of shell-plugs when the shells are filled with picric acid or derivatives thereof.

Picrates, $C_6H_2(NO_2)_3.MO$. (In which M represents some metal radical.)

Structural formula:

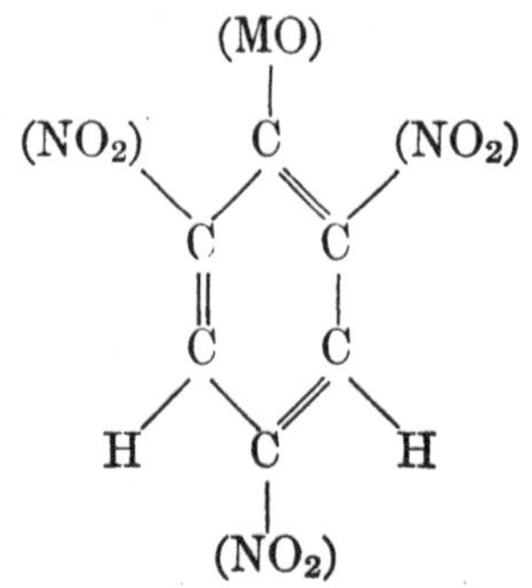

For many years attempts have been made to use the picrates of certain metals as ingredients of explosives. In 1869 a class of powders were introduced in France, known as Designolle's Powders, consisting of picrate of potassium, niter, and charcoal. Potassium picrate is, however, too sensitive to give a serviceable explosive. About the same time, Brugère in France and Abel in England suggested the use of ammonium picrate instead of potassium picrate. These powders gave excellent results.

Brugère's powder contained:

Picrate of ammonium	54 parts
Niter...............................	46 "

It was stable, safe to manufacture, burned with slower rate than black powder, was less hygroscopic, had little smoke small residue, did not attack metals. In the small-arm rifle it gave about 2½ times the effect of black powder.

Abel's powder was practically the same, the proportion being 60 parts of ammonium picrate to 40 of nitre.

Ammonium picrate appears to be the only picrate which has given satisfactory results. While the metallic picrates are very sensitive to shock, ammonium picrate is quite insensitive. It is also very stable, showing no tendency to form ammonium nitrate in the above mixtures.

It is easily made by saturating a hot solution of picric acid with a concentrated solution of ammonium hydroxide, or by passing ammonia-gas through a hot solution of picric acid. As soon as the solution is completely saturated with the new salt it is allowed to cool, when ammonium picrate separates in the form of long yellow prisms.[1]

If ignited, it burns without any tendency to explosion.

It is insensitive to shock of any kind, and can be detonated only by a very powerful primer.

Trinitrotoluol, $C_6H_2(NO_2)_3 \cdot CH_3$, also known as trotol; tritone; trinitrotoluene; trinitromethylbenzene; tolite; trilite; trinol; trotyl.

Its structural formula may be written as follows:

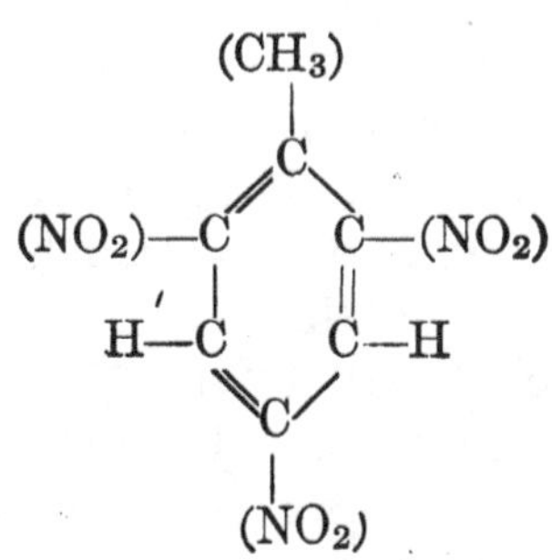

Its relation to the benzene series is clearly evidenced by its structural formula, which is similar to that of picric acid and the picrates, except that the organic radical, CH_3 (methyl), replaces the hydroxyl radical of picric acid and the metallic radical of the picrates. From this chemical similarity it might well be assumed that it would occupy a field in explosives analogous to that of picric acid and the picrates, and this is the fact. It has, however, some advantages of an important nature which suggest that it may eventually replace both picric acid and the picrates as a shell-filler, the most important of which is that it does not form sensitive compounds by combination with the metals. Its explosive force also is slightly less than either picric acid or the picrates.[1]

Manufacture.—The material is neither difficult nor dangerous to make. Commercial mononitrotoluene is made by acting on 1 part toluene[2] with 3 parts, by weight, of mixed nitric and sulphuric acids ($40HNO_3$, sp. gr. 1.495, to $60H_2SO_4$, sp. gr. 1.84); the result is a mixture of isomeric bodies. This mixture may be converted into dinitrotoluene by nitrating with 2 parts of the acid mixture to 1 part of the mononitro product; the dinitro products thus produced may be converted into trinitrotoluol by nitrating again with stronger nitric acid, but there is a considerable loss of acid and the yield is not high.

"According to Haussermann, it is more advantageous to start with the orthoparadinitrotoluene, which is prepared by allowing a mixture of 75 parts of 91 to 92 per cent nitric acid (sp. gr. 1.495) and 150 parts of 95 to 96 per cent sulphuric acid (sp. gr. 1.840) to run in a thin stream into 100 parts of paranitrotoluene, while the latter is kept at a temperature between 60° C. and 65° C., and continually stirred. When the acid has all been run in, this mixture is heated for half an hour to 80° C., and allowed to stand until cold. The excess of nitric acid is then removed. The residue after this treatment is a homogeneous crystalline mass of orthoparadinitrotoluene,

---

[1] Experiments at Picatinny Arsenal indicate that the theoretical force of picric acid is 135,800 pounds per square inch, that of Explosive D 124,600 pounds per square inch, and that of trinitrotoluol is 119,000 pounds per square inch.

[2] Toluene is a by-product in the manufacture of illuminating coal-gas.

of which the solidifying point is 69.5° C. To convert this mass into the trinitro derivative, it is dissolved by gently heating it with four times its weight of sulphuric acid (95 to 96 per cent, sp. gr. 1.540), and it is then mixed with one and one-half times its weight of nitric acid 90 to 92 per cent, sp. gr. 1.495), the mixture being kept cool. Afterwards it is digested at 90° C. to 95° C., with occasional stirring, until the evolution of gas ceases. This takes place in about four or five hours. The operation is now stopped, the product allowed to cool, and the excess of nitric acid separated from it. The residue is then washed with hot water and very dilute soda solution, and allowed to solidify without purification. The solidifying point is 79° C., and the mass is then white, with a radiating crystalline structure. Bright sparkling crystals may be obtained by recrystallization from hot alcohol. This product melts at 81.5° C."[1]

A recent modification of the manufacture permits a melting-point as low as 75.5° C., which allows second crystallization to be omitted.

It is most usual to nitrate in three stages, using a mixture of nitric and sulphuric acids (as single-stage nitration requires stronger acids, which give lower yield and have other disadvantages). In the three-stage operation, the spent acids from the last process are used in the second, etc., weaker solutions being required in the earlier nitrations and stronger for the tri-nitration. Nitration is done in vessels arranged for control of temperatures, which are about 135° F., 185° F., and 220° F. in the three stages respectively.

When nitration is complete, the stirrers are stopped and the acid settles to bottom, the nitrated material coming to the top, and the acid is drawn off.

Purification following nitration consists of boiling in water with agitation by compressed air and addition of soda to neutralize acid, followed by several neutral boilings until acid-free.

---

[1] J. C. Sanford, " Nitro Explosives."

The molten material is then run out through a jet into cold water, to prevent formation of large lumps. The material is next melted in a pot (with some water to help separation of impurities) and then cooled in pots to a finely crystallized condition. These processes eliminate impurities, but do not remove the mono- and di-nitro products.

Crystallization after a further melting with some solvent, which will remove the products of lower nitration, must follow if a higher melting-point than about 75.5° C. is required. The solvents used by manufacturers are trade secrets.

After complete purification and crystallization the product is dried, screened and boxed.

Properties.—Trinitrotoluol when pure has no odor and is a slightly yellow fine crystalline powder; it assumes a brownish appearance as the impurities increase. It melts at from 79° C. to 81.5° C., depending upon the presence of isomers. If melted and allowed to cool it crystallizes and forms a very vesiculated mass which is difficult to explode. It should be melted in a water-bath, though it is not likely to be exploded under any application of heat for melting. The sharpness of the crystallizing point is a good indication of purity. It is insoluble in water, and traces only pass into solution at a temperature of 40° C.

Its specific gravity when in the form of non-compressed crystals is from 0.8 to 1.0. It dissolves in alcohol, ether, benzene, and toluol.

Commercial trotol undergoes some change when exposed to metals, in the presence of salt water, but does not lose its explosive properties; no sensitive compounds are formed by the action.

It behaves in a very stable manner when exposed to the air under varying conditions of temperature. It is unaffected by contact with metals even in the presence of moisture, and forms no sensitive compound due to such contact.

It is a very powerful explosive when detonated, but cannot be exploded by flame or strong percussion; a rifle bullet may be fired through it without effect. Its ignition point is 180° C. When flame is applied to a mass of the explosive it melts, takes fire, and burns quietly with a heavy black smoke.

It may be detonated in its crystalline form by a mercury fulminate exploder. Heavy black smoke comes off when it is detonated in air and a characteristic black cloud is seen at the surface when it is detonated under water. A package of it in ordinary paper with a mercury fulminate exploder embedded in the powder may be detonated under water. A mound of it may be prepared, a depression made in the center and filled with water, a fulminate exploder placed in the water, and the whole mass detonated by the firing of the exploder.

Its explosive force at points some distance from the charge, when in a loose crystalline form, is about the same as that of Explosive "D," and from 50 to 75 per cent greater than that of guncotton wet with 25 per cent of moisture, but is not greater at points in close proximity to the charge. Its specific explosive force is less than that of guncotton.

As a bursting charge for projectiles it is less efficient than Explosive "D," on account of the fact that a greater weight of the latter can be inserted into a given volume. Except in the cast form, in which condition it is extremely difficult to detonate, it is more sensitive to shock than is Explosive "D," and as a bursting charge for projectiles will not withstand impact on hard-faced armor without exploding.

Chemical Specifications for Military Trinitrotoluol.

(1) The material must be chemically pure, and in the form of a slightly yellow, fine and uniform crystalline powder passing through a 12-mesh screen. No odor of any by-product or crystallizing agent must be present.

(2) Melting-point must be from 80.5° to 81.5° C., and must be sharp and distinct.

(3) Ash must not be greater than 0.10 per cent.

(4) Insoluble matter from a solution of 10 grams in 150 c.c. alcohol must not be greater than 0.15 per cent.

(5) Moisture must not be greater than 0.10 per cent.

(6) There must be no acidity.

(7) No uncoverted toluol, or dinitrotoluol or any other by-product must be present.

(8) Nitrogen must not be less than 18.30 per cent determined by Dumas' combustion method. Pure trinitrotoluol contains 18.5 per cent nitrogen.

(9) The material must stand a heat-test at 65.5° C. with K. I. starch paper of at least 30 minutes.

Tests.

(1) *Solidifying Point.*—Take a porcelain basin 15 cm. diameter and a capacity of about 500 c.c., dry the basin thoroughly, and then take about 200 to 250 grams of trinitrotoluol; melt the trinitrotoluol but do not let the temperature get above 90° C.; then, when all the trinitrotoluol is melted, remove the source of heat and stir with a thermometer that has been standardized; the temperature falls gradually until the trinitrotoluol begins to solidify, when it rises; keep on stirring until the highest point is reached, then note the degree of temperature as the solidifying point.

(2) *Softening Point, Melting-point Begins, Melting-point Ends.*—Take a small quantity of trinitrotoluol, grind it up as fine as possible, and place in a capillary tube sealed at one end (the tube having about 2½ mm. bore); attach to a standardized thermometer by means of a rubber ring, then place in a glass beaker nearly full of water and heat up slowly; note that temperature (*a*) at which it softens, (*b*) begins to melt, (*c*) ends melting. (*a*) should not be less than 79° C., (*b*) should not be less than 80.5° C. and (*c*) should not be greater than 82° C.

(3) *Specific Gravity.*—Take a quantity of trinitrotoluol and melt up in a dry basin; fill up a test-tube with the melted trinitrotoluol; suspend it in a water-bath filled with water that is kept at the boiling point. Place a hydrometer in the test-tube, leave it for a few minutes, and then note the reading of the hydrometer when the temperature of the whole is at 100° C. The specific gravity in powdered form closely packed is 1.55; in melted form, 1.65.

(4) *Softening Test.*—Take 5 grams of trinitrotoluol and grind up fine; place in a crucible 40 mm. depth and 50 mm. diameter at the top; then place in an oven and keep the temperature at 70° C. for one hour. The sample should not soften.

(5) *Staining Test.*—Grind a small quantity of trinitrotoluol finely and place on a sheet of clean white paper, keep for 24 hours; the paper should show no stain.

(6) *Ash.*—Weigh a small crucible and put in about 8 grams of trinitrotoluol; heat the trinitrotoluol gently until it has all burned away; then heat strongly for about 15 minutes, cool, weigh, and again heat till the weight is constant.

(7) *Heat-test.*—Weigh 2 grams of trinitrotoluol and grind up fine; place in a clean dry test-tube and apply the potassium-iodide heat test. No color should appear for 30 minutes.

(8) *Acidity.*—Boil in a porcelain basin some distilled water; weigh 25 grams of trinitrotoluol; place it in the basin and allow to boil for a few minutes, stirring up well. Then allow to cool and keep stirring till the trinitrotoluol is solid; then add a little phenolphthalein, and titrate with $\frac{N}{100}$ caustic soda till just neutral. From the data obtained, the sulphuric acid in 100 grams of the explosive may be calculated.

Alcohols, Ethers, Ketones.

Alcohols and alcohol derivatives are used either in the manufacture or as ingredients of modern explosives.

The alcohols may be regarded as formed from the hydrocarbons of the paraffin series by substituting the radical HO for one or more of the hydrogen atoms. They are, therefore, as already indicated, properly organic hydroxides of the paraffin series. Some authorities consider all hydroxides of the hydrocarbons as alcohols, there being a series of alcohols corresponding to each series of hydrocarbons.

Alcohols containing (HO) are monohydric;
" " $(HO)_2$ " dihydric;
" " $(HO)_3$ " trihydric;
" " etc. " etc.

There are but two alcohols proper which need be described in connection with substances used in explosives, namely, monohydric ethyl alcohol, $C_2H_5.HO$, and trihydric propenyl alcohol (glycerine), $C_3H_5(HO)_3$.

The structural formula of ethyl alcohol is

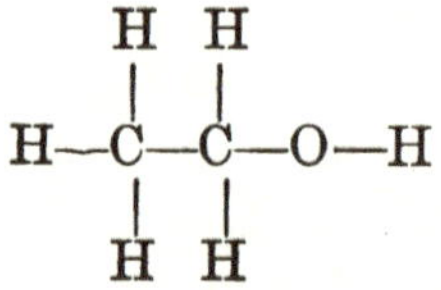

$C_2H_5.HO$;

of glycerine:

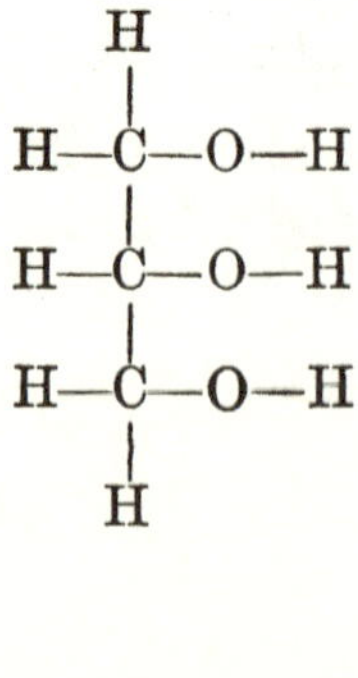

$C_3H_5.(HO)_3$.

Two other substances may be referred to here as allied in structure to the alcohols, in order to emphasize both the relation existing and the differences in structure.

1. *Ethyl ether*, which, as before explained, is the oxide of the paraffin hydrocarbon radical, C_2H_5. Its molecular formula is $(C_2H_5)_2O$ and its structural formula is

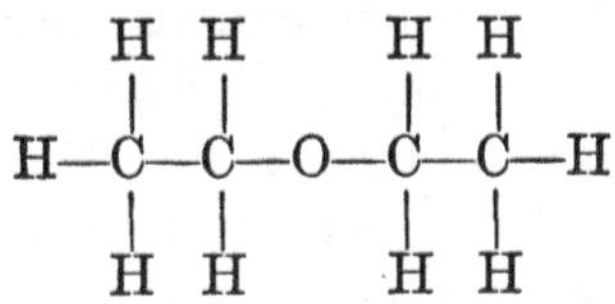

2. *Cellulose.*—This has a more complex structure. It does not fall strictly under the alcohols or ethers, but its chemical behavior leads to its classification as a hexhydric alcohol. Under this conception its structural formula, using a double grouping, may be written as follows:

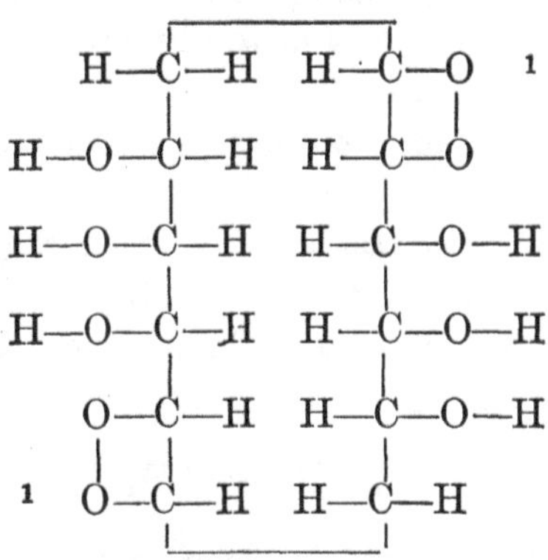

$(C_6H_{10}O_5)_2$.

Another hydrocarbon derivative closely allied to the foregoing is acetone or dimethyl ketone ($CH_3.CO.CH_3$). The relation is as follows:

Acetic acid results from the oxidation of alcohol:

$$C_2H_5.HO + O_2 = CH_3.HO.CO + H_2O.$$

[1] Quinone arrangement suggested by Dr. John W. Mallet, University of Virginia. See Walke's Lectures on Explosives, p. 205.

Acetone may be considered as derived from acetic acid by displacing the HO group by a paraffin hydrocarbon radical, thus:

acetic acid: $CH_3.CO.HO$; acetone: $CH_3.CO.CH_3$.

Acetone is the standard solvent for highly nitrated celluloses used in smokeless powders containing nitroglycerine, and ordinary guncotton for demolitions, etc.

A mixture of ethyl alcohol, $(C_2H_5)HO$, and ethyl ether, $(C_2H_5)_2O$, in the proportion by volume of 1 to 2 is used in dissolving nitrocellulose of medium nitration in the manufacture of smokeless powders that are made of pure nitrocellulose without an admixture of nitroglycerine.

Ethyl Alcohol. Vinic Alcohol. Alcohol. $C_2H_5.HO$.

When mixed with water known as spirits of wine.

As stated above this substance is one of the ingredients of the solvent used in colloiding nitrocellulose in making smokeless powder.

It is a colorless liquid having a characteristic odor and burning taste.

Pure or "absolute" alcohol has a specific gravity of 0.794 at 15° C. It freezes at −130.5° C. Its boiling-point is 78.3° C. It burns with a blue smokeless flame, the reaction of combustion being as follows:

$$C_2H_5.HO + O_6 = 2CO_2 + 3H_2O.$$

It evaporates rapidly in the open air without combining with oxygen. Exposed to the air it absorbs water. Bottles containing it should therefore be tightly corked. It mixes with water in all proportions, evolving little heat and giving a mixture rather smaller in volume than the sum of the volumes of the constituents.

Next to water it is the most universal solvent. It is especially useful as a solvent of certain resins and alkaloids which are insoluble in water.

To test for alcohol in a liquid, add HCl and enough potassium dichromate to give an orange-yellow color. Divide between two test-tubes for comparison. Heat one until the liquid boils. If alcohol is present, the color will change to green and give off odor of aldehyd.

The strength of alcohol is usually determined by its specific gravity. This may be determined by using a hydrometer for liquids lighter than water, or by weighing a few cubic centimetres carefully measured, the weight in grams per cubic centimetre will be its specific gravity (1 cubic centimetre H_2O at standard density = 1 gram).

In the commercial grades, *rectified spirit* has a specific gravity of 0.833 and contains 84% of alcohol; *proof spirit* has a specific gravity of 0.92 and contains only 49% of alcohol. This is the weakest spirit that will answer the old rough proof of firing gunpowder which has been moistened with it.

Ethyl Ether, $(C_2H_5)_2O$.

May be considered as derived from the corresponding alcohol by process of dehydration. Ethyl ether is sometimes called sulphuric ether, from the fact that it is prepared by distilling a mixture of ethyl alcohol with sulphuric acid in the proportion by volume of 2 to 1. The sulphuric acid is left unchanged by the process, the reaction being apparently as follows:

1. Production of hydro-ethyl sulphate $(H.C_2H_5.SO_4)$:

$$C_2H_5.HO + H_2SO_4 = C_2H_5.H.SO_4 + H_2O.$$

2. Production of ethyl ether heating with more alcohol at 140° C.:

$$C_2H_5.H.SO_4 + C_2H_5.HO = (C_2H_5)_2O + H_2SO_4.$$

The ethers as a class are insoluble in water and lighter and more volatile than the corresponding alcohols. They are not as easily acted upon chemically by other bodies as alcohols are.

Ethyl ether is a very mobile, colorless liquid with a characteristic odor; has specific gravity of 0.70 at 15° C.; it boils at 34.9° C.; evaporates rapidly in air at ordinary temperatures, producing great cold and yielding a heavy vapor (specific gravity 2.59) which is very inflammable and in unskilled hands is dangerous. It is very sparingly soluble in water, requiring 10 volumes of H_2O to dissolve 1 volume of ether. 34 volumes of ether are required to dissolve 1 volume of H_2O. But commercial ether contains alcohol, and this latter takes up considerable water. Ether and alcohol may be mixed in any proportion, but the addition of excess of water displaces the ether. Ether dissolves the nitro-substitution compounds and is the great solvent for fats.

Acetone, $CH_3.CO.CH_3$. Dimethyl-ketone.

Acetone is the solvent for cellulose that has been nitrated so as to contain a high percentage of nitrogen, say 12.9% or above. At ordinary temperature and pressure cellulose so nitrated is not soluble in the ether-alcohol mixture, but is soluble in acetone. Acetone is found among the products resulting from the distillation of wood. When wood is distilled, the condensed products separate into two layers: the lower is wood-tar, and the upper is a mixture of water, methyl alcohol, acetic acid, and acetone.

Acetone is a colorless liquid with characteristic, pleasant odor; specific gravity 0.81; boils at 56.3° C.

It burns with a bright flame; it evaporates readily, and its vapor is dangerous if mixed with air. It mixes with water, alcohol, and ether in all proportions. Adding KHO to its aqueous solution displaces it and it rises to the surface. It is a good solvent for resins, camphors, fats, guncotton and nitroglycerine, and freely dissolves the nitro-substitution compounds.

Glycerine. Glyceryl Hydroxide. Propenyl Alcohol.
$C_3H_5(HO)_3$.

Glycerine is a trihydric alcohol, having the structural formula

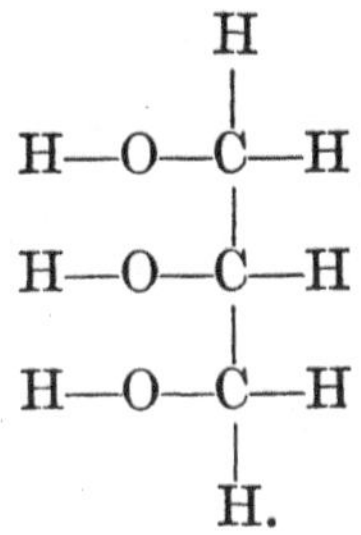

It may be obtained from all fats and is the sweet principle of them. Fats are sometimes called glycerides.

Glycerine is also formed as a by-product in the alcoholic fermentation of grape-sugar, and is present in small quantities in beer and wine. It is a by-product also in the manufacture of soaps and candles, being separated in the mother-liquor when fats are saponified by lime or superheated steam. The crude glycerine resulting from these processes is purified by distillation. A quantity of crude glycerine is placed in a copper still, and steam at 280° C. is forced through it. The pure glycerine is volatilized, passes over, and is condensed.

Glycerine is a colorless sirupy liquid, its viscosity increasing as the temperature is lowered. Although it is a viscous liquid, it has the property of working its way by capillary action through the smallest openings or fissures. Its specific gravity is 1.269 at 12° C.; it boils at 290° C., but then undergoes partial decomposition; it is slightly volatile at 100° C., but not at ordinary temperatures. Glycerine crystals may be obtained from an aqueous solution kept for some time at 0° C. Pure glycerine solidifies at −40° C., forming a gummy mass. It ignites at 150° C. in air, burning with a faint blue flame resembling that of alcohol. It absorbs water readily from the air

It mixes with water and alcohol in all proportions. It is insoluble in ether, but is soluble in ether-alcohol mixtures. It is soluble in carbon disulphide, petroleum, benzene, chloroform. Glycerine is one of the most important solvents, dissolving most substances which are soluble in water, and some others, such as some metallic oxides, which are not.

The best test of identifying glycerine is to mix it with $KHSO_4$ and heat it strongly, when the unpleasant odor of acrolein (odor of smouldering candles) is noticed.

Its importance in explosives results from the fact that it forms nitroglycerine when acted upon by nitric acid.

Cellulose, $C_6H_{10}O_5$ or $n(C_6H_{10}O_5)$.

Cellulose is by far the most important substance used in the manufacture of the new explosives. It is the source from which guncotton and most of the smokeless powders are derived.

As already stated, the most recent practice classifies cellulose as a hexhydric alcohol, although its molecule has not the simple structure of the alcohol series.

Captain Willoughby Walke, Artillery Corps, on page 205 of his Lectures on Explosives, gives the suggestion of Dr. John W. Mallet, the celebrated chemist of the University of Virginia, that only three atoms of hydrogen are grouped in the hydroxyl radical, the fourth hydrogen atom being united directly to the carbon atom, while the corresponding oxygen atom and the remaining free oxygen atom are linked together with the same carbon atoms after the manner of grouping of oxygen atoms in quinone, thus:

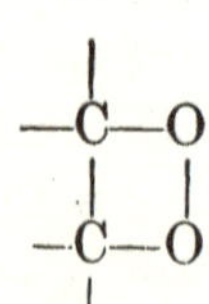

If tnis be adopted, the cellulose molecule may be wiitten as follows for the double molecule:

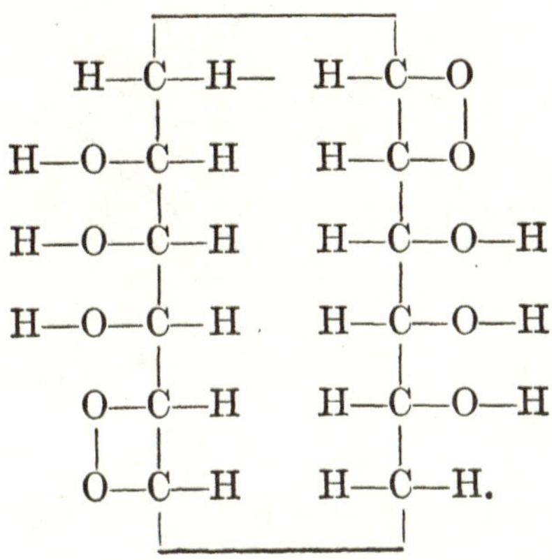

The formula for the single molecule $C_6H_{10}O_5$ may be arranged after the plan of the quinone group, thus:

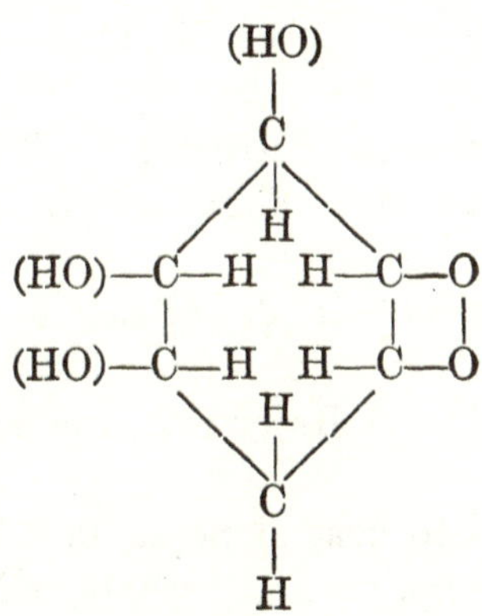

A cellulose ring may be written composed of any number of groups of this type of arrangement. It should be understood, however, that this arrangement of the cellulose molecule is theoretical and of value only in so far as it agrees with observed chemical facts.

The group $5(C_6H_{10}O_5)$ would be written structurally as follows:

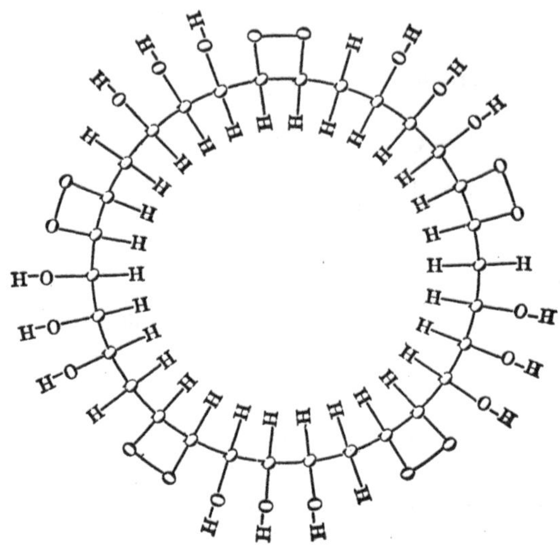

Cellulose is the constituent of the cell-walls of all plants. When the soluble ingredients of all forms of vegetable life and mineral substances are removed, cellulose remains as a white, opaque, organized structure. White filter-paper, cotton wool, and pure linen fibre are familiar examples of cellulose.

It is infusible; insoluble in all ordinary solvents. It is dissolved by "Schweitzer's Reagent" (a solution of cupric hydroxide in ammonia); it is precipitated from this solution by adding an acid.

Cellulose subjected to heat alone, as in destructive distillation, breaks up into organic volatile compounds, especially into certain organic acids, such as:

Formic: 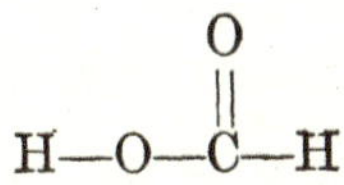

Acetic: 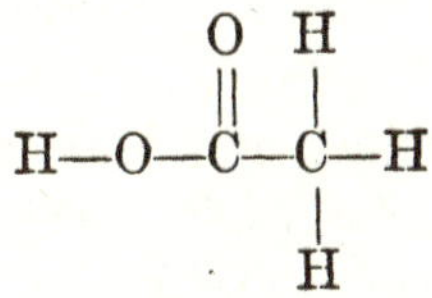

Propionic: 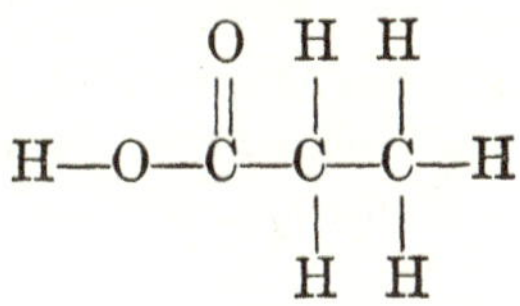

Butyric: 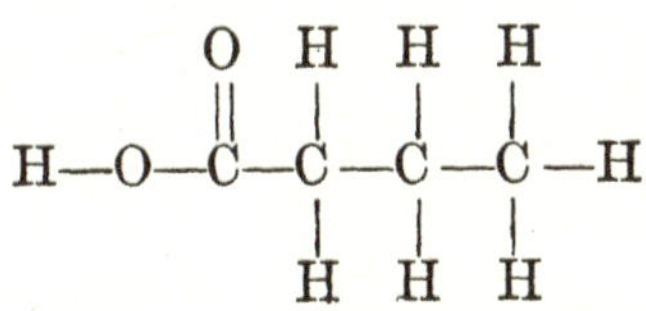

Strong sulphuric acid acts on dry cellulose, converting it into a gummy mass which dissolves in the cold in an excess of the acid with very little color.

Unsized paper immersed in a cold mixture of strong sulphuric acid with one-half its volume of water converts the cellulose into a tenacious translucent substance called amyloid. A strong solution of zinc chloride affects cellulose in the same way. This property is made use of in manufacturing vegetable parchment, shipping-tags, cartridge paper, etc.; it increases the strength of paper about five times and makes it water proof.

Cellulose left in a bath of sulphuric acid (specific gravity 1.453) or in hydrochloric acid (specific gravity 1.16) for 12 hours is converted into a brittle mass of hydrocellulose ($C_{12}H_{20}O_{10}.H_2O$) which is more easily oxidized than cellulose and is soluble in a hot solution of potassium hydroxide. This is made use of in separating cotton from old fabrics (rags) of cotton and wool mixed; the wool remaining is called shoddy.

Dry-rot in wood is supposed to be due to a similar change.

The action of nitric acid on cellulose will be described later, in connection with the manufacture of guncotton and smokeless powders.

Liquefied chlorine or bromine, or a mixture of these two elements, have been used as charges for steel shells and hand grenades in the European war. The shells on striking are exploded by percussion, and the liberated liquefied gases vaporize as soon as released from pressure, causing great pain to all exposed membranous surfaces such as the eyes and the breathing organs. (See page 370, Appendix III).

III.

GENERAL REMARKS ON EXPLOSIVES.

As a matter of practical military interest, explosives may be divided into three classes, namely:

1. Progressive or propelling explosives. (Low explosives.)
2. Detonating or disruptive explosives. (High explosives.)
3. Detonators or exploders. (Fulminates.)

The first includes all classes of gunpowders used in fire-arms of all kinds; the second, explosives used in shell, torpedoes, and for demolitions of all kinds; the third, those explosives used to originate explosive reactions[1] in the first two classes.

Each of these classes is distinguished by the character of the explosive phenomenon it produces, and it may be said that, corresponding to these respective characteristics, explosive phenomena may be divided into three classes, namely:—

1. Explosions proper: explosions of low order; progressive explosions; combustion.
2. Detonations: explosions of high order.
3. Fulminations: the characteristic type of explosion produced by the fulminates, possessing exceptional brusqueness.

---

[1] An *explosive reaction* is a chemical reaction usually involving the change of state of a substance from a solid or liquid to a gas, attended with great increase of volume, or the combination chemically of two or more gases with sudden increase of volume.

Explosion Proper. (Explosion of Low Order. Progressive Explosion.)

As the heading implies, this class of explosion is marked by more or less progression; the time element is involved as a controlling factor, the time required to complete the explosive reaction being large compared with that in the other forms of explosion. In this class of explosion the time consumed in the reaction is to some extent under control by varying the physical characteristics of the explosive. The explosion, indeed, is of the nature of an ordinary combustion. The mass is ignited at one point and the reaction proceeds progressively over the exterior surfaces and then perpendicularly to these surfaces until the entire mass is consumed.

The explosion of a charge of black, brown, or smokeless powder is not different in principle from the burning of a piece of coal, wood, or other combustible; there is a *progressive* change of state from particle to particle, from the solid state to the gaseous state, accompanied by the heat due to the chemical change.

The word "combustion" as used above has a definite meaning. It is the combination of the carbon and hydrogen of a combustible with oxygen. The *calorific value* of a combustible is the number of units of heat involved in its combustion.[1] This is independent of the time involved; it is the same whether the change takes place in a fraction of a second or is prolonged through years; a pound of wood will give the same number of units of heat whether it be burned as fine shavings or pass into the gaseous state through slow oxidation in the air. *Calorific intensity* is the maximum possible *temperature* of the products of combustion.[2] Its numerical value is determined

[1] Unit of heat is the amount of heat required to raise a pound of water from 0° C. to 1° C., or from 32° F. to 33° F.

[2] It may be defined as the temperature to which the heat generated by the burning of each portion of the fuel can raise its own products of combustion when burned in its own volume without loss of heat due to conduction or radiation.

by dividing the total number of units of heat produced by the number of units of heat required to raise the products 1° C. at the temperature of these products.

It may be represented in the form of an equation as follows:

Let H = total number of units of heat produced;

W, W', W'', etc., = weights of products of combustion;

S, S', S'', etc., = heat required to raise 1 unit of weight 1° C. at the temperature and pressure of the products = specific heats of products;

T = calorific intensity.

Then

$$T = \frac{H}{WS + W'S' + W''S'' + \text{etc.}}$$

The heat of combustion is due chiefly—

1. To H burning to H_2O.
2. To C burning to CO, in limited supply of oxygen.
3. To C burning to CO_2, in unlimited supply of oxygen.

The calorific value of most substances may be estimated approximately from the molecular composition, by determining the number of atoms of C and H that are free to combine with O. In some combustibles, as in the carbohydrates, part of the O is present in association with H in the molecule in the proportion found in the water molecule (H_2O) and no heat results from these atoms; indeed, on the contrary, heat is absorbed in the physical change of state of this water to vapor. Water held mechanically in the pores or intermolecular spaces of substances must be treated in the same way. It requires 537 units of heat to convert one pound of water at 100° C. into one pound of steam at 100° C. This is the latent heat of evaporation or condensation. To evaporate 9 pounds of water (containing 1 pound of H) at 100° C. requires $537 \times 9 = 4833$ units of heat. Eight pounds of O combine with 1 pound of H to produce 9 pounds of water-vapor at 100° C., and, in doing so, produce 29,629 units of heat. This is the calorific value of hydrogen when the products are in the state of vapor. If this water-vapor be condensed to liquid water at 100° C., the latent heat of condensation must be added to this, and the calorific value of hydrogen when the product is in the form of liquid water at 100° C. is $29,629 + 4833 = 34,462$ units of heat. In the same manner $\frac{16}{12}$ pounds of O combining with 1 pound of C produces 2481 units of heat in burning to CO; $\frac{32}{12}$ pounds of O combining with 1 pound of C produces 8080 units of heat in burning to CO_2.

If the weights and specific heats of the products of combustion are known, it is possible to compute the maximum possible temperature developed in an explosive reaction.

The specific heats of gases which constitute the chief products of combustion in explosions vary with the temperature and pressure, and their values at high temperatures and pressures are not known accurately.

Temperature and pressure are both dependent on the space in which the reaction takes place. In a restricted space, like the chamber of a gun, both the temperature and pressure rise very high, and as a result of the high temperature the phenomenon of *dissociation* may occur; that is, the elements may separate by a physical process due to the weakening of the molecular bonds by the action of the high heat.[1] The effect of this "dissociation" is to reduce temperature. The motion of the projectile in the gun, enlarging the volume, will reduce pressure and temperature. The result is, at a certain stage of the lowering of both temperature and pressure, chemical combination again takes place and the heat due to this tends

[1] Oxygen and hydrogen at atmospheric pressure separate at 2800° C.

to increase the pressure. The phenomenon of dissociation occurs in the first instants of explosions; the phenomenon of recombination in the later instants.

Heat thus may be the cause of directly resolving a substance into its component parts; if the body is reformed upon the lowering of the temperature, the phenomenon is *dissociation;* if not reformed, it is *decomposition.*

When a body is disintegrated by heat in a confined space, some of the products being gaseous, the disintegration proceeds until the gas or vapor liberated has attained a certain pressure, greater or less according to the temperature. No further disintegration then takes place, nor will the separated elements combine so long as that particular temperature and pressure are maintained. If the temperature be raised, disintegration will be resumed until some higher limiting pressure is produced; if the temperature be lowered, combination will take place until a certain lower pressure is attained; if the temperature remain constant and the pressure be increased, combination will take place; if lowered, disintegration. The amount of dissociation is definite in all cases for the same substance and the same condition of temperature and pressure.

This action is not limited to compound substances, but is believed to take place with the molecules of the elements; that is, the molecules of multi-atom molecules may be dissociated by heat into their separate atoms.

Detonation. (Explosion of High Order.)

The second class of explosion is of a different nature. The explosive reaction is not confined to the surfaces exposed, but appears to progress in all directions throughout the mass, radially, from the point of initial explosion; it appears to pass from the molecules at the initial point to those adjacent, and from these to the next adjacent, and so on, throughout the body, at a very rapid rate. Apparently the atomic bonds of the initial molecules are disrupted by the molecular energy or blow at the initial point; this breaking up of the initial molecules is transmitted by a wave-like action known as *the explosive wave*, extending throughout the body, the initial disruptive energy being transmitted from molecule to molecule, and these, in succession, giving way, the nascent atoms thereof combining according to the newly existing affinities which yield mostly gaseous substances.

The effect is to transform the explosive in an almost inappreciably brief time from the solid or liquid state to the gaseous state, the gases being greatly increased in volume and pressure by the heat of combination attending the reaction. It has been determined experimentally that the velocity of propagation of the explosive wave throughout a mass of guncotton is from 17,000 to 21,000 feet per second.

The calorific value and calorific intensity of disruptive explosives may be determined as explained for progressive explosives, the combination between oxygen and carbon and hydrogen having the same heat value regardless of the form of explosion.

The phenomena of dissociation and combination may take place in the products of this type of explosion, also, giving rise to a more prolonged explosive blow than in the case of the explosion of fulminates.

Fulmination.

This class of explosion is still more brusque than the last. It is like the last in that the initial molecule is broken up by the crushing effect of the blow due to the exciting cause, and the molecular energy thus applied is transmitted by the disruption of the first molecules to those adjacent, and these to the next, and so on throughout the mass.

The characteristic feature of this form of explosion is the *absence of dissociation*. The gases are evolved in such a simple form that there is little or no dissociation and the new affinities do not invite chemical combination. The explosive blow is thus not prolonged by these phenomena, and is therefore relatively much sharper than in the last class.

The heat of the first phase of the explosion is also very great, tending in itself to increase the sharpness and energy of the blow on the initial molecules.

A brusque explosive blow such as described is thought to have the effect of breaking up the molecular bonds of explosive molecules, and thereby initiating an explosive wave throughout the mass of the explosive. With progressive powders, it would be effective in initiating the explosion, but there would not, in ordinary cases, be an explosive wave. It is this property of initiating detonation and explosion which gives rise to the use of the detonators or exploders. They are used in caps and primers of all kinds; the abruptness of their explosion, and the consequent sharpness of the blow and the concentration of heat on the point of ignition, constituting their efficiency as originators of explosions of the first two classes.

In all cases, explosions are attended by a sudden and large increase of volume of the substances which constitute the explosive. Generally there is also evolution of heat; always so when due to chemical reaction in the first phase of the explosion, and recombination after dissociation in the later phase.

An explosion due to physical causes alone, as when compressed air is released, causes cold; the firing of the pneumatic gun produces so much cold as to cause the condensation of the water-vapor in the air of the charge as it leaves the muzzle of the gun.

IV.

PROGRESSIVE EXPLOSIVES.

Progressive explosives may be considered under the two headings:

1. Charcoal powders.
2. Nitrocellulose powders.

CHARCOAL POWDERS.

These may be divided into:

1. Black charcoal powder, or black powder.
2. Brown charcoal powder, or brown powder.

Brown charcoal powder is now obsolete. Black charcoal powder is used chiefly for saluting purposes and as an igniter for nitrocellulose powders, also in fuses and as mealed powder in primers.

Black Powder.

In the manufacture of black powder, fully charred black charcoal is used. The wood is charred at about 350° C.

Charcoal charred at this temperature contains about 76 parts by weight of pure carbon, 4 parts of hydrogen, 19 parts of oxygen, and 1 part of ash.

The ingredients of black powder are, besides pulverized charcoal, pulverized sulphur and pulverized nitre. The proportion in which these ingredients are mixed is about as follows:

75 parts by weight of nitre;
15 " " " " charcoal;
10 " " " " sulphur.

Variations from these proportions occur in different countries, but the differences are insignificant.

The ingredients are purified as a preliminary step. They each are then pulverized by grinding.

The charcoal is ground in a machine resembling a large coffee-mill. It consists essentially of a vertical metal cone, having teeth placed spirally on its surface. This revolves within a vertical cylinder, having teeth projecting inwardly and arranged spirally, inclining in the opposite direction to that of the teeth on the cone. These teeth are susceptible of adjustment, so that the clearance between the two sets may be increased or decreased. By this means the degree of fineness of the ground charcoal is regulated.

The sulphur and nitre are ground in a machine resembling a mortar-mill. It consists of a pair of circular edge-rollers, travelling around a strong, circular cast-iron bed, revolving at the same time on their axes. The rollers are about 4 feet in diameter and weigh about 3000 pounds. They are placed at different distances from the centre of motion, so that each passes over the cast-iron bed on a separate path, one being just inside of the other. The two rollers have a common horizontal shaft about which they turn. At a point on this horizontal shaft, nearer one roller than the other, is a vertical spindle, which is geared below to the driving-train of machinery, so as to give a motion to both rollers about this spindle. The nitre or sulphur is spread evenly over the bed, about 1 to 2 inches thick, and motion given to the rollers. They move over the material, and in a few minutes it is reduced to a fine powder. A scraper follows behind each roller, and is so formed as to throw the material under the next following roller.

After grinding the charcoal, nitre, and sulphur, each is passed through a separate *sifting-reel.* This sifting-reel consists of a frame cylinder covered with wire cloth, 32 meshes to the inch. The ground materials pass through the interior of this reel, which revolves slowly. The fine particles suitable for powders pass through the meshes and fall into a bin. The coarser particles pass through the reel, are received in a barrel at the lower end, and are taken back to the grinding-mill for regrinding.

The sifted materials are weighed out very carefully in 50-pound lots, in the relative proportions given above (75 parts of nitre, 15 parts of charcoal, and 10 parts of sulphur), and placed in bags. The contents of three bags constitute a charge for the *mixing-machine*.

The mixing-machine consists of a copper drum mounted on a horizontal shaft. The drum has a capacity of about 150 lbs. of the mixed materials. It revolves at about 35 revolutions per minute. The shaft of the drum is hollow, and through this passes a second shaft, which carries a series of arms or "flyers" on the interior of the drum. These arms are flat, with forked ends, and just clear the interior surface of the drum. They revolve in the opposite direction to the drum at about 70 revolutions per minute.

Three bags of the ingredients are emptied into the drum, the machine set in motion, and the mixing is completed in five minutes.

The *mixed ingredients* are allowed to fall through a chute into a tub, carefully examined to see that the mixing is regular, placed in bags, and tied very compactly. These bags are laid on their sides to prevent, in so far as possible, the tendency of the ingredients to separate in layers according to their specific gravities; when necessary to handle the bags, it should be done carefully and without jarring or shaking, for the same reason.

The mixed ingredients are next taken to the *incorporating mill*, to be put through the process of *incorporation*. This is the most important process in the manufacture of charcoal powder. Its object is to bring the ingredients into the closest possible contact, so that each particle of the resulting *cake* shall be composed of the three ingredients in proper proportion.

The incorporating mill is of the edge-roller type, like the sulphur and nitre grinding-mills, except more massive; the rollers are about 6.5 feet in diameter, 15 inches wide, and weigh about 4 tons each.

The mixed ingredients from the mixing-machine are spread evenly over the bed of the incorporating mill; it should not be thicker than 0.5 inch nor less than 0.25 inch: if thicker than 0.5 inch, the incorporation is defective; if less than 0.25 inch thick, there is danger of explosion.

After the charge has been spread over the bed, it is moistened with from 4 to 8 pints of distilled water, depending on the state of the atmosphere. Greatest care must be exercised by the attendant in regulating the water, as the nature of the product depends very much on uniformity in the amount of moisture present.

It requires from 3 to 4 hours to incorporate a charge. The incorporated mass is called *mill-cake.* It should have a uniform blackish-gray color, without any white or yellow specks. A small amount of it flashed on a plate should burn smoothly, at the proper rate, and give little residue.

The incorporated powder, in the form of soft mill-cake, is put into open tubs and placed in small magazines, where it is exposed to the action of the air, so that all workings may either absorb or give off water-vapor and come to about the same percentage of hygroscopic water present; 2 to 3 per cent of water in the mass is necessary to give good results in the subsequent pressing.

In so far as the chemical requirements for combustion are concerned, the powder is now completed. The subsequent operations have for their object the production of certain physical effects, depending upon the use to which the powder is to be put. In order that its rate of burning may be regulated, the *size* and *density* and *form* of the grains must be fixed.

Before being *pressed,* the mill-cake is broken into lumps of uniform size, in a machine called the *breaking-down machine.* This machine consists essentially of two pairs of grooved cylinders arranged one pair above the other. These cylinders are so placed on shafts, and are so geared, that they have motions downward between each pair. The clearance between the cylinders, and the dimensions of the grooves, are adjusted to the nature of the cake, and, for safety purposes, the clearance may be automatically increased by the action of a sliding-bearing of one of each set of cylinders, which allows this cylinder to move back in case the cake is fed to the rollers too rapidly, or a hard lump happens to pass through. The hopper of the breaking-down machine takes about 700 lbs. of mill-cake. It is open below, resting on a continuous canvas belt with cleats, which, as it moves, feeds the mill-cake to the top set of cylinders. After passing through these, the cake falls between the second set of cylinders and then into suitable box-cars or trucks. It requires about a half-hour to break down a charge of 700 pounds. The product of the breaking-down machine is called *powder-meal*. It is stored again for several days, so as to equalize the moisture, and is then ready for *pressing*.

In order that the powder may be granulated, it is first pressed into solid compact cakes, called the *press-cakes*. These are formed by hydraulic pressure, applied to powder-meal placed between gun-metal plates, in a large, strong, gun-metal box. The press-box is laid on its side, and the upper side removed; the metal separating-plates are inserted; the meal is filled in between the plates, the space between plates being about $\frac{5}{8}$ of an inch: the box is then placed under the head of an hydraulic ram and pressure applied. The plates are free to move under the applied pressure and compress the powder-meal to a hard, compact cake.

The press-cake is broken up into *grains* by passing through the granulating-machine. This consists of a series of pairs of gun-metal cylinders, with teeth of suitable size and suitably placed on the surfaces of the cylinders. The press-cake passes between these rollers, and is broken into grains of various sizes. There is a screen under each pair of rollers, to catch the broken press-cake and to conduct it to the next set of rollers. The sizes of the breaking-down teeth, and of the screen-meshes, are altered to suit the special requirements of any particular granulation that may be desired.

The sharp corners of grains are worn off and the dust separated from any grade of grained powder by the *dusting-machine*. This consists of horizontal cylindrical frames, covered with canvas, having 24 meshes to the inch. Several barrels of foul grain are put in the cylinders, and the latter set to revolving at about 40 revolutions per minute. In about half an hour the process is completed, the powder dust having passed through the meshes of the canvas.

At the end of the process, the powder is collected in barrels. Sometimes it is necessary to repeat the dusting once or twice before the powder is sufficiently free of dust.

Some powders are *glazed*. The grains are put into a horizontal, barrel-like receptacle, and revolved for 5 to 6 hours with a small quantity of pulverized graphite The object of this is, to make the powder less liable to form dust in storage and transportation, and to protect the grains to some extent from the effects of moisture in the air.

The final operation is to remove excess of moisture from the powder by *drying*. The powder is spread out over shallow canvas-bottom frames, arranged in tiers, over a steam radiator, and is subjected to a temperature of 130° F. for 16 to 18 hours. After standing for 2 to 3 hours to allow it to cool, it is run through a dusting-reel and then packed.

Usually black powder is packed in 100-pound packages. The receptacles are, as a rule, either wooden barrels or metal canisters. If wooden barrels are used, the wood is oak and the hoops are made of some wood, like cedar, not liable to become worm-eaten. Zinc-lined boxes have also been used. These boxes are often arranged to be hermetically sealed, or are provided with gasket covers, to protect the powder in storage from the moisture of the air.

Black gunpowder should be of even granulation, of good hardness and density, free from dust. A small quantity poured on the back of the hand should leave little or no trace of dust; when flashed in 10-grain samples on a copper plate, there should be no bead or excessive residue. It should absorb little water from the air.

Brown Powder.

The foregoing description applies in its essential features to the manufacture of all mixtures of nitre, charcoal, and sulphur. In *brown powder*, the charcoal is made from rye-straw and is under-charred. The proportions of the ingredients vary some from those given for black powder, the proportions for brown powder being,[1] approximately:

Nitre	80 parts
Charcoal	16 "
Sulphur	3 "
Moisture	1 part
	100

This mixture is slower burning than black powder. It has been discontinued as a service powder by the United States.

Granulation of Powder.

The introduction of large-grained, perforated, slow-burning brown prismatic powder marked the last phase of a long line of investigation, begun in the early sixties in the United States by the late General T. J. Rodman, Ordnance Department, U. S. Army. Some reference may well be made here to those series of experiments which, initiated by Rodman, were taken up and extended by many others, both in the United States and in Europe, especially as the principles established thereby still survive and apply to the new powders.

Rodman sought to increase the powers and endurance of large guns by controlling the combustion of powders used in them. He conceived the idea that there were certain definite relations among the elements, size, form, and density, that would give a *best powder* for a given gun, that is, a powder which would give the highest velocity for a given pressure; or, stated in other words, a *special powder* could be determined for

[1] Captain Robinson reports the following analyses at the School of Submarine Defense, Fort Totten, N. Y.:

	English.	*German.*
Nitre	79 parts	77 parts
Charcoal	18 "	20 "
Sulphur	3 "	3 "
	100	100

each piece of ordnance, and the idea came to be known as the *principle of special powders.* In carrying out this idea, he experimented with powder having much larger grains than had been used prior to his time, and with powders of varying density and forms, including those subsequently known as "mammoth," "pebble," lenticular, perforated prismatic, and perforated cylindrical cake-powders.

The Civil War put a stop to Rodman's experiments, and, after the war, although he desired to continue them, he was, for some reason difficult to appreciate, ordered to a post of duty where it was impossible to give any attention to the matter.

Knowledge of his work had, however, become known abroad, and the line of investigation was taken up there, resulting, after a time, in the adoption of the perforated prism as the standard form of grain for large guns.

The fundamental idea involved in this development may be said to be, to so control the combustion of a charge of powder in a given gun that there shall be a certain uniformly progressive evolution of gas, so that the projectile will be started from rest under a minimum pressure, with the quantity of gas evolved in consecutive instants of time, gradually increasing until the projectile reaches a certain point in the bore of the gun. The pressure in the gun increases to a maximum soon after the projectile is started, and then falls regularly: the velocity increases to a maximum at a point just beyond the muzzle.

The first step was to gain slow combustion through increasing the density and enlarging the size of the grain; the result of this was evidenced in the old "mammoth" powder that was used in the 15-inch smooth-bore Rodman guns.

The next was, while holding to the above principles, to control the rate of evolution of gas in burning a grain of powder *by perforating it,* and to have thereby a certain portion (from the interior outward) of the grain burn on *increasing surfaces,* giving *for this portion* increasing quantities of gas in succeeding intervals of time.

This same effect was later obtained in another way by the so-called *Fossano Powder,* made in Italy. The powder-grain was in itself a conglomerate of smaller grains bound together by a suitable powder matrix, the whole being compacted into large grains by pressure. As the large grain burned, it was broken up, exposing the surfaces of the smaller grains, and in this way offering successively increasing surfaces for ignition and burning, and, consequently, increasing quantities of gas.

The next was, to establish uniformity in time of burning of each grain by moulding the grains, as in the hexagonal and sphero-hexagonal powders, and, in connection with the pressure applied in forming these moulded powders, to produce a higher density of grain on the surface than in the interior of each grain, illustrating the principle of varying density of grain.

The perforated prism gave, however, the best results, and the right hexagonal prism came in time to be the standard form of grain for large guns the world over. Variation existed in the number of perforations, some prisms having but one perforation, others seven, one opposite each angle and one in the centre. The last-named is thought to be the arrangement generally adopted. That portion of a prismatic grain between the perforations is called *the web* of the grain; its thickness is the determining factor in the time of combustion.

In determining the "special powder" required for a given gun, the density and granulation (number of grains to the pound) of hexagonal and sphero-hexagonal are the data to be fixed by computation or experiment; if prismatic powder is to be used, the dimension of the prism and the number and size of the perforations must be determined.

As the ability of the powder-manufacturers to make slow-burning powders developed, the maximum pressures in the rear portion of the bores of guns fell, but the pressures in front of the trunnions was increased. At the same time, the gun-makers were able to increase the strength of the built-up gun, so as to make it possible for the gun to bear slightly higher pressure: The improvement in gun-making also made it possible *to increase the lengths of bores*; this, in turn, made it possible to burn more powder in the guns and thereby increase velocity. To receive these larger charges, and also to further control the powder-pressure over the charge, enlarged powder *chambers* were introduced. By properly adjusting the relations the volume of the powder chamber, the weight of charge and of the length of the bore, the pressure corresponding to a given velocity could be kept within the limit of the gun's elastic strength.

NITROCELLULOSE POWDERS.

Nitrocellulose may be considered as the base of all forms of smokeless powders.

The nitrocellulose molecule contains within itself the elements carbon, hydrogen, and oxygen, so that, when conditions favorable to a disruption of existing molecular bonds and to a recombination of these elements are produced, the reactions of combustion take place, producing the gaseous oxides of carbon and water-vapor. In the case of charcoal powders, these elements were brought by *mechanical* process into such intimate relations that each particle of black or brown powder should contain the elements necessary for combustion. The two classes of explosives have, therefore, a fundamental difference in this respect. In nitrocellulose, the elements to produce combination are present in the molecule in accordance with the law of fixed proportions, in great purity and in closer relations than is possible with a mechanical mixture, like charcoal powder.

It will be remembered that the structural formula of cellulose was written (p. 82) to show its analogy to the alcohols, thus:

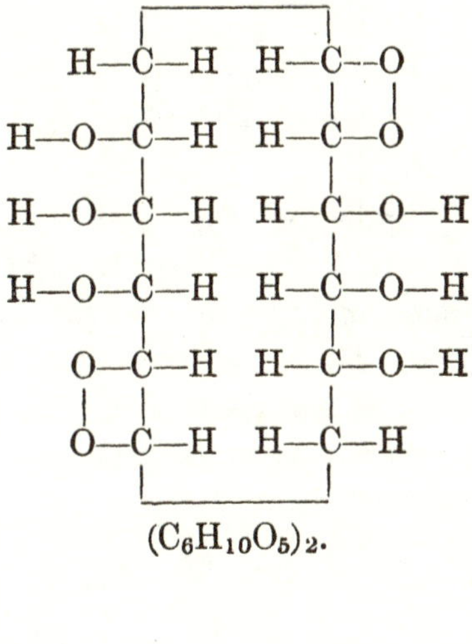

$(C_6H_{10}O_5)_2$.

It will be recalled, also, that ethers may be considered to be formed from the alcohols by substituting a suitable hydrocarbon radical for the hydrogen of the hydroxyl radical of alcohol (p. 81).

Thus, ethyl ether is derived from ethyl alcohol by substituting the ethyl radical for the hydrogen of the hydroxyl of the alcohol:

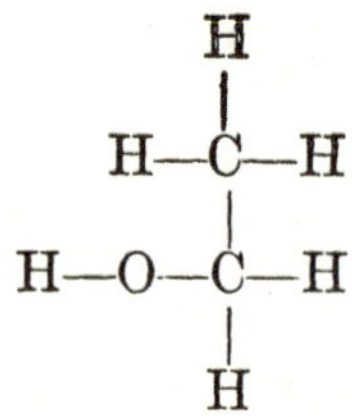

Ethyl alcohol [$C_2H_5(HO)$]

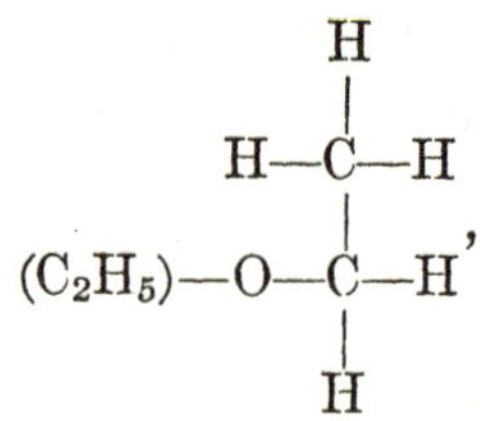

Ethyl ether [$(C_2H_5)_2O$]

In the same way nitric ether may be produced from alcohol by the action of nitric acid on alcohol, the radical nitryl, NO_2, displacing the hydroxyl hydrogen atom and giving:

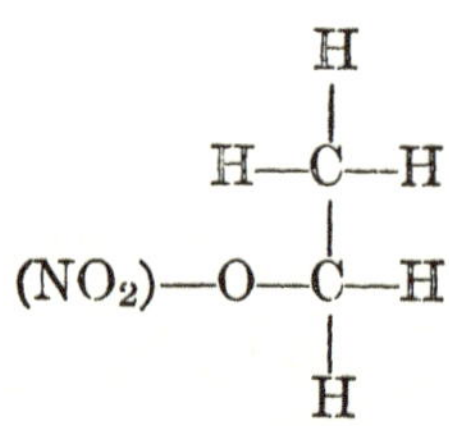

Nitric ether ($C_2H_5.O.NO_2$)

In like manner the hydrogen of the hydroxyl radicals of the cellulose molecule may be displaced by NO_2 by the action of nitric acid, giving substances which in molecular structure resemble nitric ethers. There are three hydroxyl groups in the cellulose molecule that are susceptible of this substitution; there may, therefore, be three separate displacements, as follows, using the double grouping:

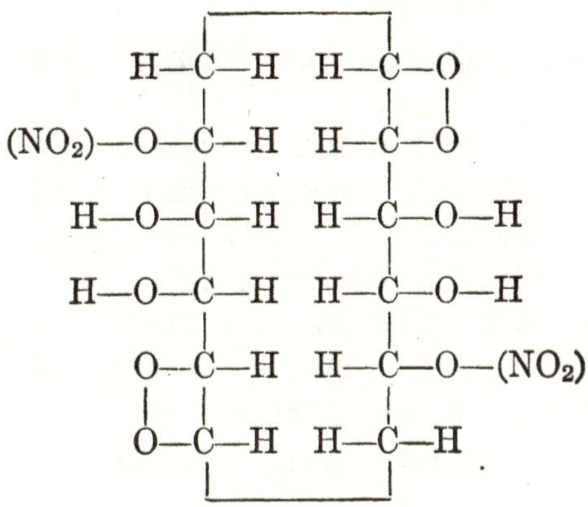

Mononitrocellulose

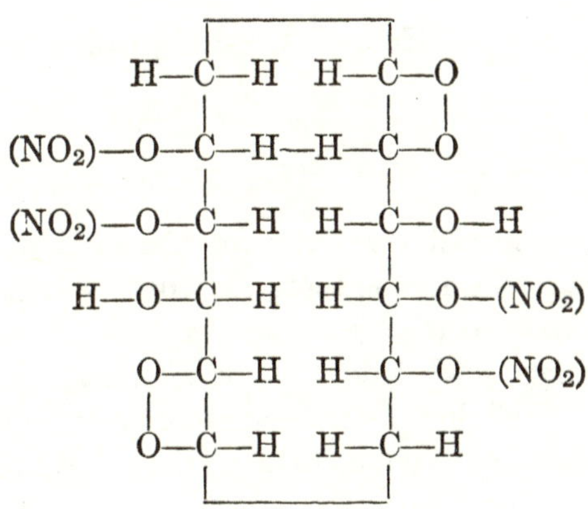

Dinitrocellulose

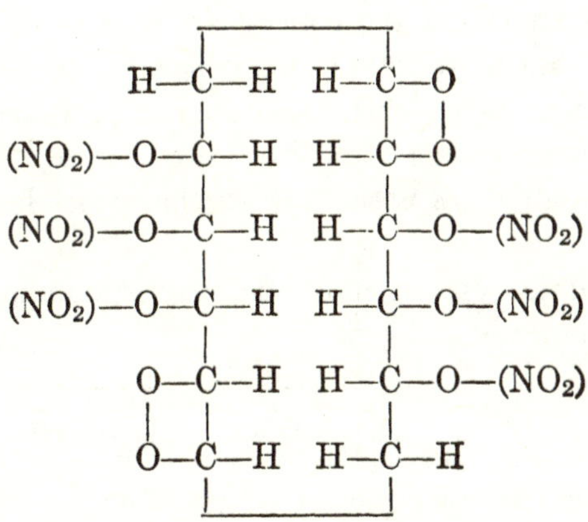

Trinitrocellulose

On account of this and other chemical analogies nitrocellulose is generally classed as a compound nitric ether of the trihydric alcohol, cellulose.

The nitryl radicals which are transferred when cellulose is acted on by nitric acid introduce weak molecular bonds, which give way under the action of heat and permit the elements to combine with great energy, according to their relative affinities for each other, and it is this feature particularly, which constitutes nitrocellulose an explosive. The result of the breaking-up of the trinitrocellulose molecule in explosion may be represented by the following reaction:

$$[C_6H_7.O_2.O_3(NO_2)_3]_2 \text{ exploded} = 7H_2O + 9CO + 3CO_2 + 3N_2.$$

The Nitration of Cellulose.

For military explosives, the cellulose used for nitration is, as a rule, the waste from cotton-spinning factories, cotton-cloth factories, or other forms of pure cotton fibre.

Within the past few years much attention has been given to the subject of nitration of cellulose by several eminent investigators and scientists, including Vieille, Bruley, Lunge, Will, and others.[1]

In 1878 Dr. J. M. Eder arrived at the conclusion, as a result of a series of experiments, that there were as many as six degrees of nitration of cellulose, three of which he was able to produce and isolate, namely, the *hexa-*, *penta-*, and *di-*; two, the *tetra-* and *tri-*, he obtained in admixture with others; the *mono-* he was unable to prepare.

Eder assumed the double type of molecule, corresponding to C_{12}, and wrote the formulas as follows:

[1] "Nitration of Cotton," by M. Bruley. "Researches upon the Nitration of Cotton," by M. Vieille. "Investigations as to the Stability of Nitrocellulose," by Dr. W. Will. G. Lunge's experiments in nitrating cellulose.

Mono-nitrocellulose	$C_{12}H_{19}O_9(NO_3)$
Di- "	$C_{12}H_{18}O_8(NO_3)_2$
Tri- "	$C_{12}H_{17}O_7(NO_3)_3$
Tetra- "	$C_{12}H_{16}O_6(NO_3)_4$
Penta- "	$C_{12}H_{15}O_5(NO_3)_5$
Hexa- "	$C_{12}H_{14}O_4(NO_3)_6$

Vieille, as a result of extended research made in 1883, arrived at the conclusion that, in order to account for the amount of NO_2 given by the products of his experiments, the formula $C_6H_{10}O_5$ must be quadrupled, and the molecular formula of cellulose written $C_{24}H_{40}O_{20}$; giving rise to eight varieties of nitrocellulose, as follows:

Cellulose tetra-nitrate	$C_{24}H_{36}O_{20}(NO_2)_4$
" penta- "	$C_{24}H_{35}O_{20}(NO_2)_5$
" hexa- "	$C_{24}H_{34}O_{20}(NO_2)_6$
" hepta- "	$C_{24}H_{33}O_{20}(NO_2)_7$
" octo- "	$C_{24}H_{32}O_{20}(NO_2)_8$
" ennea- "	$C_{24}H_{31}O_{20}(NO_2)_9$
" deca- "	$C_{24}H_{30}O_{20}(NO_2)_{10}$
" endeca- "	$C_{24}H_{29}O_{20}(NO_2)_{11}$

Of these the *deca-* and *endeca-* varieties were found to be insoluble in ether-alcohol; the *ennea-* and *octo-* were soluble and capable of being colloided; the lower nitrations gave friable products insoluble in ether-alcohol.

In Vieille's researches the present military smokeless powder may be said to have had its origin. Soon after his deductions were announced, the manufacture of smokeless powder in France was begun. The French powder was kept a secret for some time. The success of the French inaugurated activity throughout Europe, and, before long, the nitrocellulose base came to be the essential ingredient of all smokeless powders.

In Russia the development of a smokeless powder was intrusted to the celebrated chemist, Professor D. Mendeléef. His investigations resulted in the claim that he had been able to produce a definite nitrocellulose having the formula $C_{30}H_{38}O_{25}(NO_2)_{12}$, which he called "pyrocollodion," which colloided perfectly in ether-alcohol, and in combustion gave the maximum volume of gas possible from the elements represented in the molecule, since the content of oxygen, as given in the formula, is just sufficient to burn all of the C to CO, after burning the H to H_2O; the explosive reaction being as follows:

$$C_{30}H_{38}O_{25}(NO_2)_{12} \text{ exploded} = 30CO + 19H_2O + 6N_2.$$

Mendeléef's claim that his pyrocollodion is a definite compound is disputed. It is claimed by others that the substance is, rather, a mixture of nitrates of different degrees of nitration, such, for example, as the following:

$$\begin{aligned} 2[C_6H_7O_5(NO_2)_3] &= C_{12}H_{14}O_{10}(NO_2)_6 \\ 3[C_6H_8O_5(NO_2)_2] &= C_{18}H_{24}O_{15}(NO_2)_6 \\ \hline & \quad C_{30}H_{38}O_{25}(NO_2)_{12} \end{aligned}$$

Pyrocollodion, according to Mendeléef, results from the following reaction:

$$5C_6H_{10}O_5 + 12HNO_3 = C_{30}H_{38}(NO_2)_{12}O_{25} + 12H_2O.$$

Perhaps the most complete series of experiments made in connection with the nitration of cellulose are those made by the French Government chemist, M. Bruley, published in the *Memorial des Poudres et Salpêtres,* vol. viii, 1895–96, in a paper entitled "Sur la Fabrication des Cotons Nitrés," an English translation of which is to be found in Bernadou's "Smokeless Powder, Nitrocellulose, and Theory of the Cellulose Molecule."

M. Bruley points out that of recent years the various grades of nitrocellulose have given rise to many varied uses, such as photographic films, celluloid, mercerized cotton, in the mechanical arts; and guncotton and smokeless powder in military explosives. Each of these requires a special variety of nitrocellulose, and it becomes important, if possible, to fix the conditions which regulate the nature of the product.

For many years military guncotton had been manufactured from the standard mixture of acids, three parts of sulphuric acid by weight (65.5° Baumé) and one part by weight of nitric acid (48° Baumé). But, as a result, chiefly of Vieille's investigations, his classification of the nitrocelluloses and the manufacture of smokeless powders based on his deductions, it became desirable to determine some practical rules and guides for the manufacture of the new nitrocelluloses of lower nitration.

In the ordinary manufacture of nitrocellulose, the product is apt to contain a mixture of the three classes of nitrocelluloses, guncottons (endeca- and deca-nitrates), collodions (ennea-, octo-, and hepta-nitrates), and friable cottons (penta- and tetra-nitrates). The first of these is insoluble in ether-alcohol, the second is soluble in that mixture, the third not soluble.

The experiments of M. Bruley had for their object, therefore, the determination of some practical method of obtaining a certain desired product in the nitration of cellulose.

His experiments may be well explained by reference to the accompanying figure.

For the purpose of graphically representing the conditions of the experiments let O represent the origin of a set of axes, OX and OY. Let OX represent the axis of the proportion by weight of water used in the mixture, and OY the axis of the proportion by weight of nitric acid used. Let $OX' = OY'$ represent the fixed quantity of sulphuric acid used. Let OX'' represent a certain quantity of water used in a particular experiment, and OY'' represent a certain quantity of nitric acid used in the same experiment. Express OX'' as a percentage of OX' and OY'' as a percentage of OY'; that is, if OX' and $OY' = 100\%$ weight (the fixed weight of the sulphuric acid), $\frac{OX''}{OX'}$, carried out to hundredths in decimal form, will represent the percentage quantity by weight of water used in terms

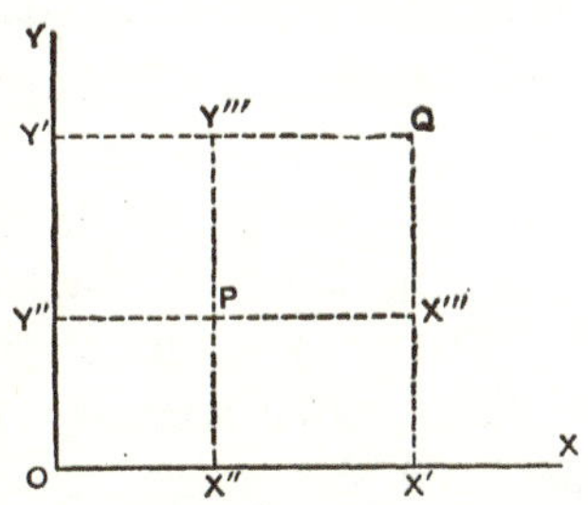

of the fixed quantity of sulphuric acid used, and, similarly, $\frac{OY''}{OY'}$ will represent the percentage quantity by weight of nitric acid used in terms of the fixed quantity of sulphuric acid. The line $X''Y'''$ represents the locus of all products, corresponding to the ratio $\frac{OX''}{OX'}$ between water and sulphuric acid.

The line $Y''X'''$ similarly represents the locus of all nitric-acid mixtures, corresponding to the ratio $\frac{OY''}{OY'}$ between nitric acid and sulphuric acid. The point P corresponds to a definite mixture of OX'' parts of water, OY'' parts of nitric acid, and $OX'=OY'$ parts of sulphuric acid. The area $OY'QX'$ includes within it all possible combinations of mixtures of water and nitric acid with sulphuric acid, when the quantities of water and nitric acid do not, either of them, exceed the quantity of sulphuric acid used.

M. Bruley assumed twenty-five points uniformly distributed throughout this area, prepared the mixtures to correspond thereto, immersed the cellulose in these mixtures, and steeped them for 6, 12, and 24 hours, thus producing three series of nitrations. He subsequently determined, by chemical analysis and physical experiment, the following data:

1. The nitrogen content, expressed in c.c. of NO_2.
2. The solubility in ether-alcohol.
3. The viscosity in ether-alcohol.

The temperature of the immersions was 12° to 13° C.

The water normally present in both nitric and sulphuric acid was determined carefully, and considered as a part of the water ingredient of the acid mixtures. These determinations were 5 to 6 per cent in the sulphuric acid, and 10 to 15 per cent in the nitric acid.

The fixed weight of sulphuric acid taken was 1.2 kilos.

A separate mixture was made for each of the twenty-five points, corresponding to a range of nitric acid of 10 to 60 per cent; and of water, 10 to 45 per cent. The inferior limit for nitric acid being fixed by the time required for nitration, and the superior limit by that cost of the acid beyond which it would not pay to go in manufacturing nitrocellulose for the trade; the lower limit of water was fixed by the quantity of water always present in the strongest acid, the higher limit by the limit of colloidable nitrocellulose.

When the quantity of nitric acid fell below 15 per cent, the time required to nitrate completely was so prolonged that it would not be practicable, commercially, to use so low a percentage.

The samples consisted of 4 grams of bleached spun-cotton waste, and were immersed in 400 grams of mixed acids.

The table on page 118 gives the results of the experiments.

Bruley divided the products into: (1) guncotton, (2) colloids, and (3) friable cottons.

In general terms it may be said that the guncottons resulted from mixtures within the following ranges of percentages, by weight of nitric acid and water, the weight of sulphuric acid being 100 per cent:

For nitric acid,	55	per cent;	water,	12	to	24	per	cent.
" " "	15	" "	"	8	"	16	"	"

In the same way the limits of mixtures for the most perfect colloids having, say, a solubility above 90 per cent, were:

For nitric acid,	55	per cent;	water,	27	to	35	per	cent.
" " "	15	" "	"	18	"	25	"	"

A fairly high degree of solubility extended beyond these water percentages to about 40 per cent of water for 55 per cent of nitric acid, and about 27 per cent of water for 15 per cent of nitric acid.

Beyond these latter water limits the products were friable cottons.

The guncottons correspond to nitrocellulose, having a nitrogen content above about 12.9 per cent; the higher colloids, a nitrogen content less than about 12.9 per cent and more than

No. of Mixture.	Composition.			General Results as Affected by the Time of Steeping.								
				6 Hours.			12 Hours.			24 Hours.		
	H_2SO_4.	HNO_3.	H_2O.	NO_2 c.c.	Solubility, %	Viscosity, Degree in Scale.	NO_2 c.c.	Solubility, %	Viscosity, Degree in Scale.	NO_2 c.c.	Solubility, %	Viscosity, Degree in Scale.
I	The weight of H_2SO_4 used (1.2 kilos) taken as the base of comparison and constituted 100/100. 65.8° Baumé; about 5% of water.	55/100	28.8/100	196.2	99.	195	197.2	97.0	206			
II		45.9/100	24.3/100	205.0	52.9	....	205.8	52.9				
III		37.2/100	18.3/100	206.8	2.6	....	209.6	2.4				
IV		18.3/100	13.1/100	295.8	5.2	....	203.2	5.2	....	205.2	2.5	
V		9.1/100	10.9/100	122.6	19.8	....	128.2	20.1	....	134.0	14.2	
VI		9.4/100	21.3/100	143.0	19.1	....	142.8	19.4	....	147.8	21.9	
VII		18.0/100	20.8/100	183.4	89.7	52	182.0	90.9	45			
VIII		17.6/100	33.2/100	Cotton	dissolved	in acid.	Cotton	dissolved	in acid.	Cotton	dissolved	in acid.
IX		28.8/100	22.0/100	196.2	95.4	117	196.2	97.2	103			
X		37.5/100	29.3/100	184.6	96.3	86	186.2	95.8	90			
XI		54.9/100	37.8/100	164.4	87.5	56	164.6	94.4	41	172.6	95.3	45
XII		47.0/100	34.6/100	171.6	92.8	....	172.6	96.8	40	177.0	96.4	74
XIII		27.5/100	33.3/100	153.4	47.0	....	154.2	61.8	....	157.2	66.0	
XIV		37.3/100	38.0/100	144.4	16.0	....	146.2	14.0	....	146.8	10.8	
XV		47.6/100	45.4/100	121.6	1.4	....	123.6	1.5	....	142.8	1.9	
XVI		54.1/100	34.5/100	174.2	94.8	33	178.6	96.5	45	178.2	91.6	151
XVII		45.1/100	28.3/100	185.0	96.9	40	186.2	97.8	51			
XVIII		45.4/100	38.4/100	155.2	26.0	....	155.6	39.8	....	152.8	21.8	
XIX		36.5/100	23.7/100	197.6	92.2	178	199.8	91.3	122			
XX		36.4/100	33.5/100	163.4	86.4	58	163.4	81.8	59	164.8	80.8	71
XXI		27.7/100	18.4/100	207.8	5.4	....	208.8	3.3				
XXII		27.8/100	28.5/100	180.4	94.5	97	182.0	96.2	107	180.2	95.7	72
XXIII		18.3/100	18.1/100	193.8	20.0	....	201.6	21.9				
XXIV		18.1/100	28.0/100	160.4	65.1	....	162.4	65.7	....	162.6	77.4	
XXV		13.9/100	17.6/100	168.0	26.2	....	175.0	28.0	....	187.0	6.2	

about 12 per cent; the inferior colloids, a nitrogen content of from just below 12 per cent to just above 10 per cent.

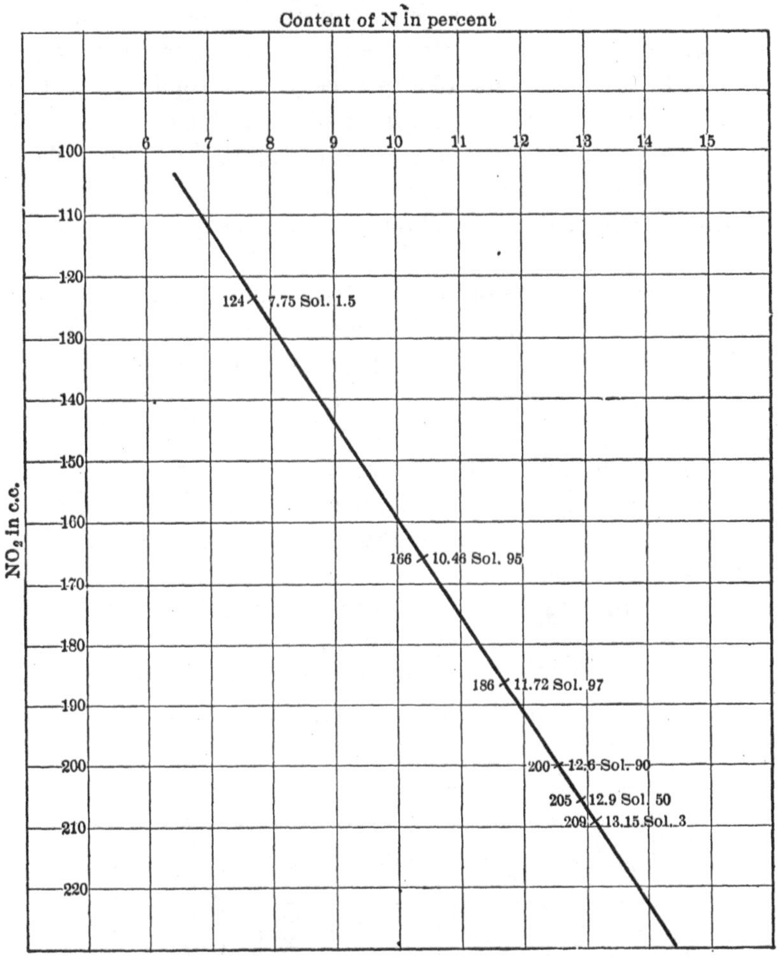

For a fixed per cent of nitric acid in a series of mixtures in which the per cent of water only varies, the nitrogen content changes slowly in passing beyond the guncotton zone as the water percentage increases, while, at the same time, the solubility changes very rapidly. A nitrogen content of about 12.5 per cent is soon reached, having a solubility of about 95 per cent, and after this has been attained considerable variation may be made in the quantity of water with little change in either the nitrogen content or in the solubility. When the increase of water for this same quantity of nitric acid causes the nitrogen content to fall to about 10.5 per cent, the solubility drops below 90. Beyond this an increase of water causes a gradual decrease in both nitrogen content and solubility to take place until the lowest recorded limit is reached; that is to say, a limit of nitrogen content of about 7.75 per cent and a solubility of about 1.5.

While the relative proportion of the ingredients of the acid mixture is the chief factor influencing the result of nitration other causes have an effect, such as (1) duration of steeping, (2) temperature of dipping and steeping, (3) subsequent steps in purification.

Cotton-wadding nitrates more readily than spun-cotton waste. Generally speaking, the more perfectly the fibres are separated and the waste freed from tangles and knots the quicker and better the nitration.

In order to obtain the same degree of nitration, the steeping should be prolonged in proportion as HNO_3 is reduced in the acid mixture. The influence of duration on the NO_2 content and solubility appears in the following table:

Duration of Steeping.	Inferior Colloid. I.		Superior Colloid. II.		Guncotton. III	
	NO_2 Content, c.c.	Solubility, per cent.	NO_2 c.c.	Solubility.	NO_2 c.c.	Solubility.
1 hr.	165.8	91.7	186.8	94.9	206.4	10.9
2 hrs.	166.8	95.5	189.	95.0	209.4	8.3
4 "	167.8	93.0	191.8	96.2	209.2	6.8
6 "	167.8	94.8	198.	94.1	210.2	6.7
8 "	166.8	95.4	191.8	96.7	210.2	5.6
12 "	.....	.....	.....	.....	210.8	7.4
24 "	166.8	98.1	194.	96.6	210.6	10.6

From which it is observed that with the colloids from 2 to 6 hours are required, and with guncotton, 8 to 10 hours. If the reaction be continued beyond 6 to 8 hours, the solubility for the same nitrogen content is materially increased.

Increase of temperature, during dipping and steeping, up to 26° C., increases both the solubility and NO_2 content of colloids, and has a tendency to the same for guncottons.

When, therefore, it is desired to produce a definite nitrocellulose it is first necessary to fix the composition of the acid mixture, following the principles set forth above, and testing the nitrogen content by the usual nitrometer method.

While the degree of nitration may be regulated by the foregoing principles, the *stability* of the product depends chiefly on the process of purification. It is found that any nitrocellulose after nitration contains certain nitro-by-products which are more or less unstable, and these are liable to spontaneous decomposition in storage; some of these nitro-products may disintegrate under comparatively low heat and often cause the condemnation of nitrocellulose which, except for their presence, is thoroughly trustworthy. Dr. W. Will, of the German Central Station for Scientific-Technical Investigation, New Babelsberg, near Berlin, has investigated this phase of the problem, and arrived at the conclusion that these nitro-by-products are produced in the nitration of cellulose, besides the nitrocellulose proper, and the nature of these by-products is such that they are not wholly soluble in cold water, and, when cold water alone is used in the purification, they are not carried off. Boiling and subsequent washing in cold water removes them, due, perhaps, to the fact that the boiling modifies the chemical nature of some of these products, rendering them soluble in cold water and, when the latter is applied after boiling, the objectionable products are removed.

Dr. Will claims that when boiling and cold washing are properly conducted, practically all of such unstable by-products are removed, and the resulting nitrocellulose proper, whatever its degree of nitration, is a safe compound and may be stored for years under normal temperatures, without change. Nitrocellulose so prepared is said by him to be in its "limit state," and such nitrocellulose, if subjected to a higher heat, say 135° C., as in the German heat-test, will evolve equal volumes of N in equal times; this Time-Nitrogen relation, when plotted, approximates closely to a right line for the limit state.

Nomenclature of Nitrocelluloses.

There are various products resulting from the nitration of cellulose to different degrees and under different conditions. These may be enumerated as follows, following the nomenclature given by Bernadou:

Nitrocellulose. A general term applied to products resulting from the action of nitric acid on cellulose, in which the organic cellular structure of the original cotton fibre has not been destroyed.

Nitrocellulose of high nitration.[1] Those in which the content of nitrogen is large, say 12.9% or greater.

Nitrocellulose of mean nitration.[1] Those in which the content of nitrogen is mean, say less than 12.9% and greater than about 11%.

Nitrocellulose of *low nitration.*[1] Those in which the content of nitrogen is less than about 11%.

Insoluble nitrocelluloses. Those insoluble in ether-alcohol mixture[2] at ordinary temperature and pressure.

Soluble nitrocelluloses. Those soluble in ether-alcohol mixture at ordinary temperature and pressure.

Hydrocellulose. The product obtained by acting on cellulose with the fumes of HCl, or by immersing cellulose in HCl, H_2SO_4, or very dilute HNO_3. It is a white, pulverulent mass which, examined under the microscope, shows that the cellular tissue of the original cotton fibre has been modified.

Nitrohydrocellulose. The product resulting by acting on hydrocellulose with HNO_3 (strong), the product still retaining the modified cellular form of the hydrocellulose.

Nitrohydrocellulose of high nitration. Contains relatively a high per cent of N.

Nitrohydrocellulose of mean nitration. Contains relatively a mean per cent of N.

Nitrohydrocellulose of low nitration. Contains relatively a low per cent of N.

Insoluble nitrohydrocellulose. Those insoluble in ether-alcohol at ordinary temperature and pressure.

---

[1] See Vieille's Classification of Nitrocelluloses (table), p. 121.

[2] In the proportion of 2 parts by volume of ether to 1 part by volume of alcohol.

Soluble nitrohydrocellulose. Those soluble in ether-alcohol at ordinary temperature and pressure.

Guncotton. Those nitrocelluloses of high nitration used for disruptive purposes in war. They consist, as a rule, of a mixture of insoluble nitrocellulose with a small quantity of soluble nitrocellulose and a very small quantity of unnitrated cellulose.

Pyrocellulose. A soluble nitrocellulose of so called definite percentage of N(12.4), corresponding to the molecular formula, $C_{30}H_{38}(NO_2)_{12}O_{25}$, claimed to have been produced by Mendeléef; it possesses just sufficient content of O to burn all of the C to CO, the H to H_2O.

Colloid, or collodion nitrocellulose. Nitrocellulose that may be colloided in ether-alcohol.

VIEILLE'S CLASSIFICATION OF NITROCELLULOSES.

Molecular Formula.	Designation.	Theoretical c.c. of NO_2.	Experimental c.c. of NO_2.	Theoretical per cent of N.	Remarks.	
$C_{24}H_{36}O_{20}(NO_2)_4$	Tetra-nitrocelluose	108	109	6.76		
$C_{24}H_{35}O_{20}(NO_2)_5$	Penta-nitrocellulose	128	132	8.02		
$C_{24}H_{34}O_{20}(NO_2)_6$	Hexa-nitrocellulose	146	143	9.15	Only slightly attacked by acetic ether and ether-alcohol.	Friable cottons.
$C_{24}H_{33}O_{20}(NO_2)_7$	Hepta-nitrocellulose	162	164	10.18	Becomes gelatinous in acetic ether and ether-alcohol.	
$C_{24}H_{32}O_{20}(NO_2)_8$	Octo-nitrocellulose	178	182	11.11	Soluble in ether-alcohol.	Inferior colloid.
$C_{24}H_{31}O_{20}(NO_2)_9$	Ennea-nitrocellulose	190	192	11.96	Highly soluble in ether-alcohol.	Superior colloid.
$C_{24}H_{30}O_{20}(NO_2)_{10}$	Deca-nitrocellulose	203	205	12.75		
$C_{24}H_{29}O_{20}(NO_2)_{11}$	Endeca-nitrocellulose	214	215	13.47	Insoluble in ether-alcohol. Soluble in acetone.	Gun-cotton.

According to Guttmann Vieille's formulas are not beyond question. Guttman himself claims to have made guncotton on a large scale, containing 13.65 per cent of nitrogen, which, according to Vieille, would be impossible.

Colloidization.

After cellulose has been dipped in nitric acid ("nitrated") and "purified" of the free acid and nitro-by-products by boiling and washing in water, it possesses a property it did not have before, namely, it is soluble in certain liquids in which it was not soluble as cellulose. The two most important of these liquids are *acetone* and a mixture of *ether* and *alcohol*, in the proportion by volume of 2 to 1.

If an excess of the liquid be used a true solution is formed, and if the liquid be evaporated off, the nitrocellulose will remain as a horn-like compact mass, called "colloid," in which all evidences of cellular structure have disappeared. If the quantity of solvent be reduced sufficiently, the solid nitrocellulose will soften and take the form of a paste-like mass, one of the states passed through from the true solution to the compact, horn-like solid in evaporating the solvent.

This process of dissolving nitrocellulose and producing the colloid form of it is called colloidization.

In connection with nitro-explosives there are two important series of colloids: one, the acetone series; the other, the ether-alcohol series.

Acetone dissolves the nitrocelluloses of highest nitration, and gives colloids which are characterized by brittleness. Under pressure or shock they break up. This fact renders such colloids dangerous when used alone for powder; the shocks due to handling and the pressure in the bore of a gun would cause grains to be disintegrated, the rate of combustion to be enormously increased, and excessive pressures.

The ether-alcohol colloids, on the other hand, are tough and elastic. It is from this class of colloids that most smokeless powders now in use are made.

The several physical states of the two series of colloids, as a result of evaporation from the solution, may be described as follows:

Acetone series: Liquid, slime, plastic mass, brittle colloid.

Ether-alcohol series: Liquid, jelly, elastic mass, tough colloid.

Manufacture of Smokeless Powder.

While there are minor differences in the manufacture of smokeless powder as conducted at the different factories, the essential steps are the same, and are performed in practically the same manner. The following description of the commercial method of manufacture gives these steps in sufficient detail.

1. Cleaning.

(a) *Washing-house.* The *base*,[1] in the form of cotton-waste or cotton rags, is brought to the washing-house in large bales. These are broken open and the waste put into the

(b) *The Washer.* *washer*. This consists of a large iron cylinder mounted on a horizontal axis, with pipes running through the centre, which carry steam for heating the charge. The cylinder is filled with a solution of caustic soda and the cotton-waste is added to this. The washer revolves very slowly. Its motion keeps

(c) *First Washing.* the mass constantly agitated, and accomplishes the removal of oil and grease. A temperature of 120° to 130° F. is maintained during the washing, which lasts about 4 hours.

(d) *Centrifugal Wringer.* From the washing-house the cotton is taken to a centrifugal wringer, and wrung as dry as possible.

(e) *Second Washing.* It is then returned to the washer and washed a second time in clear, pure water.

(f) *Second Wringing.* It is then wrung out a second time in the centrifugal wringer.

(g) *The Picker.* After the second wringing it is taken to the *picker*. The cleaned cotton-waste, or rags, is placed on the

[1] The Germans have found that wood pulp gives a higher nitrification and forms a better base for smokeless powder than cotton. The pulp is prepared in a way similar to that of the manufacture of paper, and paper scraps, after suitable mechanical and chemical treatment, are also available. Judging by the results obtained by the Germans, it is thought to be probable that in the course of time wood pulp will displace cotton fibre generally in the manufacture of nitrocellulose powder.

apron of the machine, which conducts it between two horizontal toothed cylinders which revolve in opposite directions, pulling in between them the cotton, tearing apart the knots and tangled lumps of waste, or the cotton rags, into shredded strips, about 1 inch to 1½ inches long, and about ¼ inch wide. After passing through the picker it is collected in boxes and taken to the *drying-house.*

2. Drying.

Drying-house. This house has large wooden bins with perforated bottoms. Hot air circulates under the bottoms, and is forced up through the bins and through the *cleaned, dried,* and *picked* cotton placed therein. The temperature of the air is from 90° to 105° F. The cotton is turned over by hand from time to time. It is kept in the bins about 8 hours. It then contains about 0.5% of water. As soon as the cotton is thus dried it is placed in air-tight cans. This is necessary, as it absorbs from 1½% to 2% of water by mere exposure to the air. It is then taken to the *nitrating-house.*

3. Nitrating.

(*a*) *Nitrating-house.* The cotton, as now prepared, is nitrated in earthen pots containing the acid mixture, or by placing it in a centrifugal machine, so arranged as to allow the acid mixture to be admitted and the spent acids to be withdrawn through suitable pipes with stop-cocks. In case the nitration takes place in a centrifugal machine it is conducted as follows: One can of dried cotton, containing about 16 pounds, is placed at one time in the machine with about 900 pounds of mixed acids, consisting of 3 parts sulphuric acid and 1 part nitric acid, both very strong, 98% and 95% respectively. The mixed acids are drawn from a large tank, called the mixed-acid tank. The spent acids, after "revivifying" by additions of "fortifying" acids of concentrated strength, are let into the mixed-acid tank.

(*b*) *The Acids.*

(c) Nitration. The charge is kept in the centrifugal machine about 30 minutes. In becoming nitrated the cotton increases in weight about one-half; the 16 pounds of cellulose giving about 24 pounds of nitrocellulose. The degree of nitration is about 12.6% of N. During the 30 minutes the charge is turned over and over by iron hooks.

(d) Drawing off Spent Acid. After 30 minutes the drain-cocks of the machine are opened, the machine is started, and the spent acids are forced out by centrifugal action.

4. PURIFICATION.

The remainder of the process has for its object getting rid of the free acids remaining in the nitrated cotton and of the nitro-by-products.

(a) Drowning. The nitrated cotton is taken at once from the nitrating machine, and immersed or drowned in a large quantity of pure cold water. It is kept immersed in this water for 8 hours, two changes of water being made.

(b) First Washing. From the drowning-tanks the cotton is taken to another centrifugal machine. The machine is started as soon as the charge is in it, and while it is revolving cold water is played on it from a hose. After about ten minutes the washing is discontinued, and the machine then revolved at its highest capacity and the cotton wrung as dry as possible.

About 1000 pounds are allowed to accumulate from the foregoing operations, and this constitutes a factory "lot."[1] This lot receives a definite number which attaches to it throughout its existence. In connection with this number all subsequent purification operations, stability- and ballistic-tests are recorded.

(c) Purifying-tanks. These are large wooden tanks, having steam-pipes arranged over the bottom. Steam circulates through these pipes and keeps the cotton and water at the desired temperature. Pure water is put in

---

[1] The size of the "lot" of different factories varies.

(d) First Purification. the tanks and one lot added. The lot is kept in the purifying-tank for two days, the temperature being maintained at 80° C., except that the water is renewed three times during this period, and at each renewal the temperature is raised to 100° C. for two hours. The mass is kept agitated by revolving arms set at different angles.

In some factories the purification consists of alternate two-hour washings at 80° and 100° C., with renewal of water each time to include five renewals.

(e) Second Washing. From the purifying-tanks the nitrated cotton is taken to a centrifugal machine, where it is washed with pure cold water from a hose for a few minutes. It then goes to the *pulper*.

(f) Pulper. This is the ordinary pulping-machine used in paper-mills. It consists of an oval-shaped vat or tank, with a horizontal shaft across its narrowest dimension. On one end of this shaft is a drum, which has on its outer surface a series of parallel knife-edges. Directly below the drum is a concentric surface, with a second series of knife-edges. The clearance between these edges can be regulated. Pure water circulates slowly through the vat, running in at one point and overflowing at another. About 1000 pounds of cotton from the purifying-tank is placed at one time in the pulper. The contents of the vat are submitted to an acid color-test from time to time, and sufficient sodium carbonate is added to neutralize any free acid that may be liberated as the pulping proceeds. The drum revolving pulls the cotton down and forces it between the two series of knife-edges, cutting it finer and finer until the whole mass is a smooth, even, fine pulp, about the consistency of corn meal; this requires about six hours. From the pulper the cotton goes to the *poacher*.

(g) Poacher. This is a vat similar to the pulper in form, but it has no knife-edges. The horizontal shaft across its narrow part carries only wooden paddles. The object of the machine is simply to continue the washing, with a view to removing all free acid or

alkali. The contents are tested for both acid and alkali as the poaching proceeds. The operation is continued until the lot is shown to be free from acid and alkali. A chemical stability-test is now made. Further treatment depends on its result. Another form of poacher consists of large, deep, cylindrical vats, with a propeller-shaped wheel on a vertical axis near its bottom. Steam-pipes may be placed over the bottom, and the mass subjected to the

(*h*) *Second Purification.* action of boiling water and rewashing with cold water, as in the purifying-vats. The propeller keeps the mass circulating. The process should continue for three days, having twelve changes of water and two hours' boiling with each change.

(*i*) *Third and Final Washing.* From the poacher, as just described, the cotton is dumped into a large volume of pure cold water, which is contained in a large trough. Through the trough circulates an endless belt of coarse cotton cloth, which passes between two rollers at some distance outside of the trough. As the belt moves through the mass of suspended cotton a certain quantity adheres to it, and the belt carries this up through the rollers, which squeeze out the surplus water, and a scraper detaches the squeezed cotton from the belt and it falls into receptacles placed to receive it on the other side of the rollers. It is

(*j*) now in the form of small thin flakes. It contains about 4% of water. This is submitted to careful laboratory tests.

5. Colloidization.

(*a*) *Dehydrating.* This product is taken to the *dehydrating-press.* The water is extracted by means of alcohol; the latter displacing the water. The alcohol thus mixed with the cotton is sufficient to accomplish its colloidization when mixed with ether in the next operation. In extracting the water, 15 pounds of nitrocotton is placed in the cylinder of the dehydrating-press, and submitted to a pressure of 3000 pounds per square inch, which forms it into a cylindrical "cheese." A large quantity of water is pressed out by this pressure, but some still remains. A quantity of alcohol is let into the press cylinder. Air is admitted over the alcohol, and a pressure of 100 pounds per square inch put on. This forces the alcohol through the mass of the cheese, and the liquid flows out through a pipe below; first water comes, then a mixture of water and alcohol, and, finally, alcohol of full strength. A pressure of 3000 lbs. per square inch is again put on the cheese, and this forces out surplus alcohol. Enough remains for colloiding. The cheese now weighs about 17 lbs., 15 lbs. of cotton and 2 lbs. of alcohol.

(b) *Colloiding.* From the dehydrating-presses the product is taken to the colloiding-machine. This consists of an ordinary bread-dough kneading-machine, as used in large bakeries. Three cheeses from the dehydrating-press are broken up and put into the kneader with about one-half the weight of ether. The kneader is started, and the mixing continues until all of the ether is absorbed, which, as a rule, requires about two hours. When the colloiding is finished, the charge from the mixing-machine is pressed into a cake by hydraulic pressure. This cake is a cylinder about 9″×14″. The product should now be a smooth, compact colloid, with a clear amber or light brown color. Some few white spots seen in the colloid cake are air-bubbles. To get rid of these air-bubbles and to blend better the colloid, the cake is put through the *macaroni press.*

6. Granulation.

(a) *Macaroni Press.* This is an hydraulic press, having small holes in the bottom of its cylinder. The colloid is forced by the pressure through these small holes, and falls in a receptacle below in macaroni-like strings. These are collected and put into the final press, and pressed into the final *powder-cake.*

(b) *Die-press.* The powder-cake is put through the *die-press,* from which it emerges in the form of a continuous cord- or rope-like cylinder, of the diameter of the powder-grain being made, and with the requisite perforations. This result is accomplished by having the end of the press a cone, and fitted into the apex of the cone is a die, with needles of proper size for the perforations. The press is horizontal. The head forces the colloid to fill the cone and surround the needles. Continued pressure forces the colloid out through the die; it is received on rollers, carried thereon to the end of a long table, at which point a revolving disk-cutter cuts the rope into grains of proper lengths. The die can be changed so that one press may turn out many sizes of grain.

7. Drying.

(*a*) *Solvent Recovery.*

The grains from the powder-press are collected in suitable cases and taken to the *solvent-recovery house.* At this house the grains are placed in certain receptacles, and hot air forced up through them. This hot air carries off the greater part of the solvent, the grains shrinking and shrivelling in the process. The air, laden with the vapors of alcohol and ether, passes to an elaborate refrigerating-apparatus, in which the two vapors are separately condensed and collected. The process takes about 8 hours. About 60% of the solvent should be recovered, but this degree of efficiency is rarely attained.

(*b*) *Dry-house.*

The powder is then taken to the *dry-house,* where it is kept from two to four months in a drying temperature of 100° to 105° F.[1]

---

[1] The methods heretofore employed for drying smokeless powders,—that is, the removal of moisture and of excess solvent, have required long periods, from one to four months, after the powder is grained and before it can be issued or used. Various methods of shortening this period have been investigated. Abnormally elevated temperature air-drying has a tendency to injure the grains and may destroy them.

Of the other methods proposed, " water-drying " has been the one giving best promise, and since the outbreak of the European war, this process has been largely used by many manufacturers, because of the urgent need for prompt deliveries.

This process consists of immersing the powder,—either with or without prior solvent-recovery, in water, where it remains for a period varying from a few days to two weeks, depending upon the degree of urgency of requirements, etc. The water is generally cold at first and raised more or less gradually to a maximum of 55° to 60° C. Naturally, the shorter the period, the

All powder is doubly blended before being formed in acceptance lots.

The delivery of a lot of powder dates from the completion of the blending and boxing it, at which time the powder inspector of the Government selects samples for chemical analysis and for ballistic test. Its acceptance depends on the passage of these tests.

Powder is shipped in zinc-lined boxes containing, approximately, 100 pounds. Each box is marked with the number of the lot, maker's initials, year, gun intended for, muzzle-velocity, pressure, and granulation.

General Remarks on Smokeless Powders.

Powder, such as that just described, is a pure cellulose or colloid powder. Sometimes nitroglycerine or certain metallic nitrates are added to the colloid in the mixing, with a view to giving a better ballistic effect. These substances when added are to be considered as distributed throughout the mass of

---

more heating is required for powder with an equal amount of solvent. Too severe heating of water injures the grains, especially large ones. The action of the hot water is to dilute the solvent, which by this action and subsequent wringing in a centrifugal, is mostly removed from the powder. Thereafter, a period of air-drying is needed to remove the moisture, and this period, while preferably about two weeks with moderate temperature, is sometimes made quite short by using higher temperature.

By what may be called a conservative water-drying process, therefore, cannon powders may be completely dried in about one month, while some private manufacturers are now drying small-grain powders in a few days. Powders dried in water have a whitish or milky appearance, but do not differ markedly in other physical aspects from air-dried powders. Tests applied are of the same nature and for powders conservatively dried in water, give about the same results as for air-dried powders.

Some samples of considerable age are still stable, but not sufficiently long observation of such powders has been had to demonstrate fully their keeping qualities.

This process is therefore still regarded as one to be used for quantity manufacture only in time of need, and when the powder may be expected to be consumed within a reasonable period,—for which conditions its value is proven.

the colloid: the nitroglycerine like water in a sponge, the metallic nitrates like particles of sand, or earth, in ice made from muddy water. They are not *essentials;* they are added to modify the character of the explosion in the bore of a gun.

The ballistic efficiency of a powder may be represented by the ratio:

$$\frac{\text{Velocity given to projectile in f. s.}}{\text{Pressure in tons per sq. in. in bore}}.$$

It is desirable that this ratio should have a maximum value.

The strength of guns now in use limits the denominator to about 16 to 18.

With this limitation muzzle-velocity for a given projectile is dependent on the rate of burning of the powder, its quantity, and the length of bore.

Under existing conditions, including kind of powder and capacity of powder chambers, a muzzle-velocity of about 2300 f. s. is had in guns having bores about 35 calibers long, about 2600 f. s. in guns having bores about 40 calibers long, and 2800 to 3000 f. s. in guns having bores about 50 calibers long.

It has been universally thought desirable, heretofore, to so design a nitro-powder that the carbon would all burn to carbon dioxide. Lately this has been questioned by Mendeléef, who advances the claim that the best results with progressive explosives are to be had when the carbon is burned to CO instead of CO_2, for the reason that a given weight of carbon will give double volume of CO compared with CO_2 at same pressure and temperature, and this will be more efficient in a gun than the increase of volume due to the increased temperature in burning to CO_2. Furthermore, the higher temperature of the products of explosion when C is burned to CO_2 is so destructive to the metal of the bore of guns by erosion as to make such explosives less desirable.

For example, military guncotton has insufficient oxygen to burn all of its C to CO_2, and nitroglycerine has an excess of oxygen. By mixing these two substances in proper proportions the excess of oxygen in the explosion of the latter supplies the deficiency of oxygen in the explosion of the former, and the products of explosion of the mixture are CO_2, H_2O, and N.

The English smokeless powder, cordite, is an illustration of such a combination; it is composed of

Guncotton (acetone colloided)........	37	parts
Nitroglycerine........................	58	"
Vaselin...............................	5	"
	100	

It gives high muzzle-velocities with low pressures, but the temperature of its explosion is very high comparatively, and has caused thereby such rapid erosion of the bores of English guns as to cause it to be discarded in favor of a powder with less nitroglycerine, about 38 per cent.

The celebrated French BN powder had barium and potassium nitrate.

The explosion of such powders containing an oxygen carrier disseminated throughout the mass of the colloided nitrocellulose, appears to be more prolonged and increasing in its effect than that of the pure colloid powders.

Based on the foregoing considerations nitrocellulose powders may be classified as follows:

I. Pure Colloids.

(*a*) *Acetone colloids.* Composed of nitrocellulose of high nitration colloided in acetone. Such colloids are brittle, and apt to disintegrate under pressure in the bores of gun, giving excessive pressures. They are dangerous.

(*b*) *Ether-alcohol colloids.* Composed of nitrocellulose of mean nitration colloided in ether-alcohol. Such colloids are tough and elastic, and do not break up under pressure in the bores of guns.

II. Composite Colloids.

(*a*) Acetone colloid for matrix with nitroglycerine.
(*b*) Acetone colloid for matrix with metallic nitrate.
(*c*) Ether-alcohol colloid for matrix with nitroglycerine.
(*d*) Ether-alcohol colloid for matrix with metallic nitrate.
(*e*) Acetone colloid for matrix with organic nitrate.
(*f*) Ether-alcohol colloid for matrix with organic nitrate.

Some examples of these types of powders are given in the following table:

PURE COLLOIDS.

ACETONE.

Maxim-Schupphaus.	
Guncotton	80.
Nitrocellulose (sol.)	19.5
Urea	0.5
	100.0
Poudre B (Vieille's Powder).	
Guncotton	68.21
Nitrocellulose (sol.)	29.79
Paraffin	2.00
	100.00
Rifleite.	
Guncotton	75.
Nitrocellulose (sol.)	22.48
Nitrobenzene	2.52
	100.00
Indurite.	
Guncotton	40.
Nitrobenzene	60.
	100.
Swiss Normal Powder.	
Guncotton	96.21
Nitrocellulose (sol.)	1.80
Resin	1.99
	100.00

ETHER-ALCOHOL.

Mendeléef.
Pyrocellulose, $C_{30}H_{38}(NO_2)_{12}O_{25}$, contains 12.4 per cent of nitrogen.

Powder for U. S. Army and Navy (Cannon).
Nitrocellulose containing not less than 12.60 per cent of nitrogen ± 0.1 per cent.

U. S. Army and Navy (Small Arms).
Pyrocellulose.

COMPOSITE COLLOIDS.

METALLIC NITRATES AND NITRO-DERIVATIVES OF THE AROMATIC SERIES.

Plastomenite.	
Nitrocellulose (N = 12.33%)	67.18
Barium nitrate	9.76
Di-nitro-toluene	22.06
Poudre BN.	
Guncotton	38.67
Nitrocellulose	33.23
Barium nitrate	18.74
Potassium nitrate	4.54
Sodium carbonate	3.65
Solvents, etc.	1.29
Schultze.	
Guncotton	32.66
Nitrocellulose	27.71
Cellulose	1.63
Barium nitrate	27.62
Potassium nitrate	2.47
Paraffin	4.20
Solvents, etc.	1.48
E. C. Powder.	
Guncotton	28.35
Nitrocellulose	27.95
Cellulose	3.15
Potassium and barium nitrates	37.80
Resins, etc.	2.75

NITROGLYCERINE.

Ballistite.	
Guncotton	50
Nitroglycerine	49
Diphenylamine	1
Cordite.	
Guncotton	37
Nitroglycerine	58
Vaselin	5
Maxim-Schupphaus.	
Guncotton	80
Nitrocellulose	10
Nitroglycerine	9
Urea	1

V.

DETONATING EXPLOSIVES.

(a) Guncotton.

As already explained guncotton is nitrocellulose of high nitration, containing above 12.9 per cent of nitrogen. Its manufacture has been described in connection with the manufacture of smokeless powder. The degree of nitration is regulated by the relative quantities of water, sulphuric acid, and nitric acid used in the nitrating bath, the time and temperature of the steeping. The purification of guncotton for disruptive military uses is accomplished in the same manner as described for nitrocellulose used in the manufacture of smokeless powder.

In manufacturing guncotton for military purposes, purified pulp, produced as explained under the head of manufacture of smokeless powder is taken from the poacher to a *stuff-chest* by suction. This consists of a large vat with air-tight top. Through the centre of the vat passes a vertical shaft, on which are mounted a number of feathered paddles. After the purified pulp has been sucked up into the stuff-chest it is kept agitated by these paddles, so that the pulp will be kept evenly distributed in suspension throughout the liquid.

From the stuff-chest the pulp is drawn into the *moulding-press*. This is an hydraulic press made of bronze and containing moulds. The pulp is run into these moulds, and the pressure applied for about four minutes. The mould-press blocks are taken to the *final press*, placed in the moulds of the final press, and the pressure applied, increasing from a minimum of 6000 to a maximum of 7000 pounds per square inch, through an interval of about three minutes; the highest pressure is maintained for one minute.

The blocks as they come from the final press contain about 15 per cent of water. While in the press they are stamped with the name of the factory, the lot and year. Before being issued for storage or service they should be soaked in pure water until they contain about 35 per cent of water.

In order to get dry guncotton for primers a block of wet guncotton is split up into one-half inch sections; these are strung on a copper or brass wire or tube separating the sections from each other, and exposed to a drying atmosphere out of direct rays of the sun. The sections should be weighed from time to time, and the drying should continue until the weights are constant.

While, theoretically, 183.3 pounds of guncotton (trinitro-cellulose) or 176 pounds of endeca-nitrocellulose (Vieille's) should be obtained from 100 pounds of cellulose, in practice the yield is about 105 pounds of guncotton to 100 pounds of unnitrated cotton; this makes about 230 blocks.

After nitrating and before pulping, guncotton retains the complete cotton structure; even under the microscope no difference is to be detected between nitrated and unnitrated cotton. The only outward evidences of the change is the rough feeling it has, the crackling sound when rubbed between the fingers, and its electrical properties, sticking to the fingers if rubbed between them. Rubbed in the dark, dry guncotton is to some extent phosphorescent.

It may easily be distinguished from unnitrated cotton by treating with solution of iodine in potassium iodide, and subsequently moistening with dilute sulphuric acid. Unnitrated cotton, when so treated, gives a blue color; nitrated cotton, a yellow.

Dry guncotton varies in color from white to light yellow. The yellow is often an indication of sodium carbonate. Sometimes, there is a brownish or reddish shade; this is due, as a rule, to iron, from the washing-water.

When pure, it is without color, odor, or taste, and free from either alkaline or acid reaction.

The density of unpulped dry cotton is about 0.1; after pulping, about 0.8; and in the block form after compression, about 1.2. The absolute specific gravity of guncotton is 1.5.

It is insoluble in both hot and cold water and in alcohol, ether, and ether-alcohol, at ordinary temperatures.

It is soluble in acetone, acetic ether, and in a number of the nitro-derivatives of the aromatic hydrocarbons.

It is insoluble in nitroglycerine; but both guncotton and nitroglycerine dissolve in acetone, and a combined colloid may be obtained by dissolving them in this solvent and then evaporating the common solvent. Soluble nitrocellulose is partly soluble in nitroglycerine, and explosive gelatine is based on this property.

Guncotton is completely decomposed by boiling in a solution of any alkaline sulphide, while unnitrated cotton is not; this principle is used in analyzing guncotton.

Caustic potash solution, with alcohol added, decomposes guncotton almost instantly.

For disruptive purposes, guncotton is used to fill the cavities of shell, to charge torpedoes, and for demolitions of all kinds. For these purposes, it is pressed by hydraulic pressure while in the wet state, in the form of purified pulp, into suitable disks, blocks, or special forms. It is not colloided.

Its value as a disruptive agent rests upon its great force, and its safety in handling, storage, and manufacture.

While many disastrous explosions have occurred with it in the past, none have of late years; and the fact that it is kept in storage in the wet state in which it is non-explosive, except with a powerful detonator or a small piece of dry guncotton, makes it less likely to accidental or spontaneous explosion than any other explosive now used.

If properly purified, guncotton may be kept for years,

even in the dry state, without the slightest deterioration. If not purified completely, some of the nitro-by-products may decompose, and these initiate a progressive decomposition of a mass of guncotton. If, however, the gases generated in such decomposition are free to pass off, the mass will quietly disintegrate. The first evidences of decomposition are acid fumes. These may be recognized by their pungent odor, or, if a piece of moist, blue litmus paper be *confined* with a mass of guncotton thought to be in the state of incipient decomposition, it will soon be reddened. As its decomposition progresses, the fumes become more copious and may be seen as the reddish-brown gas, NO_2 (nitric peroxide). At the same time, the mass begins to show soft, pasty, yellow spots, which extend and coalesce until the whole mass is soft and pasty; and, in connection with the escape of gas, the mass shrinks in volume. As the process proceeds, other gases than NO_2 pass off. The residue is an amorphous, porous, sugar-like substance, almost entirely soluble in water. As long as the gases escape and the heat developed by the reactions is carried off with them, there is no danger of explosion; but if the gases cannot escape, the pressure increases, the heat is retained, the reaction is accelerated, the temperature rises, and ultimately an explosion may result from these causes.

In the case of a mass of decomposing guncotton, it should be spread out, exposed to the air, out of the sun, and wetted with water.

While nitrocellulose is one of the safest explosives known and, when carefully purified, is not liable to decomposition, still it should be kept in mind that it is an explosive, and due care in handling and storing it should be observed.

Some authorities claim that strong light will act slowly to originate the decomposition of nitrocellulose; but Abel, who made searching investigation of this matter, says that "guncotton produced from properly purified cotton may be exposed to *diffused* daylight, either in the open air or in closed vessels, for very long periods, without undergoing any change. The preservation under these conditions has been perfect after three and one-half years." But long-continued exposure of dry guncotton to the direct rays of strong sunlight produces a very gradual change. If moist guncotton be exposed to sunlight, it is affected to some greater extent than dry guncotton, but the change is very small even after several months' exposure to sunlight in a glass bottle.

It has been found that guncotton, exposed to the sunlight without confinement, has had its stability, as determined by the heat-test, improved. This would seem to suggest that the action of sunlight decomposes the unstable nitro-by-products, and the escape of these into the air slowly leaves the nitrocellulose proper in a purer and more stable state. Indeed, the evolution of acid fumes from a nitrocellulose exposed to strong, *diffused* light would be evidence of incomplete purification.

Instructions for blending or drying smokeless powders require that the operation be performed out of the direct rays of the sun.

Heat of sufficient degree, of course, disintegrates the nitrocellulose molecule; but nitrocellulose of either high or low nitration, that has been properly purified, will stand a temperature approaching 130° F. without change.[1]

Water or a damp atmosphere serves to protect nitrocellulose from the disintegrating effect of heat (not light). A guncotton stored in water or in damp magazines is able to withstand,

[1] "In general, it may be said that no nitro-compound will stand heating to temperatures above 160° F. for any prolonged period. At 194° F., even the best and purest product is sure to decompose within a few hours, and even pure guncotton cannot be exposed to a temperature above 122° F., without impairing its capability of subsequently standing the heat-test. It is true, decomposition may not take place at this temperature, and that the product may be kept indefinitely without decomposition under favorable conditions; but whenever it is again subjected to the heat-test at 160° F., it will at once give a distinct reaction. In general, it would appear that only the most perfect products will stand a temperature of 113° to 122° F. for some months without impairing their capability of standing the heat-test."—GUTTMANN.

without change, temperatures as high as 200° F. for long periods. This property renders guncotton a desirable explosive in hot, damp climates. Water not only protects the nitrocellulose proper from the disintegrating action of heat, but also the nitro-by-products present in incompletely purified nitrocellulose.

To be non-explosive, it is only necessary that the guncotton be damp; nitrocellulose, with only the water left in it after coming from the centrifugal wringer, is not to be exploded by fire or ordinary shock.

Guncotton made for disruptive purposes contains, as a rule, a small amount of carbonate of sodium; this, disseminated through the mass of the cotton, tends to neutralize any free acid that may be formed in storage. It would not be desirable to have it in finished smokeless powder, as it would increase the solid residue in guns and cause some smoke.

Guttmann is opposed to the use of sodium carbonate in guncotton even to neutralize free acid due to incomplete purification or incipient decomposition. If the guncotton is properly purified, there is no reason why there should be free acid, or why decomposition should take place; and if decomposition should begin, the action of the carbonate would only neutralize the first gases given off, it would not arrest the process: indeed, alkalies have a tendency to decompose nitrocellulose at temperatures above 86° F. It is not desirable to check incipient decomposition by sodium carbonate; on the contrary, if incipient decomposition takes place, it is desirable that the gases given off should pass off and serve themselves to give evidence of the condition existing. "At ordinary temperatures—that is, those occurring under normal circumstances of storage and carriage—decomposition of guncotton, so far as present experience goes, is out of question."

Cold has no effect on dry guncotton. The compressed cakes and disks are caused to flake off on the surfaces if wet and exposed to freezing and thawing, and the freezing also causes the mass of the cake or disk to open out and be less compact.

Variations of temperature between 105° F. and 32° F. have no effect on either the physical or chemical conditions of guncotton.

Guncotton, even when dry, is not liable to explode by blow or friction, unless very closely confined and compressed. For example, in order to explode by a blow a piece of guncotton, it is necessary to take a small piece, wrap it tightly in tinfoil, place on an anvil, tap it two or three times lightly to compress it, then strike it a heavy blow. Shells filled with disks of dry compressed guncotton have been fired from guns into masonry at fifty yards from the gun without explosion.

Flame, or metal heated to red or white heat, will ignite guncotton. Its rate of burning is affected by the degree of confinement and physical state of the mass: if woven into wicks or compact cloth, the rate is much reduced; if compressed while in the pulped state into compact blocks, its rate is also reduced. Burning guncotton may be extinguished by water; but if a mass of considerable size be burning, it may be quenched on the exterior and continue to burn in the interior. Wet guncotton in any form cannot be ignited by flame. A wet disk of guncotton thrown into a fire will first dry out on the outer surface and burn there, and continue this progressively until the whole disk is consumed. As much as a ton of wet guncotton has been consumed in this way without the slightest evidence of explosion.

The igniting-point for nitrocellulose is about 186° C.

The specific heat of the gases composing the products of explosion may be taken approximately at 0.28.

The experiments of Roux and Sarrou indicate 1056.3 centigrade units of heat given off by the explosion of guncotton. This indicates a temperature of 3700° C. Nobel and Abel fixed the temperature, as a result of their experiments, at 4400° C. Sarrou and Vieille found that water was dissociated at the temperature of the explosion, all of the carbon burning to CO_2. Berthelot estimates that guncotton of density 1.1, exploded in its own volume, will give a pressure of 160 tons per square inch. The rate of propagation of the explosive wave of guncotton in rigid tubes has been found to be 5000 to 6000 metres per second.

Experiments of Professor C. E. Munroe, at the naval guncotton factory at Newport, R. I., have shown that thoroughly dry guncotton can be detonated by three grains of mercury fulminate; air-dry guncotton, by five grains, if the fulminate be confined in copper tubes and the tubes are *in close contact with the cotton.* The Navy primers, however, have 35 grains of mercury fulminate in order to have a liberal certainty factor.

A disk of guncotton detonated on an iron plate reproduces on the surface of the plate the reliefs and depressions on the surface of the disk; a depression on the surface of the disk will be reproduced as a depression on the surface of the plate. The explanation of this is to be found in the erosive effect of the rushing gas at those points where there is no contact; it is the same effect as is to be noted near the bands of projectiles in the bores of guns: the enormous velocities of the gaseous molecules impinging on the metal at these points, in connection with the weakening of bonds of cohesion and affinity by the high heat, is thought to be sufficient to account for the phenomenon.

The violence of explosion is greater in proportion as the confinement is greater; the maximum being when confined rigidly in its own volume, and, in accordance with this principle, tamping increases the violence of an explosion. Even the amount of air-pressure will have its effect on the character of an explosion; the same explosive and detonator would give a more mild explosion on a mountain-top than at the seashore.

Wet guncotton gives a more brusque explosion than dry guncotton; and Professor Munroe explains this by supposing that the water in its pores, being nearly incompressible and highly elastic, increases the rate of propagation of the explosive wave or disturbance and diminishes thus the time of explosion.

The energy of the explosive wave may be sufficient to initiate explosion in a mass placed at a certain distance from an exploding mass. This is called explosion by influence. Two theories are advanced to account for the phenomenon: One, that the explosion is due to certain *synchronous* relations of the motions of the molecules of gas and the molecules of guncotton,—that a wave of certain amplitude and length passing over the guncotton causes "sympathetic" motion to be taken up by the latter, and this in turn accomplishes the disruption of the guncotton molecule; just as a string of a musical instrument may be set to vibrating by sounding near it the note which gives the wave of sound that correspond to the string, or as certain glass beads under strain may be shattered by musical notes of certain pitch. The other considers that in all cases the explosion is initiated by the *energy of the impact of the molecules in motion*,—that there is a definite product of molecular mass into molecular velocity, which, if it be delivered against a molecule of guncotton, will disrupt the molecule, and the disruption of one molecule will disrupt all adjacent molecules, and so on. As temperature varies directly with molecular velocities, an explosive molecule, for a given pressure, requires a given temperature to disrupt it.

While both theories have advocates, the latter is thought to be more generally accepted at the present time. Some authorities claim that in all cases heat initiates explosions.

Explosion by influence may be illustrated by placing guncotton disks side by side at varying distances apart ($\frac{1}{4}''$, $\frac{1}{2}''$, $\frac{3}{4}''$, $1''$) and noting the effect.

Berthelot fixes the heat of combustion of guncotton at 12 calories for each nitryl radical, and, accepting the products of explosion as determined by Sarrou and Vieille, gives the total heat of combustion at 633 calories per molecular weight proportion.

Sarrou and Vieille conducted a series of experiments in which guncotton was exploded in a closed vessel. They found the volume of gases reduced to 0° C. and 760 mm. pressure, to vary with the density of the charge, both as to proportion of each kind and total volume. Some of their results are given in the following table:

Density of charge		0.01	0.023	0.2
Volume of gases (reduced) per material		658.5	670.8	682.4
Composition of gases per 100 volumes:	CO	49.3	43.3	37.6
	CO_2	21.7	24.6	27.7
	H	12.7	17.2	18.4
	N	16.3	15.9	15.7
	CH_4	0.0	trace	0.6 and N.

From this it appears that the proportion of CO and N decrease, and CO_2 and H increase, as the density of charging increases; also, that for the higher charging a little CH_4 appears.[1]

From these results Berthelot writes out the following reaction for the explosion of guncotton exploded in closed vessels, under the ordinary conditions of charging for disruptive purposes, as in torpedoes:

$$C_{24}H_{18}O_9(HNO_3)_{11} \text{ exploded} = 24CO + 24CO_2 + 17H_2 + 12H_2O + 11N_2.$$

When guncotton is burned in the open air there is some nitric oxide in the products of combustion, amounting to about 24 per cent of the total volumes of the product.

(b) Nitroglycerine.

Nitroglycerine is a nitric ether of propenyl alcohol (commonly termed glycerine).

Propenyl alcohol is trihydric and may be written structurally:

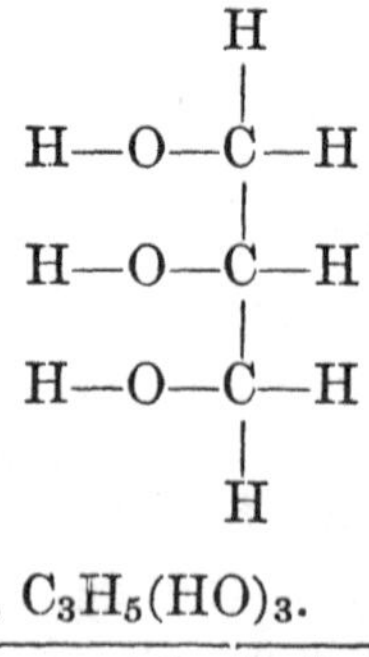

or, in ordinary symbols, $C_3H_5(HO)_3$.

[1] This CH_4 may, perhaps, account for the flare-backs from cannon in using smokeless powder.

The nitric ether is formed by replacing the H of the HO radicals by NO_2, and theoretically there may be three ethers, corresponding to one, two or three replacements of H, forming:

Mononitroglycerine, $C_3H_5(HO)_2ONO_2$,
Dinitroglycerine,[1] $C_3H_5(HO)O_2(NO_2)_2$,
Trinitroglycerine, $C_3H_5O_3(NO_2)_3$.

Only the last is of interest.

The process of manufacture of nitroglycerine follows in a general way the operations performed in the manufacture of guncotton, and consist essentially of:

1. Nitrating the glycerine; and,
2. Purifying the product of free acid and other nitro-compounds.

In nitrating, it is not possible to place a large amount of glycerine in the acid, for the reason that the action would be too energetic and the temperature would rise too high. Therefore the process is so modified as to bring small amounts of pure glycerine (free from lime, iron, aluminum, chlorides, fatty acids, glucose, or other adulterants and having a specific gravity of 1.26) in succession into the presence of the mixed acids.

Sulphuric acid is used in the acid bath for the same reason as in making nitrocellulose.

The acids must be of the highest possible concentration, in the proportion of 1 part by weight of nitric acid (93% to 95%) to 2 parts by weight of sulphuric acid (96%).

According to the chemical formula, 227 parts of nitroglycerine should be obtained from 92 parts of glycerine and 189 parts of nitric acid; in practice it is necessary to take a much higher proportion of acid. As a rule, 1 part of glycerine is taken to 8 parts of nitric acid and about 14 parts of sulphuric acid.

In order to keep down the heat developed by the reaction, the glycerine must be kept between 68° and 77° F. This is regulated by the amount of glycerine injected into the mixed acids. The heat is caused by the water combining with the sulphuric

[1] It is understood that dinitroglycerine has been used with very promising results.

acid. A rise in temperature may explode the nitroglycerine or cause a loss of product, converting it into oxalic acid and other products; these are difficult to remove, and make the nitroglycerine unstable if not removed. This excessive heating and accompanying NO_2 fumes is called "firing." If the temperature rises above 86° F. and cannot be controlled by stopping admission of glycerine, compressed air is forced through pipes into the mixture, and the acid bath cooled by the expansion of this air and the agitation it causes. If the temperature still continues to rise, the whole charge is run out into safety-tanks. These safety-tanks are large leaden chambers or vats, situated at some distance from the nitrating apparatus, into which, in case of firing, decomposing mixtures may be run directly at any stage of the manufacture and "drowned" in a large mass of cold water, which is kept agitated and cooled by compressed air escaping through the mixed liquids.

It requires about one hour to charge, nitrate, and discharge the contents. During the nitration copious fumes of NO_2 are given off from the surface of the acid mixture. The condition of the charge and the degree of reaction are judged by inspection.

When the nitration is completed the contents are permitted to run out into the separating apparatus, which consists of a large leaden tank. The nitroglycerine, having less specific gravity than the waste acids and mixed by-products, collects on top. It is drawn off through a stop-cock into a second tank containing water. While the nitroglycerine is being run into this latter tank, compressed air is forced from below through the water, keeping it agitated. The effect of this is to "wash" the nitroglycerine and to keep the temperature between 60° and 86° F., which is of first importance. Its specific gravity being greater than water (1.6), it settles to the bottom of the tank as soon as the compressed air is shut off, and is drawn off from it for further purification. A small amount of nitroglycerine will be left in the wash-water; this is partially recovered by mixing with other washings and subsequent separations.

There remain also some slight traces of free acid; these

are removed by adding a small quantity of sodium carbonate in solution. The washing process is then repeated in a washing-tank of similar construction, agitating the liquid in a warm dilute solution of sodium carbonate by compressed air, repeating the washings and renewing the solution until the desired degree of purity is attained.

After it is thoroughly washed, it is filtered through flannel or felt, stretched on suitable frames, two frames being used, to remove all slimy and foreign particles which may have gotten into the liquid during the manufacture. A layer of dried salt is placed on the filters, to remove small quantities of water still in the liquid and to favor the rate of filtering.

The nitroglycerine is allowed to stand in a warm room for several days, and still a small quantity of water will rise to the top, and may be removed by skimming or absorption.

The waste acids and wash-waters are subjected to special treatment to recover the small quantities of nitroglycerine carried off in them, and to place the acids in such condition that, after properly "fortifying" them, they may be used again.

Physical Properties.—Nitroglycerine, made from chemically pure ingredients and at a temperature between 60° and 80° F., is a water-white oily liquid, without odor at ordinary temperature. Commercial nitroglycerine has a yellow color, more or less deep. When free from water it is transparent; the presence of water makes it milky and translucent.

It has a slightly sweet taste, and gives a burning sensation. It is very poisonous, and a very small quantity absorbed through the mouth, nostrils, or skin gives characteristic symptoms of giddiness, faintness, and severe headache; if the quantity be increased, these symptoms become more aggravated, producing rigor and unconsciousness. Robust and highly nervous persons appear to be specially susceptible to the effects described. Sometimes one never becomes immune to these effects, but, as a rule, the human system little by little adjusts itself so that workmen experience no unpleasant effects. The headache effect is most often experienced by those not accustomed to handling nitroglycerine.

Nitroglycerine contracts about .08 of its volume in freezing, which it does at 3° to 8° C. (37° to 46° F.);[1] it does not melt from the frozen state until at about 11° C., or about 51° F.

Nitroglycerine is soluble in alcohol of above 90 per cent strength, ether, chloroform, benzene, concentrated sulphuric acid, glacial acetic acid, warm turpentine, methyl and amyl alcohols, carbolic acid, nitrobenzene, toluene, acetic ether, acetone, olive-oil, stearine oil, hot nitric acid.

It is insoluble in cold water, 50 per cent alcohol, carbon disulphide, cold turpentine, kerosene, caustic-soda solution, borax solution.

It is decomposed by cold hydrochloric acid, specific gravity, 1.2, slowly; hot ammonium sulphydrate, hot iron chloride, 1.4 grams of $FeCl_2$ to 10 c.c.

The presence of nitroglycerine may be detected by acting on the suspected liquid with a solution[2] of aniline in concentrated sulphuric acid. This gives a *purple color*, which turns green on the addition of water.

Another simple test is to absorb the suspected drop or quantity with blotting-paper. If it is nitroglycerine, it will not dry, and, when struck on an anvil with a hammer, it will explode. Lighted, it burns with a yellowish flame; placed on a hot metal plate, it explodes.

In the frozen state, nitroglycerine is less sensitive to shock than in the liquid state; but the process of thawing frozen nitroglycerine is a very dangerous one, and many accidents have resulted therefrom. It should never be attempted over

---

[1] According to Walke, at 3° to 4° C. (37° to 40° F.); according to Bloxam, at about 4° C. (40° F.); according to Munroe, at 39° to 40° F.; according to Guttmann, freezes at 8° C. and melts from the frozen state at 11° C.

A small per cent (0.5 to 3.) of nitro-benzene reduces the freezing-point very much, but diminishes the explosive effect also.

[2] 1 volume aniline to 40 volumes H_2SO_4, specific gravity 1.84.

a naked flame, or by direct contact with a solid in contact with a flame. The only safe way is to thaw over steam-pipes heated not higher than 50° C. (122° F.), or immersed in a water-tight vessel itself immersed in a vessel of water heated not higher than 50° C. (122° F.).

Nitroglycerine can be completely evaporated at a temperature of about 70° C. (158° F.). It evaporates slowly at lower temperatures; at 40° C. (104° F.) 10 per cent has evaporated in a few days. Washing for two hours with water at 50° C. (122° F.), with agitation by compressed air, 0.15 per cent of nitroglycerine is lost.

Although *frozen* nitroglycerine is very liable to explosion if brought over a naked flame or hot metal, *liquid* nitroglycerine is insensitive to flame. A lighted match plunged into liquid nitroglycerine will be extinguished without causing explosion; an incandescent platinum wire will be cooled down, the nitroglycerine only volatilizing.

If the liquid is ignited in the open air, it will burn quietly provided the mass is small; if it is large and the temperature is increased by a failure of the heat of the burning surface to be conducted off, explosion will take place when the temperature of the surrounding medium rises to 180° C. (356° F.).

Formerly, nitroglycerine was thought to be liable to undergo spontaneous decomposition, but, as now manufactured, such danger is very remote. If properly purified, there should be no tendency to decompose. When decomposition starts, it proceeds slowly and quietly, giving off NO_2 and CO_2 and forming crystals of oxalic acid; the escaping gases, some of which are held in the liquid, color it green. As the decomposition proceeds, the entire mass, after some months, is converted into a greenish, gelatinous substance, composed chiefly of oxalic acid, ammonia, and water. Decomposing nitroglycerine is, therefore, characterized by a *greenish* color. While in this state, it is more liable to explosion than when normal, and every care should be taken not to subject it to jar, blow, or shock; decomposing nitroglycerine should be exposed to the open air, so that the heat of chemical action may be carried off.

All nitroglycerine should be tested from time to time for free acid with blue litmus-paper.

If heated above 45° C. (113° F.), decomposition will ensue, but below this temperature it may be kept in storage indefinitely without change.

A mass of nitroglycerine heated above 180° C. will explode. It will explode by shock under certain conditions. If pinched between two rigid surfaces like metal or rock, it explodes; e.g., a small piece of blotting-paper saturated with a drop of nitroglycerine, struck by a hammer on an anvil, will explode at the point struck, but, as a rule, not beyond. A thin thread or sheet of nitroglycerine on a metal surface will detonate if struck with a piece of metal. A bullet fired into a mass of nitroglycerine will detonate it.

Shock, friction, and heating of all kinds must be carefully guarded against in handling and keeping nitroglycerine.

Nobel, in 1863, discovered that the highest type of explosion could be initiated in nitroglycerine by a small cap of fulminate of mercury. This marked an epoch in explosives, in that it for the first time established the fact that the *character of the explosion* is dependent upon the *character of the initial disturbance.* Nitroglycerine should, therefore, be fired by a cap of mercury fulminate if its full explosive force is to be developed, and for this purpose the cap should be in direct contact with the liquid. In the frozen state, it requires a powerful cap to detonate it.

Great care should be taken of cans or other receptacles which have contained nitroglycerine. The film of nitroglycerine left on the surface of such empty receptacles has caused disastrous explosions. All such receptacles should be immersed in an alkaline sulphide solution before being used for other purposes.

The explosive reaction for nitroglycerine may be given as follows:

$$2C_3H_5O_3(NO_2)_3 \text{ exploded} = 6CO_2 + 5H_2O + 3N_2 + O.$$

One kilogram of nitroglycerine should give 1135 litres of gaseous products.

The temperature of explosion has been ascertained by experiment to be about 3000° C. The theoretical temperature, exploded in its own volume, is 6980° C.

The energy represented by 1 kilogram is about 6000 kilogram-meters. It is about eight times more powerful than gunpowder, weight for weight. Exploded in its own volume, it gives a pressure of about 164 tons per square inch.

Nitroglycerine was first used by Nobel in 1864 for blasting purposes. It proved to be a very dangerous explosive, on account of its liquid state and its "creeping" and "sweating" properties. Small masses could not be distinguished from water, and the detonation of a drop might explode huge masses, causing great destruction of life and property.

To avoid these dangers, Mowbray, of North Adams, Mass., made use of it in the *frozen state* in the construction of the Hoosac Tunnel. Not only were the dangers due to the liquid state avoided, but in the frozen state it is less sensitive to shock.

Nobel next resorted to the device of dissolving nitroglycerine in wood alcohol (15 to 20 per cent) for shipment. While in this state it is absolutely non-explosive, and can be recovered in its explosive form by adding 6 to 8 times its volume of water to the solution. While this made shipment safe, the danger of handling it in blasting, and for disruptive purposes generally, remained, and its use in the liquid state was discontinued abroad some years ago, and recently in America.

At present, nitroglycerine is only used in explosives as an ingredient of dynamites of various types and of smokeless powders.

If at any time it is necessary to store liquid nitroglycerine, it should be kept in earthen crocks, standing in copper vessels, and a layer of water should be kept on the nitroglycerine.

If the liquid show a green color at any time, the mass should be destroyed by explosion or by chemical action, any alkaline sulphide solution being efficient for this purpose. "Sulphur solution," made by dissolving flowers of sulphur in a solution of sodium carbonate, is the solution used, as a rule, for this purpose. Whenever nitroglycerine is stored either in the liquid form or as dynamite, a sulphide solution should be kept on hand to pour over particles that may get on shelves or floor.

(c) Dynamites.

"Dynamite" is a term that has both a general and a specific meaning. As a general term, it includes all mixtures of nitroglycerine with solid substances, in which the latter hold the liquid nitroglycerine in absorption. The mixing may be done directly or indirectly through the medium of a solvent. The solid substance is called the *base* or *dope*. The base may be itself an explosive, or a combustible material, or entirely inert in the chemical reaction of explosion. In this sense, smokeless powders that have nitroglycerine as an ingredient partake of the nature of dynamite, but the name is used with reference to explosives designed for disruptive purposes only.

Berthelot divides dynamites into several classes:

1. Those having an inert base of silica, magnesium carbonate, brick-dust, tripoli, sand, etc., having little or no chemical action in the explosion, and acting along physical lines to render the mixture safer by checking the transmission of molecular shock-waves, the harmonious propagation of which, through a homogeneous mass, gives rise to the explosive wave.
2. Those having an active base, which may be
    (*a*) An explosive compound.
    (*b*) A combustible base.
    (*c*) A mixed base, consisting of a combustible and an oxygen-carrier.

The bases are modified to suit the work in hand; the nature of the explosion may be either *shattering* (local) or *propulsive*, the latter grading off into the slow-burning powders.

In a more special sense, the term refers to the first practical form of dynamite, namely, that in which liquid nitroglycerine was mixed with the infusorial earth called kieselguhr as the absorbent base.

The difficulties and dangers attending the use of liquid nitroglycerine have been referred to. In 1866 Mr. Alfred Nobel, in attempting to avoid these, hit upon the means of absorbing the liquid explosive into the mass of pulverized kieselguhr, an earth found in beds in various parts of the world, consisting of the silicious remains of infusorial life. This earth has marked absorptive properties, due to the cellular nature of the particles which constitute it, and, having absorbed a liquid-like nitroglycerine, it holds it tenaciously. Nobel, by making use of this property of kieselguhr, avoided the difficulties and dangers of transporting, handling and exploding nitroglycerine without materially impairing its power, and to this mixture he gave the name "*dynamite*." When carefully calcined the best kieselguhr will absorb over four times its own weight of nitroglycerine. The amount of nitroglycerine present in any case is regulated by the character of the work to be done; the highest commercial percentage is 75 per cent, and this is called "Dynamite No. 1." Dynamite "No. 2" has 50 per cent of nitroglycerine; "No. 3," 30 per cent of nitroglycerine. A little sodium carbonate is usually present to neutralize any free acids that may form.

The following is a summary of the steps taken in the manufacture of dynamite:

1. The kieselguhr is calcined in a reverberatory furnace.
2. It is ground between rollers.
3. It is passed through fine sieves.
4. It is dried.
5. It is packed in bags and stored in a *dry* atmosphere.
6. It is dried until it does not contain more than 0.5 per cent of water.
7. Dry guhr is spread over the bottom of lead-lined troughs.
8. Nitroglycerine is poured over it and mixed thoroughly.

9. It is rubbed through sieves: 1st, 3 meshes to the inch; 2d, 7 meshes to the inch.
10. The bulk dynamite is pressed into cylinders about 1 inch diameter and 8 inches long. These cylinders are called "sticks" or "cartridges."
11. The cartridges are carefully wrapped in paraffined paper.

The sensitiveness of dynamite is increased very much by heat. According to Eissler, "at 350° the fall of a dime upon it will explode it."

It ignites at 180° C. (356° F.); at this temperature it will burn quietly, if free from pressure and not affected by jar, vibration, or extraneous force of any kind, otherwise it explodes.

If a thin layer be spread over a tin plate and the plate be heated over a burner the nitroglycerine will evaporate, but if the layer be more than one-quarter of an inch deep the dynamite is liable to explode.

At a temperature less than 180° C. the sensitiveness increases with the temperature and time exposed.

Exposed to gentle heat, dynamite undergoes no change. Heated at 100° C. for one hour, no change should take place. Heated rapidly to 220° C., it ignites and burns. If ignited it burns quietly when free, but if confined will explode. If a large mass of dynamite is ignited, the interior portion may be heated high enough to explode, being *confined* by the surrounding mass.

If exposed to high storage temperature for a considerable time the nitroglycerine is liable to "leak." Dynamite should be tested for "leaking" at the highest temperature to which it is liable to be exposed in storage.

When dynamite is exposed to a temperature below 12° C. the nitroglycerine has a tendency to freeze; and if it be lowered much below this, down, say, to 4° C., the nitroglycerine freezes, and in doing so separates from its base to a certain extent and does not always become absorbed again on melting. If solidly frozen it is very insensitive to shock. A frozen stick of dynamite may, however, be exploded by attempting to cut it or chop it in two. It is dangerous to ram a frozen cartridge; forcing the frozen crystals over each other is apt to initiate an explosion. The violence of the explosion is much reduced in the frozen state.

While a stick of unfrozen dynamite may be ignited without danger, it is very dangerous to bring a frozen stick in contact with a naked flame or highly heated surface. It is only safe to thaw it in a covered vessel itself immersed in a water-bath.

Dynamite as a disruptive explosive is most efficient with hard, rigid material. With soft, yielding material it gives only a local effect. With such materials a slower acting explosive, like black powder or the modified dynamites described later, should be used.

A dynamite with an inert base containing less than 30 per cent of nitroglycerine will not explode. When the proportion of nitroglycerine is reduced below 30 per cent it is necessary to use an active base, either a mixture or compound (see Judson Powder and Blasting Gelatin).

Kieselguhr dynamite usually has a light-brown to reddish-brown color, and looks like brown sugar. It should not feel greasy to the touch, and the wrappers of dynamite sticks should show no evidences of liquid nitroglycerine on the inside. The outside of the stick when the wrapper is removed should be smooth, even, and compact; there should be no evidences of a pasty condition, or greenish spots. Broken across, the stick should present an even, granular surface on the cross-section, with no evidence of exuded globules of nitroglycerine.

The white deposit often seen on the outside of a stick of dynamite is not necessarily an indication of deterioration.

Kieselguhr dynamite is used at present chiefly in America. Dynamite No. 1 is used to charge submarine mines, and for military demolitions. According to Munroe, dynamite No. 1, exploded in its own volume, gives a pressure of 125 tons per square inch.

For rock-quarrying, tunnel-making, and blasting generally

there are many varieties of dynamites, particularly those having active bases, either explosive or combustible.

The following are some examples of American commercial dynamites:

GIANT POWDER (Dynamite No. 1).

Nitroglycerine	75	parts
Kieselguhr	25	"
Sodium carbonate	0.5	"

ATLAS POWDER (A).

Nitroglycerine	75	parts
Sodium nitrate	2	"
Wood-fiber	21	"
Magnesium carbonate	2	"

ATLAS POWDER (B).[1]

Nitroglycerine	50	"
Sodium nitrate	34	"
Magnesium carbonate	2	"
Wood-fiber	14	"

SAFETY NITRO-POWDER.

Nitroglycerine	68.81	parts
Sodium nitrate	18.35	"
Wood-pulp	12.84	"

GIANT POWDER (No. 2).

Nitroglycerine	40	parts
Sodium nitrate	40	"
Sulphur	6	"
Resin	8	"
Kieselguhr	8	"

RENDROCK.

Nitroglycerine	40	parts
Potassium nitrate	40	"

[1] Dr. Thos. B. Stillman of the Department of Engineering Chemistry, Stevens Institute of Technology, reports the following analyses of Atlas Powders, which are being used in the construction of the Panama Canal:

Atlas Powder B+

	Per cent.
Moisture	0.2
Nitroglycerine	61.1
Wood pulp	14.1
Magnesium monoxide	3.0
Saltpetre	21.6
	100.0

Atlas Powder C+

	Per cent.
Moisture	1.0
Nitroglycerine	45.7
Wood pulp	10.5
Chalk	1.9
Saltpetre	40.9
	100.0

Wood-pulp	13 parts
Pitch	7 "

VULCAN POWDER.

Nitroglycerine	30.0 parts
Sodium nitrate	52.5 "
Sulphur	7.0 "
Charcoal	10.5 "

JUDSON POWDER.

Nitroglycerine	5 parts
Sodium nitrate	64 "
Sulphur	16 "
Cannel-coal dust	15 "

There is a dynamite made in England, in which the base is charcoal made from cork. This has a remarkable absorptive power, taking up as much as 90 per cent of nitroglycerine, and retaining it even if kept under water for a prolonged period. It is known in the market as "cork-dynamite."

Like all nitro-compounds, dynamites are more sensitive to shock at the higher temperatures.

Direct rays of the sun have the same effect as with other nitro-compounds, tending to decompose them.

Dynamite made from properly purified nitroglycerine should, however, keep indefinitely at ordinary storage temperature.

Water in contact with dynamite displaces the nitroglycerine. This principle is made use of in collecting nitroglycerine from dynamite for test. All dynamite which has been exposed to water is dangerous.

Dynamite requires a much more violent shock than nitroglycerine to explode it. Iron on iron, or iron on stone, will explode it, but wood on wood will not. It is more sensitive to shock in proportion as the percentage of nitroglycerine increases.

A small-arm bullet fired at short range into dynamite will explode it.

TABLE SHOWING THE PERCENTAGE OF NITROGLYCERINE CONTAINED IN THE DIFFERENT GRADES OF "ATLAS" POWDER, AND THE DISTINGUISHING MARKS OF CORRESPONDING GRADES OF "HERCULES," "GIANT," "ÆTNA," "HECLA," "JUDSON," AND "FORCITE" POWDERS.

Atlas (Standard).		Hercules.		Giant.		Ætna.		Hecla.		Judson.		Forcite.	
Brand.	Per cent, N. G.	Brand.	Per cent, N. G.	Brand.	Per cent, N. G.	Brand.	Per cent, N. G.	Brand.	Per cent, N. G.	Brand.	Per cent, N. G.	Brand.	Per cent, N. G.
										R. R. P.	5 and under		
										F.	10		
				Extra	20	No. 5	15	....	.....	F. F.	15		
E	20	No. 4	20	M	20	No. 4	20	No. 3	20	F. F. F.	20	No. 3 E	20
				F. F. F.F.	20	No. 4 X	25	....	.....	....	.....	No. 3 D	25
E+	27	No. 4 S	27	X X X	27	No. 3 B	27	No. 3 X	27	Dbl. Ex.	27		
D	30	....	.....	....	.....	No. 3	30	No. 2	30	....	.....	No. 3 C	30
D+	33	No. 3 S	33	No. 2 C	33	No. 3 A	33	No. 2 X	33	Tpl. Ex.	33	No. 3 B	32
						No. 3 X	35						
C	40	No. 2	40	No. 2	40	No. 2	40	No. 1	40	....	.....	No. 3	40
C+	45	No. 2 S	45	No. 2 Ex.	45	No. 2 X	45	....	.....	....	.....	No. 3 X	45
B	50	No. 2 SS	50	New No. 1	50	No. 2 XX	50	No. 1 X	50	....	.....	No. 2	50
B+	60	No. 1	60	....	.....	No. 1	60	....	.....	....	.....	No. 2 X	60
A	75	No. 1 XX	75	No. 1	75	....	.....	No. 1 XX	75	....	.....	No. 1	75

(d) Explosive Gelatin.

In 1875 Nobel introduced a new type of explosive—a mixture of collodion cotton and nitroglycerine—under the name of "blasting gelatin." The chemical principle involved in this explosive is that a more complete combustion takes place with the mixture than with either ingredient; the excess of oxygen in the products of explosion of nitroglycerine supplying the deficiency in the explosion of nitrocellulose, causing the C to burn to CO_2, instead of partly to CO; this additional chemical action greatly increasing the *heat*, hence the *volume* and *force* of the explosion. The proportions vary, but may be taken at about 90 per cent of nitroglycerine with about 10 per cent of nitrocellulose completely soluble in ether-alcohol. According to Berthelot the proportions of the mixture are 93 to 95 parts of nitroglycerine, 7 to 5 parts of collodion cotton. The mixing is done in troughs at a temperature of 122° F. with wooden spades, and when the mass is so gelatinized as to make it difficult to work with spades it is kneaded by hand, like bread-dough, until it has a smooth, even consistency. It is then removed and allowed to cool, finally the mass becomes a rather firm, compact, jelly-like substance, soft enough to be easily cut by a knife. The finished product is worked into cylindrical cartridges, but this cannot be done in presses as with ordinary dynamite. As a rule, the gelatinous mass is placed in an inclined cylinder in which an Archimedean screw revolves. The action of the screw is to force the gelatin to the upper end of the cylinder and out through a circular orifice in that end, forming a continuous "cable" or "rope" of the explosive. This cable is cut by a bronze knife into the lengths desired. These cartridges are wrapped with paraffined paper, the same as ordinary dynamite-cartridges.

This substance is the most powerful explosive known, having, according to Abbot's experiments, 17 per cent greater intensity of action than Dynamite No. 1. According to Berthelot, by theory, it should have 30 per cent more power.

Explosive gelatin is a yellow or light brown, gelatinous, elastic mixture, more stable than ordinary dynamite. It differs from ordinary dynamite, also, in that ordinary pressure does not cause the nitroglycerine to exude, and it is not affected by the action of water, except at the surface. One gram of mercury fulminate is required to detonate uncamphorated explosive gelatin. By adding to the mixture a small quantity of benzene, or, better, camphor (1 to 4 per cent), it is rendered insensitive to ordinary shock and friction, but at the same time it requires a more powerful primer to detonate it; the addition of camphor also raises the temperature of explosion to above 300° C.; if mixed with 10 per cent of camphor it fuses without explosion. The special primer required to detonate camphorated blasting gelatin consists of 60 parts of nitroglycerine and 40 parts nitrohydrocellulose. The initial shock required to detonate blasting gelatin is six times greater than that required to detonate ordinary dynamite. Owing to these causes blasting gelatin is far less sensitive to explosion by influence; the sensitiveness of blasting gelatin varies in a general way inversely as the quantity of nitrocellulose used in the mixture.

A cartridge of blasting gelatin placed in water will turn white on its surface, owing to the fact that nitroglycerine in the outer layer is displaced. The nitrocellulose remaining forms a protective coating to the rest of the mass.

At ordinary temperature it is much less sensitive than ordinary dynamite; in the frozen state it is very sensitive to shock, and, in this respect, the opposite of ordinary dynamite.

It burns in the open air without exploding, when small quantities are used.

It is very stable under the action of heat, keeping for days unchanged at 70° C., but its sensitiveness is increased by heat. Slowly heated to 204° C. it explodes.

The potassium-iodide-starch stability test is for 10 minutes, but it will often stand the heat of the test for from 40 to 60 minutes.

The action of blasting gelatin is too violent for many purposes, and modifications of it have been introduced. The explosive gelatin, as made above, or a little thinner (using less nitrocellulose), is mixed with other substances, with a view to deaden the violence or prolong the duration of the explosive force, in the same manner as is done in ordinary dynamites. A large variety of these mixtures have been suggested; one of them is known as *Gelatin Dynamite*, and has the following composition:

Explosive gelatin.	65 per cent
Base.	35 " "

The explosive gelatin has only 3 per cent of nitrocellulose.

The base is a powder made up of 75 parts of sodium nitrate, 24 parts of wood-pulp, and 1 part of sodium carbonate.

Analogous to these is a class of explosives, in which nitrocellulose is mixed with oxidizers like the metallic nitrates, and the mass held together by some cementing matrix, such as paraffin, gums, resins, etc.[1] There is a great variety of this class of explosives; they are of interest only historically, and in the sense only that it was through them that the present composite smokeless powders were evolved. Originally they were used largely for blasting purposes.

(e) Picric-acid Derivatives.

Picric Acid and the Picrates have been considered (see pp. 76 and 79). Recently certain derivatives of these compounds have come into use as charges for shells. The compositions of these explosives are kept secret and cannot be given. They were first exploited in 1869 by Brugère and Designolles in France, and Abel in England. The powders proposed by these consisted of a mixture of ammonium picrate and saltpetre.

Melinite, used by the French as a shell-filler, is essentially picric acid alone or with other substances. Originally it was a mixture of picric acid and colloided nitrocellulose, later only fused picric acid was used, and Cundill says "there is some

[1] An excellent scheme for the chemical analysis of this class of explosives, suggested by Thos. B. Stillman, Ph.D., and Peter T. Austen, Ph.D., will be found in the *Bulletin de la Société Chimique de Paris*, April 20, 1906, and an English translation thereof in *The Chemical Engineer* for July, 1906.

reason to believe that nitrobenzol or a similar material is employed as well."

Lyddite is the English equivalent of melinite.

The later forms of these shell-filler explosives, such as were used by the Japanese and Russians in the recent war, are thought to be either pure picric acid or a mixture of it with a nitro-compound, of the aromatic series, as suggested by Cundill.

Much attention has been given to the use of high explosives in shell by the U. S. Army Ordnance Board, with results superior, it is thought, to those attained elsewhere.

The most successful explosive of this type in the United States is Explosive D.

Explosive D is not fusible; it is used as a shell-filler by compression; this is considered a disadvantage, both because the density of charging is less and because application of pressure of such magnitude as is necessary to properly charge shell introduces a source of danger. Explosive D is, however, the least sensitive to shock of all the explosives named, and this is a very great advantage.

A high explosive for charging shell must fulfil many conditions, some almost contradictory, in order to be thoroughly serviceable.

The Ordnance Board, U. S. Army, enumerates the following requirements for high explosives for shell:

I. Information to be Furnished by the Inventor.

1. Form of material (mealed, crystalline, plastic, or molded).
2. Chemical composition.
3. Facility of manufacture and time required for manufacture.
4. Relative strength or force as compared with black rifle powder or picric acid. If estimated, explain method; if determined by experiment, state how.
5. Commercial purposes for which the material is or may be used.

6. Results of tests, if any, that have been made to show:
   (*a*) Safety in handling.
   (*b*) Sensitiveness to friction and shock.
   (*c*) Means required to produce good detonation.
   (*d*) Keeping qualities under exposure to moisture, heat, cold, and continued storage.
7. Actual firing tests from guns, if any.
8. Proposed method of loading in shell.

II. Laboratory Examination.

1. Complete quantitative chemical analysis and determine calculated force and temperature of explosion.

2. Test in Trauzel lead blocks to determine strength in comparison with that of known explosives.

3. Tests with standard impact testing machine to determine sensitiveness to blow in comparison with that of known explosives.

4. Determine relative ease of detonation by exploding lead or tin tubes, using mercuric fulminate detonators of varying strengths.

5. Determine the solubility of the explosives or any of its components in water and the effect of water on the ease of detonation.

6. Note hygroscopic qualities.

7. Determine chemical action on metals, especially iron, and determine sensitiveness of salts formed, if any.

8. Determine probable ease of loading into projectiles as compared with that of known explosives.

9. Determine maximum practicable gravimetric density when loaded into projectiles.

10. Determine "residue from flash" and "mineral ash" as approximate measures of the production of smoke.

11. Determine the stability by noting rate of decomposition at 65.5° C. of dry and wet samples of the explosive.

12. Determine the temperature of ignition and the character of burning in open air.

13. Determine the melting point, if any, or if the explosive is a mechanical mixture of chemical compounds, some of which have melting points, determine them.

III. General Requirements

A satisfactory high explosive for shell should fulfil the fellowing requirements:

Safety and Insensitiveness.

1. Should be reasonably safe in the manufacture and free from very injurious effects on the operatives.

2. Must show a safe degree of insensitiveness in the impact machine.

3. Must withstand the maximum shock of discharge under repeated firings in the shell for which it is intended.

4. Must withstand the shock of impact without explosion when fired in unfused shells against the strongest target that the shell alone will perforate without breaking up as follows:

(*a*) *Field Shell.*—With maximum velocity, against 3 feet of oak timber backed by sand. With remaining velocity, that of full-service charge at 1000 yards, against seasoned brick wall.

(*b*) *Siege Shell.*—With remaining velocity, that of full-service charge at 500 yards, against seasoned concrete thicker than shell will perforate.

(*c*) *Armor-piercing Shell.*—Against a 7-inch tempered steel plate in the case of a 12-inch A.P. shell, with striking velocity just sufficient to perforate the plate.

As a preliminary test the explosive is fired as a charge of a 6-pounder shell against steel plates of varying thickness to determine the limit of sensitiveness of the explosive. The limit of perforation for a 6-pounder shell is, approximately, a 3-inch steel plate.

Detonation and Strength.

1. Must be uniformly and completely detonated with the service-detonating fuse.

2. Should possess the greatest strength compatible with other necessary requirements.

Stability.

1. Must not decompose when a dry or wet sample is hermetically sealed and subjected to a temperature of 65.5° C. for one week.

2. Should be preferably non-hygroscopic, and must not have its facility for detonation affected by moisture that can be absorbed under ordinary atmospheric conditions of storage and handling.

3. Must not attack ordinary metals used in projectiles and fuses, especially iron, to an extent that cannot be prevented by simple means, or, if the explosive forms iron salts, they should be relatively insensitive.

4. Must not deteriorate or undergo chemical change in storage.

Charging Shell.

1. *Safety.*—Charging must not be attended with unusual danger, and should not require exceptional skill or tedious methods.

2. *Efficiency.*—It is very desirable that shell may be charged by pouring in the explosive in fused state, or by inserting the charge in the form of densely compressed blocks.

Supply.

It should be possible to obtain large quantities of it quickly and at reasonable cost.

The following table gives the data of tests made with the explosives that were favorably considered by the Army Ordnance Board:

Nature of Property.	Rend-rock, No. 400	Picric Acid.	Maximite, 20 per cent.	Explosive D.	Gun-cotton Pellets.
Relative force for actual density of loading in shell referred to guncotton as unity	2.12	2.87	1.91	1.81	1.00
Specific gravity	3.62	1.70	1.55	1.64	1.40
Density of loading in shell	2.54	1.66	1.58	1.31	0.73
Charge in pounds contained in 100 cubic inches	9.00	5.84	5.67	4.75	2.60
Cost of same charge (estimated)	$3.60	$2.04	$2.83	$1.80	$1.69
Method of charging	Melted	Melted	Melted	Bulk compression	Pellets, wax matrix
Requirements.					
1. Safety in manufacture	Yes	Yes	Yes	Yes	Yes
2. Impact test	Yes	Yes	Yes	Yes	Yes
3. Shock of discharge in gun	Tested	Tested	Tested	Tested	Tested
4. Shock of impact (*a*), (*b*), (*c*)[1]	(*a*) " (*b*) "	(*a*) " (*b*) " (*c*) "	(*b*) " (*c*) "	(*b*) " (*c*) " Yes	(*b*) " (*c*) " Yes
5. Facility of detonation	Yes	Yes	Yes	Tested	Yes
6. Relative strength in shell	2.12	2.87	1.91	1.81	1.00
7. Stability (heat) test	Yes	Yes	Yes	Yes	Yes
8. Non-hygroscopic	Yes	Yes	Yes	Yes	15% H_2O
9. Non-action on metals	Yes	Metals must be protected	Metals must be protected	Yes	Yes
10. Storage stability	Tested	Not tested	Not tested	Tested	Tested

[1] See page 164, Safety and Insensitiveness.

VI.

EXPLODERS.

As a rule the active ingredients of all exploders is fulminate of mercury.[1] The explosive used may require some adjustment of the quantity of fulminate in order to obtain an explosion of the proper order, or it may require some other ingredient to be mixed with the fulminate of mercury, such as chlorate or nitrate of potassium, sulphide of antimony, etc., but, as a rule, fulminate of mercury is present.

The ingredients of cap and primer composition vary with the kind of explosive that is to be exploded. Dynamite, guncotton, picric acid, and progressive explosives each require a different cap or primer composition. Especially is the nature of the initial blow important in progressive explosives. If the primer's flame be lacking in kind or amount some of the powder may not be burned in the gun; if it be excessive, it will be burned too soon and give too great pressures. Much attention has been given to this question. Experiments have been conducted to determine the primers best adapted to different explosives—to ascertain for each explosive the proper energy, heating effect, shape, size, and duration of the flame of the cap composition. Photography has been introduced, and it has been found that the photographs of cap and primer flames are characteristic in each case.

[1] Fulminate of mercury was discarded in 1899 from small-arm percussion composition in the United States Army service, owing to the danger of handling and to the fact that the vapor of mercury liberated when the fulminate primer is exploded attacks, the metal of the cartridge case, rendering it brittle, and thus making it impracticable to use the case in reloading. A chlorate mixture has been substituted therefor, having the following composition :

Chlorate of potassium	47.2
Sulphide of antimony	30.8
Sulphur	22.0
	100.0

There are fulminates of silver and gold, but they are too sensitive to have any uses in military explosives.

Mercury fulminate is formed by treating metallic mercury with nitric acid and alcohol.

The chemical reactions which take place are not fully agreed on among chemists. Bloxam gives the following explanation:

> When nitric acid acts on alcohol several products are obtained, among which are *nitrous acid* and some *hydrocyanic acid* (HCN). The formation of the CN group in this reaction may be explained by the tendency of nitrous acid to substitute N for H_3 in organic compounds, and it might be expected that the action of nitrous acid on alcohol would be

$$CH_3.CH_2.HO + 2HNO_2 = CN.CN.(HO)_2 + 3H_2O.$$

The group $CN.CN.(HO)_2$ is too unstable to exist separately. This is the hypothetical *fulminic acid.* If it be assumed to exist in the course of the reaction, its production in the presence of mercury would, under the usual laws governing chemical changes, exchange its hydrogen for mercury, in accordance with the following reaction:

$$CN.CN.HO.HO + Hg = CN.CN.O.HgO + H_2.$$

The structural formula (following Bloxam) may be represented as follows:

$$\begin{array}{l} H—O—C = N \\ \qquad\quad\;\; | \quad\;\; | \\ H—O—C = N \end{array}$$

According to Guttmann, Kekulé demonstrated, with a fair amount of accuracy, that fulminate of mercury should have the following rational formula: $C(NO_2)(CN)Hg$. He bases his conclusions on the reactions of fulminate of mercury with chlorine, bromine, and hydrogen sulphide. This would suggest the following structural formulas for fulminic acid and mercury fulminate:

Fulminic acid: 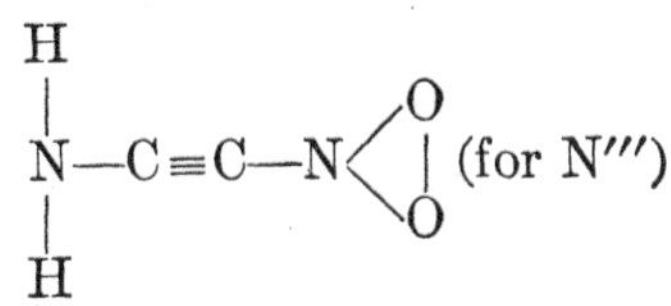

Mercury fulminate: 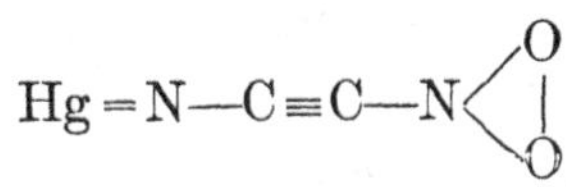

The assumption of a fulminic acid is supported by the actual existence of a mono- and tri-hydroxide of CN. The mono, CNHO, *cyanic acid,* a colorless liquid, specific gravity 1.4, and $(CN)_3(HO)_3$, *cyanuric acid,* a crystalline solid, a tribasic acid, forming salts with metals, corresponding respectively to the structural formulas

H—O—C≡N,

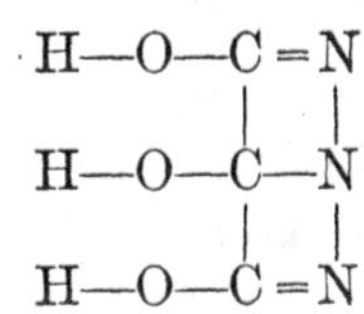

It is reasonable to assume that an intermediate hydroxide exists having two HO groups. Moreover, while the acid has not been separately produced salts, double, acid, and normal, corresponding to a bibasic acid have been produced, some of which are the following:

Mercury fulminate: 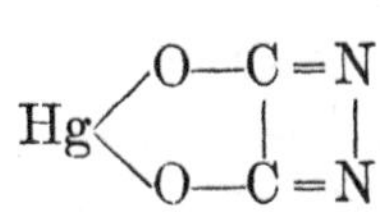

Silver fulminate: 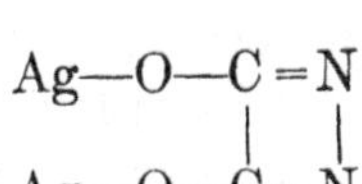

Silver-ammonium fulminate:

$$\begin{array}{l} NH_4—O—C=N \\ \qquad\quad\;\; | \quad\; | \\ \;\;\; Ag—O—C=N \end{array}$$

Silver-potassium fulminate:

$$\begin{array}{l} \;\; K—O—C=N \\ \qquad\quad | \quad\; | \\ Ag—O—C=N \end{array}$$

The manufacture of fulminate of mercury is conducted as follows:

Mercury and nitric acid (specific gravity 1.38) are mixed in a glass carboy in equal parts by weight. The mercury dissolves in the nitric acid and, when completely dissolved, the contents are allowed to cool; it is well shaken to secure uniformity of product, and then this solution is emptied into a second carboy which contains 10 parts of ethyl alcohol.

The second carboy is kept at a temperature above 60° F., and is connected with a series of receivers which stand in a trough through which water circulates. The pipe from the last receiver leads into a condensing chimney or tower.

After a few minutes the reaction begins in the second carboy, the liquid boils, and white vapors of nitric and acetic ether, aldehyde, carbonic acid, hydrocyanic acid, and some volatile compounds of mercury rise and pass off through the series of conducting pipes and receivers to the condensing tower. As the action proceeds the color of the vapors change from white to the red fumes of nitric peroxide.

In about fifteen minutes the crystals of fulminate of mercury separate from the solution in the second carboy in the form of small gray-colored needles. As soon as the reaction is completed the contents are allowed to cool, and are then poured out on a cloth filter stretched on a wooden form. These contents are then washed with pure water, until the washings show no trace of acid when tested with blue litmus. The filter is then placed in a drying atmosphere, out of the direct rays of the sun, and allowed to dry until the mass of fulminate contains only 10 to 15 per cent of water. The yield is about 125 parts of fulminate of mercury to 100 parts of mercury. Theoretically there should be a yield of 142 parts per 100 parts of mercury. Great care must be exercised that no particles of fulminate are scattered about; any suspicious particles should be treated with sodium-sulphide solution.

The principal product is usually made up in packages containing 120 grains. It is put up with about 15 per cent of water and hermetically sealed, to prevent evaporation, as it is much more sensitive to shock and friction in the dry state.

When it is necessary to dry it for use in caps and detonators great care must be exercised. The temperature must be kept below 104° F., and the dry fulminate handled with the greatest care.

Pure crystals of fulminate of mercury have a yellowish-white shade. The gray color of the commercial fulminate is due to small particles of unconverted mercury. The pure fulminate is obtained by boiling in a large volume of distilled water, drawing off the hot liquid from which the pure fulminate crystallizes on cooling in the form of a yellowish-white, silky mass. This, examined under the microscope, appears as groups of crystals. The fine crystals are more desirable for use in detonators than the coarser ones. Mercury fulminate should not be kept in a stoppered bottle, especially not one having a glass stopper, as the friction of removing and inserting the stopper might detonate a particle of fulminate caught in the neck of the bottle and transmit the explosion to the whole mass. A moderate blow of a hammer causes it to explode with a bright flash and gray fumes of mercury. It is detonated if touched with a wire heated to 195° C. or by an electric spark, by contact with strong sulphuric or nitric acid, or sparks from metals or flint. Its specific gravity is 4.42. The volume of the gases evolved is 1340 times the volume of the solid fulminate at ordinary temperatures and pressures; this would be greatly increased by the temperature of the explosion. The explosive nature of the fulminate is due to the fact that the molecule contains an oxidizing group (HgO_2) and a cyanogen (combustible) group $(CN)_2$.

Heated slowly it explodes at 305° F. (152° C.); heated rapidly it explodes at 368° F. (187° C.).

The nature of the surfaces between which the fulminate is confined when struck has an effect on its explosion; between hard rigid surfaces, like iron or steel, the explosion is certain; between soft metal surfaces, like lead, not so certain; between wooden surfaces, doubtful.

The slower the crystallization the larger the crystals, and the larger the crystals the more sensitive is the product.

When moistened with 5 to 30 per cent of water the sensitiveness is greatly reduced; if struck in this state by a hammer on iron, only that portion directly between the surfaces will explode. The explosion of a quantity of dry fulminate in contact with wet fulminate will explode the latter, even if immersed in water.

Fulminate may be subjected to high pressure without explosion, *if pure;* if sand or grit be present, the slightest pressure may explode it.

Fulminate of mercury is used very little, except in caps and primers. It often has mixed with it other substances, such as potassium chlorate, sulphide of antimony, powdered glass, etc., to modify the nature of the explosive blow, producing a prolonged action and a penetrating heat which enters deep into the mass of the explosive. The addition of oxidizing substances, like potassium chlorate, serves to increase the heat, both because the latter is an endothermic substance and because the oxygen it supplies serves to burn the CO of the products of combustion of mercury fulminate to CO_2, and thus still further increases the heat. Powdered glass is often added to increase the sensitiveness to percussion. Sulphide of antimony also increases sensitiveness, and it combines with potassium chlorate, producing heat and prolonging the action of the fulminating mixture.

The heat of formation for one equivalent, that is, a weight proportional to the weight of a molecule of mercury fulminate,

(284 grams, the molugram) is −62,900 cals. Its heat of combustion in an inert atmosphere is +116,000 cals. for constant volume and 114,500 cals. for constant pressure. This would raise the products of explosion to 4200° C.

The explosive reaction is

$$HgO_2(CN)_2 \text{ (exploded)} = Hg + 2CO + N_2.$$

One gram of it should yield 235.8 cubic centimetres of gas at 0° C. and barometer of 76 centimetres. One molugram (284 grams) should yield 66.7 litres of gas.

It is to be noted, particularly, that the products of explosion are simple gases, except CO, and therefore dissociation does not take place in a marked degree.

The effect of mixing mercury fulminate with an oxidizer, as is done in some cap compositions, is noted in the following reaction:

$$3HgO_2(CN)_2 + 2KClO_3 \text{ (exploded)} = 3Hg + 6CO_2 + 2KCl + 3N_2.$$

The heat evolved is +258,000 cals. for one molugram, almost twice that for pure fulminate, but the initial blow is greatly prolonged, due to dissociation and recombination of CO_2 and KCl.

With nitre the explosive reaction is as follows:

$$5HgO_2(CN)_2 + 4KNO_3 = 5Hg + 8CO_2 + 7N_2 + 2K_2CO_3,$$

corresponding to +227,400 cals.

Exploding in its own volume mercury fulminate gives a pressure of 28,750 kgm., as compared with 12,376 kgm. for nitroglycerine and 9825 kgm. for guncotton.

The great value of mercury fulminate as an exploder is due to this enormous pressure, and to the fact of its suddenness, owing to the absence of dissociation; the pressure is, therefore, nearly that due to explosion in the volume of the original solid, which, relatively, is very small on account of the high specific gravity of mercury fulminate (4.42). The crushing effect on the molecules of an explosive in contact with mercury fulminate is overpowering, and accomplishes the disruption of the bonds holding the atoms in the molecules; the atoms, once thus released, enter into new combinations, according to their affinities under the new conditions.

Caps and primers for progressive explosives require a more prolonged blow than that given by pure fulminate. It is, therefore, the usual practice to mix nitre or potassium chlorate for this purpose. Munroe gives the following directions for making composition for percussion caps:

100 parts of dry fulminate are rubbed to a powder with 30 parts of distilled water, 50 to 60 parts of potassium nitrate, and 29 parts of sulphur.

The rubbing is done on a marble slab, using a wooden spatula.

This mixture is dried sufficiently to admit its granulation.

It is then forced by pressure into copper caps and covered with a layer of varnish or of tinfoil, to protect it from dampness. The varnish used may be a solution of gum mastic in turpentine.

The caps are finally dried by a gentle heat and packed in boxes.

Primers for detonating explosives, for purposes of demolition or destruction, are made of pure fulminate of mercury. Such primers, as a rule, are electric, although there is one type made for use with time-fuses.

The United States Navy electric primer, according to Munroe, consists of a copper case made in two parts. The lower part is a No. 36 metallic cartridge-case. The upper part is a copper tube, open at both ends, which has been cut from a No. 38 metallic cartridge-case. A thread is pressed on each of these parts, so that the upper part or cap screws nicely on the lower part. The lower part is filled with fulminate of mercury up to the lowest thread of the screw. The top part is filled with a cement plug made of sulphur and glass, through which the lead-wires or primer legs pass to connect the bridge with the wires leading to the battery. When the fulminate is dry the spaces in the lower case and the cap are filled with pulverulent dry guncotton, and then the parts are screwed together. The lead-wires should be long enough to protect the ends of the main conductor wires from destruction by the explosion, say 6 to 10 feet in length.

The bridge is practically the same for all primers. It consists of a piece of platinum-iridium alloy, about one-quarter inch long and .001 to .003 inch in diameter. Its resistance should be (bridge and short leads), cold, 0.3 to 1 ohm; hot, 0.45 to 2 ohms; insulation resistance between conductor and case, 1 megohm; strength of current to fire, 0.3 to 0.8 ampere. Usually a small wisp of dry guncotton is placed about the bridge; next to this is placed fine gunpowder for firing progressive powder-charges, or mercury fulminate for high explosive charges. The bridge is soldered to the bared ends of the lead-wires.

Commercial detonating-primers are made on the same general principle. A drawn copper tube, closed at one end, is used for the lower part of the primer. The upper tube contains a wooden plug sealed with sulphur, which carries the legs connecting the bridge with the leading wires.

A modification of these electric primers is made in which the wooden plug is omitted, leaving the mouth open for inserting a time-fuse train. In using a time-fuse insert the end so as to *touch the fulminate in the lower tube,* then crimp upper tube tightly down on time-fuse with pincers or crimpers.

The electric primer is the safest, simplest, cheapest, and most effective means of firing charges of high explosives; it is the only means used of firing separate charges simultaneously, or a single charge at a distant point, or at a required moment, or under water.

Different grades of commercial primers or blasting-caps are known to the trade. They are specified as single, double, triple, quadruple strength caps. These are charged with detonating composition as follows:

Single strength........	0.80 grams (12.3 grains)
Double strength.......	1.00 grams (15.4 grains)
Triple strength........	1.50 grams (23.1 grains)
Quadruplé strength....	2.00 grams (30.9 grains)

The detonating composition varies according to the character of the work to be done, but as a rule consists of 75 parts of fulminate of mercury and 25 parts of potassium chloraté pressed tightly into the lower tube; sométimes a little gum dissolved in alcohol is added to make the mass more coherent. The function of potassium chlorate, sulphur, nitrates, etc., in exploders has already been explained.

Blasting-caps are tested by inserting the cap in a cork with the base of the cap flush with the end of the cork, placing the cap with base resting on a piece of wrought iron, No. 14 A. W. G., supported on block under its four corners. An efficient cap should blow a clean hole through the iron.[1]

The standard army electrical primer for high explosives consists of the following details:

1. A wooden plug grooved longitudinally on opposite sides to receive the lead-wires, and cannelured around the middle.

[1] A recent and improved " sand test " method of testing blasting powder is that which has been developed by Messrs. C. G. Storm and W. C. Cope. In this test the detonator is buried in the center of 100 grams of " Ottawa standard sand " contained in a cylindrical chamber, approximately 15 cm. deep and 3.1 cm. in diameter, bored out of a steel block. The sand is practically pure quartz, passes entirely through a 20-mesh screen, and is held on a 30-mesh screen. The grains from different lots are of remarkably uniform size. After the detonator has been fired the sand is screened, and the quantity passing through a 30-mesh screen is regarded as a measure of the strength of the detonator. In a long series of tests Storm and Cope found that the result given by the " sand test " was a definite function of the weight of the charge and that the quantities of sand crushed by mercury fulminate and its chlorate mixtures was comparable to their relative efficiencies in causing complete detonation of nitro substitution compounds. The test is an accurate indication of the grade of commercial detonators. Most other direct tests, such as the lead-plate test, are not quantitative, or depend upon effects not easily or accurately measured. (Technical Paper 125, Department of Interior, Bureau of Mines, 1916.)

2. The lead-wires (of No. 18 A. W. G. copper wire, with braided and paraffined cotton insulation) are pressed into the grooves, half-way in one groove, then in the circumferential cut around half-way to the opposite groove, then longitudinally to the end of the plug, each wire leaving the plug in the side opposite to that on which it entered. The inside ends of the wires are bared, scraped, cut to a length of about 0.1″, tinned and resined, soldered to the fine wire bridge, and bent slightly toward each other.

3. This plug is covered with a cylindrical cap with a stout shoulder at one end and having a small hole for the passage of the lead-wires. The cap fits the plug closely. The plug smeared with glue is forced into the cap until the end of the plug abuts firmly against the shoulder, leaving a chamber around the bridge to receive the priming.

4. The priming-chamber filled with mercury fulminate (4 grs.) is closed by a paper disk held in position by a drop of collodion.

5. The bridge is made of fine platinum wire (.0025″ diameter, electrical resistance 3 ohms to the inch). This bridge will carry 0.1 to 0.15 ampere without heating, and this current may be used for testing; for firing, a current of about 0.5 ampere should be used; the length of the bridge is $\frac{7}{32}$-inch.

6. The body of the primer is made of a second copper cylinder closed at one end. It contains 20 grs. of fulminate of mercury, held in place by a paper disk secured by a drop of collodion. The body fits over the cap and is pushed up over it and crimped into the wood near the top.

The completed primer is 1.4-inch long. As soon as finished it is dipped into melted Japan wax, which gives an even waterproof coating.

The electrical resistance of the completed primer is between 0.7 and 0.8 ohm.

On account of the high cost of fulminate of mercury, experiments have been recently conducted by the Bureau of Mines looking to the replacement of a portion of the fulminate of mercury in primers by other explosives. A compound primer has given satisfactory results made up of a mixture of a base charge of 0.40 gram T.N.T. and tetryl[1] in the proportions of 80 per cent of the former to 20 per cent of the latter, combined with 0.32 gram of a mixture of fulminate of mercury and chlorate of potassium in the proportion of 90 to 10. This combination corresponds to a No. 6 primer in the series experimented with by Storm, a full description of which will be found in Technical Paper 145, Department of Interior, Bureau of Mines.

---

[1] Trinitromethylnitramine, which has the structural formula:

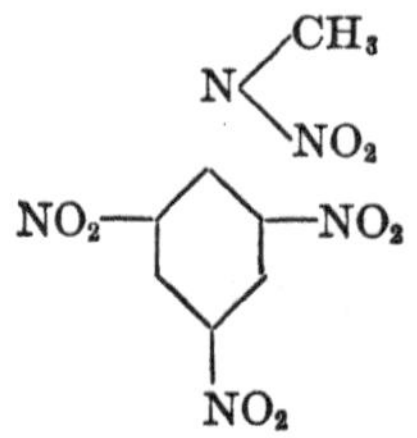

and is generally known as tetranitromethylanilin.

VII.

SERVICE TESTS OF EXPLOSIVES.

General Remarks on Tests.

From what has gone before it will be understood that it is of great importance that all explosives made by the action of nitric acid should be free of all impurities, especially of free acids used in their manufacture and of by-nitro-substances resulting from the action of nitric acid on the raw material which may not itself have been absolutely pure. If any of these remain in a nitro-explosive it is liable to decompose in course of time, especially if it be exposed to temperatures above 90° F.

All high explosives are, therefore, subjected to certain standard tests with a view to determine their stability, and especially the probability that they will not decompose in storage.

Heat of sufficient degree will decompose all nitro-compounds, and even when the heat is comparatively low it will decompose nitro-explosives if they be subjected to it for a long enough time, the time required to initiate decomposition being shorter as the temperature is higher.

It is assumed that the time required to cause incipient decomposition of a nitro-explosive is a measure of its stability in storage. Experience has shown that if a nitro-explosive will withstand the action of a certain temperature for a certain time its stability in storage may be assumed. These temperatures and times have come to be accepted as standard tests.

There are many different stability heat-tests which have been suggested by different experimenters (see Journal American Chemical Society, March and June, 1903; and Journal U. S. Artillery, September–October, 1903), but only three will be described. One known as the potassium-iodide-starch test, another as the litmus test, or 135° C. test, or German test; and the third as the U. S. Ordnance 115° C. powder-test. The first is used with all nitro-explosives, the 135° C. German test is used with nitrocellulose explosives, the 115° C. U. S. Ordnance test is at present used only with nitrocellulose powders.

In the potassium-iodide test the length of time is noted that is required to discolor a small test starch-paper saturated with potassium iodide by the nitric oxide liberated from the explosive by heat. In the litmus test the time is noted that is required to redden a litmus-test paper by fumes of NO_2. In the Army 115° C. test the rate of loss of weight of the sample is noted.

Before the heat-test is begun, preliminary tests for free acids should be made with blue litmus paper. The explosive in pulverulent state is placed in a test-tube (about 25 c.c.), the tube is then half filled with distilled water, closed with cork and shaken well; the liquid is allowed to settle; the supernatant liquid is decanted and tested with blue litmus or methyl-orange.

Nitrocellulose manufactured for use in making smokeless powder must also be examined for the presence of free alkali in the same way, using phenolphthalein as the indicator, and all nitrocelluloses are tested for the presence of mercury chloride in small quantity.

Apparatus Required for the Potassium-iodide-starch Test.

The apparatus required for making the potassium-iodide-starch test consists of a glass or copper globe or cylinder water-bath about 8 inches in diameter, with an aperture of about 5 inches; the bath is filled with water to within a quarter of an inch of the top edge. The aperture is closed by a loose cover of sheet copper about 6 inches in diameter. The globe rests on an ordinary iron tripod, so that the bottom of the globe is about 10 inches above the plane of the feet of the tripod.

A Berzelius alcohol-lamp[1] is placed under the globe. The cover has four to eleven holes: one in the center for a thermometer fitted into a rubber stopper; five to ten at equal distances around the circumference to receive test-tubes, each containing a sample of the explosive to be tested. The test-tubes after being carefully cleaned and dried are closed by clean corks, each carrying, through a hole bored in, it a glass rod with platinum-wire hook on the lower end; this hook during the test supports the potassium-iodide-starch test-paper. The test-tube corks are discarded after one test. The test-papers should be obtained from a standard source, as the value of the test depends chiefly on the uniformity and proper degree of sensitiveness of the test-papers. In case of emergency the potassium-iodide-starch test-paper may be made as follows:

> Forty-five grains of white maize starch (corn flour), previously washed with cold water, are added to 8½ ounces of distilled water, the mixture is stirred, and boiled for 10 minutes.
>
> Fifteen grains of pure potassium iodide (crystallized from alcohol) are dissolved in 8½ ounces of distilled water.
>
> The two solutions are thoroughly mixed and allowed to cool.
>
> Strips or sheets of white filter-paper, previously washed with water and redried, are dipped into the solution and allowed to remain in it for at least 10 seconds; they are then allowed to drain and dry in a place free from laboratory fumes and dust.
>
> The upper and lower margins of the strips are cut off.
>
> The paper is preserved in well-stoppered bottles *and in the dark.*

Freshly made and suitable paper should give no discoloration if touched with a glass rod holding a drop of acetic acid. When a brownish or bluish spot appears from acetic acid so applied the paper should be rejected. Often an exposure of

[1] Any convenient source of heat may be used.

one hour to bright light will destroy a set of test-papers. Papers over a month old are apt to be untrustworthy.

Owing to the differences obtained by different operators in making the KI-starch stability test, the joint Army and Navy Board on Smokeless Powder has recommended that a color scale of tiles, having standard tints, be used by operators, in all cases, to determine when the red line of proper tint appears on the KI-starch paper marking the completion of the test.

(a) Dynamite, Nitroglycerine, and Explosive Gelatin.

Dynamite.—If dynamite is to be tested the nitroglycerine must be extracted from the base. To accomplish this, advantage is taken of the fact that water will displace nitroglycerine from such mechanical mixtures as kieselguhr dynamite. The further test then becomes one simply of the nitroglycerine.

To Extract Nitroglycerine from Kieselguhr Dynamite.

A funnel, about 2 inches across, is arranged so as to filter into a small beaker. About 300 to 600 grains of dynamite finely divided are placed in the funnel, which has previously been loosely plugged by some asbestos wool. The latter should have been recently heated to white heat and allowed to cool.

The surface of the dynamite is smoothed off carefully by means of a flat-headed glass rod or stopper and some clean, washed and dried kieselguhr is spread over it to the depth of about one-eighth of an inch. This top layer of kieselguhr is then carefully and evenly saturated with distilled water by a fine jet from a water bottle. As soon as the first water has been absorbed into the mass of dynamite more is added. This is continued. The displaced nitroglycerine will, after some time, begin to drop into the measure below the funnel. The operation is discontinued when enough nitroglycerine has been collected to allow 50 grains for each test tube.

The Potassium-iodide-starch Heat-test.

Nitroglycerine.—The water-bath of the potassium-iodide-starch testing-apparatus is brought to 160° F. (71° C.) and maintained at that temperature, being regulated by the thermometer which should be immersed about 2¾ inches in the water. The source of heat should be carefully watched, and at no time should the temperature of the bath rise or fall more than 1° F. from 160° F. Fifty grains of nitroglycerine are placed in each test-tube and carefully weighed, being careful not to get any on the sides of the test-tube; this may be done by using a suitable dropper or glass tube.

A piece of test-paper is taken with the pincers and laid down on a piece of clean filter-paper. The test-paper is held in place by the end of a glass rod which has been thoroughly cleaned, heated, and cooled. A small hole is made in the test-paper with the point of the pincers opposite the middle of one end of the paper and about 0.2 inch from the edge. The test-paper is taken up with the pincers, the platinum hook inserted through the hole just made, the hook bent with the pincers until the throat of the hook is closed tightly on the paper, so that it will stand stiffly up when the paper is held vertically above the glass rod. The glass rod with test-paper is placed carefully aside under a bell glass or other protecting cover, where it will be protected from fumes and dust. In the same way the other test-papers are prepared.

A solution of pure glycerine and distilled water, in the proportion of 1 to 1, is prepared.

One of the test-papers is taken, held with the paper up, and a drop of the glycerine solution is placed on each of the lower corners of the test-paper, as held; the paper should absorb this evenly about half-way to the opposite upper edge, as held, leaving a distinct line about midway between the moistened and the unmoistened parts. One of the test tubes is placed in the bath through one of the apertures in the cover and is immersed until the sample is below the surface of the water. The test-paper moistened with glycerine is placed in the test-tube, and the glass rod is moved through the cork until the line between the moistened and unmoistened parts of the test-paper is about five-eighths of an inch above the upper surface of the cover. This time is recorded. The same is done with each of the other two test-papers. The line between the moistened and unmoistened parts of each test-paper is watched carefully, and the exact instant that a *faint brown color*[1] appears on this line of demarkation on each test-paper is recorded. This completes the test.

The nitroglycerine under examination will not be considered "thoroughly purified" unless the time elapsed between the insertion of the test-paper and the appearance of the brown color is at least fifteen minutes. The average of the records of all the tubes will be taken.

Explosive Gelatin.—If explosive gelatin is under examination a sample of 50 grains is intimately incorporated with 100 grains of French chalk, using a wooden pestle in a wooden mortar. The French chalk should be of good commercial quality; it should be thoroughly washed with distilled water, dried in a water-oven, and then exposed to moist air under a bell jar until it has taken up about 0.5 per cent of moisture. It should then be placed in a glass-stoppered jar for use.

Each test-tube is filled with this mixture to a depth of $1\frac{3}{4}$ inches, the tube being gently tapped on a table to insure a proper degree of settling.

The heat-test is then conducted as explained for nitroglycerine. Explosive gelatin will not be considered as serviceable unless the average time of the test is at least ten minutes.

Explosive gelatin is subjected also to a liquefaction and exudation test as follows:

Liquefaction Test of Explosive Gelatin.

A cylinder is cut from the cartridge having its height equal to its diameter, care being taken to have the ends cut flat and true.

This cylinder is placed on a piece of filter-paper on a smooth, clean board, and secured to the board by an ordinary pin forced through it along its axis into the board.

---

[1] In order to detect this color promptly, the water-bath should be so placed that a bright reflected light shall fall on the papers.

It is exposed in this condition for 144 consecutive hours to a temperature ranging from 85° to 90° F.

The original height of the cylinder should not decrease more than one-fourth, and the upper cut surface should retain its flatness and sharpness of edge.

Exudation Test of Explosive Gelatin.

There should be no separation of nitroglycerine in the liquefaction test or under any conditions of storage, transport, or use, or when the explosive is subjected three times in succession to alternate freezing and thawing.

(b) Guncotton.

Loose-fiber Guncotton.—The material is dried at a temperature not greater than 40° C. to constant weight; then exposed on trays to the air in a room free from fumes, until from 1 to 2 per cent of moisture has been absorbed. It is then gently rubbed through a ten-mesh sieve to insure uniformity of division, being careful that it does not come in contact with the hands or any piece of apparatus not perfectly free from any trace of acid or alkali. 1.3 grams are weighed out and placed in a test-tube 5½ to 6 inches long and not less than ½ inch internal diameter.

The potassium-iodide-starch test is conducted as explained for nitroglycerine, except that the water-bath is heated to 150° F. (65.5° C.). The test-papers, prepared as already explained, are inserted in the test-tubes,[1] and the papers adjusted in the tubes so that the line dividing the dry and moist portions of the test-paper *is on a level with the lower edge of the film of moisture which is deposited on the side of the tube soon after inserting it in the bath.*

---

[1] The standard water-bath for nitrocellulose holds ten tubes; it is long and narrow to prevent heating the upper part of the tubes as much as possible. Tubes are immersed 2¾ inches in the bath.

Nitrocellulose intended for the manufacture of smokeless powder for the Army and Navy must not show a brown color in less than 35 minutes at 65.5° C.

Blocks or Disks.—Guncotton for demolition purposes is issued in the form of compressed pulp, in disks or blocks. This form of guncotton is prepared for the heat-test as follows:

Sufficient material to serve for two or more tests is removed from the center of a block or disk by scraping, and reduced to a fine powder by rubbing between pieces of clean, dry filter-paper. This is spread out in a thin layer upon a paper tray about 6 by 4½ inches, which is then placed inside a water-oven, kept as nearly as possible at 120° F. for 15 minutes, the door of the oven being left wide open. The tray is then removed and exposed to the air of the room for two hours; during this time the material is rubbed on the paper tray with a clean glass rod and reduced to a fine and uniform state of division.

The temperature of the water-bath is the same as for fiber guncotton (150° F.).

There should be no brown color within 10 minutes.

Poacher Sample.—In case the sample is taken during the manufacture of nitrocellulose, it is taken after the poaching and after having been thoroughly washed in pure, cold water. The sample is pressed dry in a hand-press and rubbed in a clean cloth until finely divided, being careful not to let it come in contact with the hands.

(c) Smokeless Powder.

The sample should be prepared by cutting into slices 0.02 inch thick. These slices are exposed to the air for at least 12 hours.

The test-tube sample consists of 1.3 grams.

The usual potassium-iodide test is followed, except that the temperature is considerably higher for simple nitrocellulose powders, being 100° C. (212° F.) instead of 65.5° C. (150° F.).

Each sample must stand this temperature without showing a brown line for 10 minutes.

Powders containing nitroglycerine should stand the test at 65.5° C. for 20 minutes.

The British Government specifications prescribe the following times and temperatures for the potassium-iodide test:

1.	Nitroglycerine..........	15	minutes at	160° F. (71° C.).
2.	Dynamite..............	15	" "	160° F. (71° C.).
3.	Explosive gelatin.......	10	" "	160° F. (71° C.).
4.	Smokeless powders with nitroglycerine.........	15	" "	180° F. (82° C.).
5.	Guncotton.............	10	" "	170° F. (76.6° C.).
6.	Colloided pyrocellulose...	15	" "	180° F. (82° C.).

The German 135° C. Test.

Two and five-tenths grams of the sample to be tested are dried at the ordinary temperature of the laboratory for 12 hours and placed in a strong test-tube. A piece of blue litmus is placed in the tube about a half-inch above the sample, the paper being folded lightly so as to give the folds sufficient elastic power to hold the paper in place by pressure against the sides of the tube. The tube is lightly closed by a cork with a hole 0.15 of an inch in diameter bored through it, and so placed in a bath of boiling xylol (the boiling-point of which is 135°) that only 6 or 7 mm. project above the surface.

Examination of each tube is made each five minutes after twenty minutes have elapsed. In making this examination the tube should be withdrawn only half its length and quickly replaced.

Two tubes are used in each test, and there must be no failure in either tube.

Three observations are made: (1) Time of *complete* reddening of the litmus-paper; (2) time of appearance of brown nitric-peroxide fumes; (3) time at which the sample exploded.

Stable explosives should give the following times:

	Litmus not Reddened in	No Nitric Fumes in	No Explosion in
Uncolloided nitrocellulose.........	30 min.	45 min.	5 hrs.
Pure nitrocellulose powder.........	1 hr. 15 min.	2 hrs.	5 hrs.
Nitroglycerine powders............	30 min.	45 min.	5 hrs.

The substitution of methyl violet paper for litmus paper in the 135° German test has been recommended. When this substitution is made, the time of the test should be less than that with litmus paper by 5 minutes for uncolloided nitrocellulose and 15 minutes for finished powder.

For the results to have value they should be compared with that of a known stable explosive of the same kind, under the same test by the same operator, using the same test-paper.

Uncolloided nitrocellulose should be well shaken down in the tube by tapping, or lightly pressed down.

The U. S. Army Ordnance 115° C. Test.

(For nitrocellulose powders.)

Whole pieces of powder are carefully weighed on watch-glasses and then heated in an air-bath kept at 115° C. + or $-\frac{1}{2}$° for 8 hours. The sample is then removed, allowed to cool in a desiccator, and reweighed. This is repeated six times on six separate days. The oven is brought to the required temperature each day *before* inserting the samples. At the end of the 8-hour period the samples are removed and allowed to stand over night. At the end of the 6-day period the samples are allowed to cool in a desiccator, after which they are again weighed. The loss of weight must not exceed the limit shown by the test-curve, p. 187.

The air-bath may be maintained at 115° by filling the walls of the oven with a properly proportioned mixture of xylol and toluol. A reflux condenser prevents loss of the liquid by evaporation.

The temperature, 115° C., is the one that most clearly differentiates the decomposition of good powders from bad ones in a reasonable time limit. If a lower temperature is used, it requires too long a time to establish trustworthy data; if a higher temperature is used, the curves plotted to show the rate of loss of weight of good powders are not so clearly separated from those plotted to show the same for bad powders.

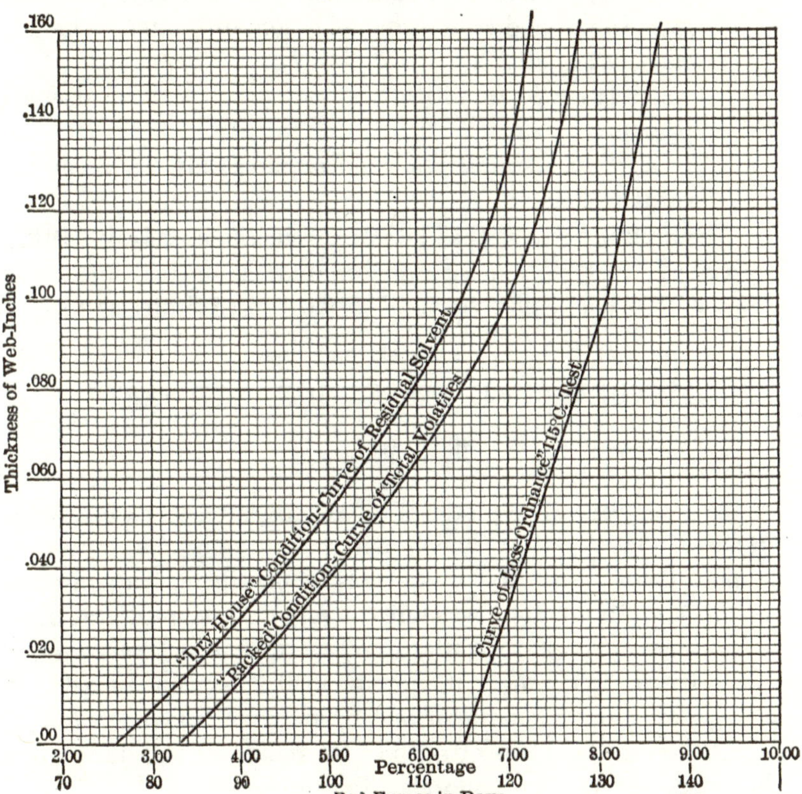

The following advantages are claimed for this test:

1. The powder is tested in its natural condition; the same in which it is stored or used.

2. It shows *all* products of decomposition; others show only *acid* or nitrogen losses by decomposition.

3. It shows the decomposition of other nitro-compounds than nitrocellulose which are often present in

powders, and shows the effect of these on the decomposition of the powder.

4. It shows the effect on the stability of powder of added substances, placed there to mask stability tests; the effect of volatiles which may set up local decomposition; traces of nitric acid; decomposition of the nitrocellulose due to saponification by water, alkalies, carbonates, etc.

5. It shows quantitatively the progress of all decompositions.

6. It is a simple test, and requires only simple apparatus to make it.

The following are the latest specifications (July, 1910) prescribed for powders for cannon, and for nitrocellulose for powders or other explosives used in the United States service.

MANUFACTURE, INSPECTION, AND TEST.

1. Raw Materials.

(*a*) *Cellulose.*—The material to be used is bleached cellulose, prepared for nitrating, which will be obtained by purifying unspun cotton wastes, or suitable short-fibered commercial cotton, and thoroughly washing to remove the purifying material or salts; it is to contain not more than 0.4 per cent of extractive matter, and not more than 0.8 per cent of ash; it is to be of uniform character, clean, and free from such lumps as will prevent uniform nitration. It should not contain more than "traces" of lime, chlorides, or sulphates.

The extractive matter will be determined by extracting not less than 1.5 grams of cotton in a Wiley or Soxhlet extractor with ethyl ether and weighing the extracted matter, after drying at 100° C.; the percentage is to be calculated on dry cotton. Ash will be determined by digesting about 1.5 grams of cotton with a little pure nitric acid, incinerating at a red heat, and weighing the residue, the percentage to be calculated on dry cotton.

Moisture will be determined by drying not less than 3 grams of cotton, at 105° C., to constant weight.

(*b*) *Acids.*—A mixture of sulphuric and nitric acids will be used, containing no metallic salts other than salts of iron, and not more than a trace of chlorine compounds.

(*c*) *Ether.*—Ethyl ether will be used, containing no impurities other than small quantities of water and ethyl alcohol. The ether to be clear and colorless, with characteristic pure odor, having less than 0.006 per cent acidity, calculated as acetic acid, and less than 0.002 per cent residue after evaporation and drying at 100° C.; and specific gravity, at 20° C., to be from 0.717 to 0.723.

(*d*) *Alcohol.*—Ethyl alcohol 92.3 per cent absolute (by weight) will be used; it is to be of the best quality, clear, and colorless, with characteristic pure odor, having less than 0.006 per cent residue after evaporation and drying at 100° C., and acidity less than 0.01 per cent, calculated as acetic acid. It shall be subjected to the silver nitrate test, as follows:

3 grams $AgNO_3$, c. p.	Make up to 100 c.c.
3 grams NaOH, c. p.	
20 grams NH_4OH, c. p. (sp. gr. 0.90)	

Ten cubic centimeters of the sample, diluted with 10 cubic centimeters of water, to be placed in a tight bottle and 1 cubic centimeter of the silver nitrate solution added. Allow to stand one hour in the dark and examine for unreduced silver salts in clear solution, after filtering; if such are found, the alcohol contains less than the allowable amount of aldehyde.

The strength of alcohol is calculated by the use of the alcohol tables published by the United States Bureau of Standards.

(*e*) Ether and alcohol obtained from any of the manufacturing processes as recovered solvent are, before use for colloiding, to be put in condition for fulfilling the requirements of (*c*) and (*d*).

(*f*) *Graphite.*—If graphite is used on the surface of powder grains, or is incorporated in the powder, it shall be dry, ground very fine, and shall contain not more than a trace of silicates or compounds of sulphur, and shall be free from sulphur and acids.

(*g*) *Carbonate of soda.*—The best quality of refined alkali, free from sulphides, containing not less than 96 per cent of Na_2CO_3, calculated on dry samples, will be used.

2. Nitrocellulose—Manufacturing Processes.

(*a*) *Quality.*—

(1) The nitrocellulose in finished poacher lots shall have a nitration of 12.60 per cent, ± 0.1 per cent. These lots may be made up by blending nitrocellulose which contains from 12.45 to 12.75 per cent nitrogen and at least 95 per cent solubility.

(2) It shall have a solubility of at least 95 per cent at 15.5° C. in a mixture of two volumes of ether and one volume of alcohol, both of the standard quality prescribed by these specifications.

(3) It shall contain less than 0.4 per cent of material insoluble in acetone.

(4) It shall leave, after ignition, less than 0.4 per cent of ash.

(5) It shall give a heat test, at 65.5° C., with potassium iodide starch paper, of at least thirty-five minutes.

(6) It shall give a "German" test, at 135° C., with litmus paper, of at least thirty minutes.

(7) It shall contain no alkali, mercuric chloride, or other substance which will mask the heat tests in any way.

(8) It shall be uniformly pulped, free from lumps, strings, or material of such consistency as to affect proper colloiding in the mixers.

(*b*) *Nitrating.*—Cellulose of standard quality shall be thoroughly dried at a temperature not exceeding 110° C. When cold, this cellulose shall be nitrated in mixed nitric and sulphuric acids. After nitrating, the nitrocellulose shall be washed in water before boiling.

(*c*) *Preliminary boiling.*—The nitrocellulose shall next be boiled for at least forty hours, with not less than four changes of water, in tubs so constructed that the nitrocellulose shall not come into direct contact with the heating coils or with the steam from the coils. There shall be complete ebullition, or boiling, over the entire surface of the tubs. No alkali shall be used in the preliminary boiling.

(*d*) *Pulping.*—The nitrocellulose shall next be pulped in fresh water, to which may be added just enough sodium-carbonate solution to preserve a slight alkaline reaction to phenolphthalein solution, the process to continue until the material is thoroughly and evenly pulped to a satisfactory degree of fineness, and shows a clean break when a handful is squeezed and broken into parts. During this process the water shall be changed to such an extent as may be necessary to remove impurities.

(*e*) *Poaching.*—After pulping, the nitrocellulose pulp shall be run into the poachers, settled, and the water decanted. The nitrocellulose shall then be boiled for six hours in fresh water, during which time a total of not more than 10 gallons of carbonate of soda solution for each 2,000 pounds of dry nitrocellulose may be added at intervals; this solution shall contain 1 pound of carbonate of soda to the gallon. During this and all other boiling in the poachers the pulp shall be thoroughly agitated by mechanical stirrers. After boiling, the nitrocellulose shall be allowed to settle and the clear water decanted as completely as possible. The tub shall then be refilled with fresh water, boiled for two hours, settled, decanted, and refilled with fresh water. The boiling shall then be continued for one hour, and this process repeated three times, making a total boiling treatment in the poachers as follows:

Six hours' boiling, with or without sodium carbonate, settle, change water.

Two hours' boiling, no soda, settle, change water.

One hours' boiling, no soda, settle, change water.
One hours' boiling, no soda, settle, change water.
One hours' boiling, no soda, settle, change water.
One hours' boiling, no soda, settle, change water.
Total, twelve hours' boiling with five changes of water.

After boiling, the nitrocellulose shall have ten cold-water washings, each washing to consist of agitation, by mechanical means, for one-half hour in a sufficient amount of fresh water, thorough settling, and decanting the clear water; in decanting, at least 40 per cent of the total contents of the poacher shall be drawn off. A sample shall then be taken for subjection to the various tests prescribed for nitrocellulose. Should the nitrocellulose fail to meet the required heat tests, it must be boiled again with two changes of water, the time of actual boiling being five hours, without the use of alkali, and must then be given the ten cold-water washings in the manner prescribed for the regular treatment.

The utmost cleanliness shall be observed in manufacture. All machinery, tools, and appliances shall be kept in the condition necessary to prevent the incorporation in the nitrocellulose of foreign matter of any kind. At all stages of the process the water used shall be clean and free from deleterious matter.

3. Testing Nitrocellulose.

(*a*) *Sampling.*—Each poacher lot or blend will be given a designating number. A sample of about 150 grams dry weight shall be selected by the inspector from each poacher charge after purification is complete, and this shall be properly marked and sent for examination. If a lot or blend is made up by blending various weights of nitrocellulose of nitration of 12.45 to 12.75 per cent, each nitrocellulose included therein shall be similarly sampled for analysis.

(*b*) *65.5° C. heat test, with potassium iodide starch paper.*—The sample shall be pressed in a clean cloth or wrung in a wringer, if it contains a large excess of water. The cake shall be rubbed up in a cloth until fine (taking care that it does not come in contact with the hands), spread out on clean paper trays, and dried in an air bath at 35° to 43° C. for a sufficient length of time to reduce the moisture to the amount required to give a minimum heat test; this amount being from 1.5 to 2 per cent. If, as sometimes happens in dry weather, the moisture has been reduced to less than 1.5 per cent, the sample shall be placed in a moist atmosphere for a time not exceeding two hours, until the required moisture percentage is obtained. The whole time of drying and making the test shall not exceed eight hours.

The dried sample for the heat test shall be weighed out in five test-tubes, 1.3 grams (20 grains) to each tube, so that a series is obtained covering the widest variation allowed for moisture. These tubes are standard, 5½ inches long, one-half inch internal diameter, and five-eighths inch external diameter, closed by a clean cork stopper, fitting tightly, through which passes a tight glass rod with platinum holder for the paper; corks are discarded after one test. The nitrocellulose is pressed or shaken down in the tube until it occupies a space in the tube of 1⅝ inches. The test papers, about 1 inch in length and three-eighths inch wide, are hung on the platinum holders and moistened on the upper half with a 50 per cent solution of pure glycerin in water. The heating bath, carefully regulated at 65.5° C. $\pm$1° C., is placed so that a bright, reflected light is obtained, and tubes placed in the bath. Time is marked when tubes enter bath. As test continues, a slight film of moisture condenses on inside of tubes, and the line of demarcation between wet and dry test paper is kept abreast the lower edge of the moisture film. The first appearance of discoloration of the damp portion of the test paper marks the end of the test for each separate tube, the minimum test of any one of the five tubes being the heat test of the nitrocellulose. The discoloration is to be greater than that obtained at the same time by a blank test.

Standard test papers will be used and will be furnished by the department to manufacturers. The standard water bath holds ten tubes and is made long and narrow, to reduce to a minimum the heating of the upper portions of the tubes. These tubes are immersed in the bath to a standard depth of 2.25 inches.

(*c*) "*German*" *test at 135° C.*—A sample of nitrocellulose shall be dried at ordinary laboratory temperature over night, or, as for the heat test; 2.5 grams of the material are to be pressed into the lower 2 inches in each of two tubes, of heavy glass about 290 millimeters long, 18 millimeters outside diameter, and 15 millimeters inside diameter, closed with a cork stopper through which a hole 4 millimeters in diameter has been bored. A piece of standard blue litmus paper 70 millimeters long and 20 millimeters wide is placed in each tube, its lower edge 25 millimeters above the cotton. When the constant temperature bath has been carefully regulated at 134.5° C. ±0.5° C., these tubes are placed in the bath so that not more than 6 or 7 millimeters of length projects from bath. Examination of the tube is made by withdrawing about one-half its length and replacing quickly, each five minutes, after twenty minutes have elapsed.

The bath must be placed in a good light and with a suitable background. The standard litmus papers will be furnished by the department. The test shall be considered completed when the litmus paper is completely reddened, and the minimum test of either tube shall be taken as the test of the lot. The standard red color must not be obtained in less than thirty minutes.

(*d*) *Nitration.*—The nitration will be determined on a 1-gram sample of nitrocellulose after drying for one hour and a half at 95° to 100° C. or in a vacuum drier after a thorough air drying. The nitrocellulose is to be washed into a Du Pont nitrometer by 20 cubic centimeters of H_2SO_4 and the per cent of nitrogen determined by comparison of the gas given off with a standard volume. The acid used shall be chemically pure sulphuric acid containing 94 to 96 per cent H_2SO_4.

The nitrometer is standardized by preparation of a calculated standard volume of dry air at a temperature of 20° C. and 760 millimeters pressure at 20° C. in the comparison tube. When pure potassium nitrate is tested with the standard sulphuric acid against such comparison tube, the nitrogen figure should invariably be 13.85 per cent.

(*e*) *Ash and organic residue.*—Ash will be determined by decomposing the nitrocellulose with nitric acid, igniting and weighing the residue. The per cent of organic residue will be obtained by dissolving 1 gram of nitrocellulose in pure acetone, filtering by decantation, and, finally, on an asbestos filter, drying and then determining the loss by ignition.

(*f*) *Insoluble nitrocellulose.*—The amount of insoluble nitrocellulose will be determined by soaking 1 gram of the dry sample over night in 95 per cent alcohol. On the following morning the mixture will be brought to 15.5° C. and a sufficient amount of ethyl ether at the same temperature added to make the mixture of ether and alcohol 2 to 1 by volume; the mixture is to be kept at 15.5° C. for one hour. The insoluble nitrocellulose is then to be filtered off and weighed, a correction for the ash and organic material insoluble in acetone being made.

When the amounts of insoluble nitrocellulose and organic residue are very small, comparative volumetric readings may be made in long tubes, allowing the insoluble material to settle after regular treatment for solution in the solvents. The lower portions of these tubes are constricted to one-half inch in diameter, cylindrical shape, and graduated by direct weighing of residue.

(*g*) *Solubility.*—The amount of soluble nitrocellulose will be found by subtracting the sum of the ash, organic residue, and insoluble nitrocellulose from 100 per cent.

(*h*) *Acceptance.*—Lots of nitrocellulose which, after chemical examination, have been found satisfactory shall be provisionally accepted, subject to the powder made therefrom successfully passing the specified ballistic and chemical tests.

4. Smokeless Powder Manufacturing Processes.

(*a*) *Quality.*—

(1) Finished smokeless powder shall be a uniform ether-alcohol colloid of nitrocellulose of standard quality. No substance whatever, except as herein specified, and in the manner and at the time specified, shall be incorporated into the powder or its component parts during manufacture, storage, or use.

(2) The powder shall be granulated, except when otherwise permitted, into cylindrical grains with seven longitudinal perforations—one in the center of the grain and six at the vertices of a hexagon—so placed as to make the outer and inner web thicknesses equal, within the limits hereinafter specified. The grains shall be carefully cut and thoroughly sorted, so as to remove cracked, distorted, or spotted grains, lumps, uncolloided material, air cavities, butt ends, and long grains. Any grains developing marked discoloration before going to the dry houses shall be removed.

(3) The total percentage of volatiles contained shall not be greater than the limits shown by the "Curve of Volatiles," "dry house" and "packed" condition, forming part of the specifications.

(4) The powder shall have physical toughness sufficient to pass the prescribed test.

(5) It shall pass the 65.5° C. "surveillance" test prescribed for its particular web thickness.

(6) It shall give a "German" test at 135° C., with litmus paper, of at least one hour and fifteen minutes and shall not explode in less than five hours.

(7) The loss of weight in the 115° C. Ordnance Department test must not exceed the limit shown by the curve forming a part of these specifications.

(8) The powder shall be stable under any or all of the above tests. It must not show, by chemical analysis or test, the

presence of any unauthorized ingredients, or that the nitrocellulose and other material employed in the manufacture did not conform to the specifications.

(9) It shall successfully pass the ballistic requirements specified.

(*b*) *Dehydrating.*—Nitrocellulose which has been accepted for the manufacture of powder shall be dehydrated with standard alcohol to thoroughly remove water, and the excess alcohol shall be removed by pressure, leaving no more alcohol than that required for mixing. At least 1 pound of alcohol shall be used for each pound of dry nitrocellulose in each pressing.

(*c*) *Mixing.*—After dehydrating, the blocks of nitrocellulose shall be broken up, placed in suitable mechanical mixers, and the necessary amount of standard ether added. The amount of ether is determined by the climatic conditions, the number and character of operations after mixing, and the caliber of the powder being made; it shall not be less than 64 per cent of the total amount of solvent used in the mixing. An amount of diphenylamine, of approved purity, equal in weight to 0.4 per cent of the weight of the dry nitrocellulose in the mixer charge, shall be dissolved in the ether and added to the charge with it.

The mixing shall be continued until the solvent is uniformly distributed throughout the mass. Clean scrap may be reworked in the mixers, but all dirty scrap and foreign material must be removed.

The manufacturer is enjoined as to the necessity for taking measures to insure that the diphenylamine shall be thoroughly incorporated in each mixer charge. The following test is suggested:

From each mixer charge at the end of the process place a pinch of the colloid in a test-tube containing ether-alcohol, shake until dissolved, then add an equal quantity of reagent consisting of a water solution of 5 per cent potassium chromate and 25 per cent strong sulphuric acid. Diphenylamine present to the extent of 0.4 per cent by weight of dry nitrocellulose will give a deep violet color; 0.04 per cent will give a pale green color.

(*d*) *Pressing.*—The material coming from the mixers shall be strained for the removal of lumps before going through the graining press. The colloid shall be pressed through dies with such uniformity as will produce the standard grain required. The area of the screen holes of the die must be at least one and one-quarter times the area of the cross-section of the die. One die, or dies of exactly the same dimensions, must be used in graining any lot of powder.

(*e*) *Drying.*—Smokeless powder shall be dried at temperatures not exceeding 44° C. until the solvent has been removed to within the limits fixed by the curve of volatiles. In all the drying operations due care is to be taken to prevent deforming the grains. Drying shall be carried out in " solvent recovery " or " dry " houses, or in both, which are operated on the " close-circuit " or " dead-air " systems, with as little circulation of air through or around the powder as is consistent with the maintenance of even temperatures throughout the powder; in any case, the quantity of new air admitted shall be a minimum.

A recording thermometer shall be suitably located in each dryhouse. At least two maximum and minimum thermometers shall be placed in the powder in the hottest parts of each dry house and daily temperature records shall be kept.

(*f*) *Blending.*—When powder is removed from the dryhouse it shall be exposed to atmospheric conditions for from twenty-four to sixty hours, in order that the absorbed moisture shall be as nearly a fixed quantity as possible. The blending shall be uniformly done in lots of such size as may be prescribed.

(*g*) *Packing.*—After blending, powder shall be packed in air-tight boxes of standard type, which shall be marked and stenciled as required.

Powder, or its standard ingredients, shall at all times be protected from the action of direct sunlight and acid fumes.

5. Tests of Smokeless Powder.

(*a*) *Sampling.*—After a lot of powder is blended, packed, and submitted for acceptance, a firing sample of the required weight and five chemical samples shall be selected by the inspector and shall be shipped to the point designated for chemical and ballistic tests. For powders granulated for 5-inch or larger guns each chemical sample shall fill a 16-ounce tight glass-stoppered bottle; for other powders, each sample shall fill an 8-ounce bottle. The chemical and ballistic samples selected must not be opened until they have arrived at the point or points designated for test.

With every lot of powder submitted for acceptance the contractor shall furnish, in quadruplicate, on official blanks, a descriptive sheet giving a complete history of its manufacture.

On receipt of the five chemical samples of a lot, a blend is made in a tight bottle of equal portions of each sample, which blend will be used in making all stability tests other than the "German" test. Each of the five portions for the "German" test must represent different samples; measurements, physical tests, and chemical tests other than stability tests will be made from portions taken at random from the various samples.

(*b*) *Measurements.*—Thirty grains will be selected at random and measured for length, diameter, perforation, and inner and outer wall thicknesses.

The outside diameter (D) of the grain shall be about ten times the diameter (*d*) of the perforations, and the length (L) about 2.25 times the outside diameter. The dimensions (L and D) of at least the 30 grains specified must comply with the requirements for uniformity as follows :

Dimensions.	Mean Variation of Individual Dimensions from Mean Dimensions, Expressed in Per cent of Mean Dimensions.	
	Permitted.	Desired Less than
L	5.0	1.0
D	2.5	0.5

Six measurements of the outside web thickness (W_0) and of the inside web thickness (W_i) will be made from the six outside holes, for each of the 30 grains, and the two sets of 180 measurements averaged to obtain the mean outside and inside webs. The difference between these means shall not exceed 15 per cent of the average web thickness.

(*c*) *Physical test.*—Ten normal grains will be taken and both ends cut off at right angles to the length until $\frac{\text{length}}{\text{diameter}}=1$. These pieces will be accurately measured for length, and then slowly compressed between parallel surfaces until the first crack appears. The pressure is then removed and the grain again measured. The decrease in length necessary to crack the grain is calculated to per cent of original length. The average compression must not be below 35 per cent. In case of failure in this test 20 more grains are tested, and if the average compression of the total 30 grains is below 35 per cent the powder will be rejected. Grains accidentally abnormal in shape, or containing obvious flaws, will not be used for this test.

(*d*) *Volatiles.*—Slices will be cut from at least five average grains for the samples of large-caliber powders. In small-caliber powder, where a large number of grains insure a proper average, whole grains may be taken, or slices may be cut from a number. The slices, or grains, having been thoroughly mixed, a sample of approximately 1 gram is taken and accurately weighed.

The sample is to be dissolved in 150 cubic centimeters of ether-alcohol mixture (2 to 1, by volume). When the solution is complete, the nitrocellulose is precipitated by the gradual addition of a suitable amount of water and the mixture evaporated to dryness on a steam or water bath. When the evaporation is apparently complete, the precipitate is dried for one hour at 95° to 100° C., or in a vacuum drier at 50° C., and weighed. Correction for residue in water, or solvent used, having been applied, the difference between the weight of the sample and that of the nitrocellulose found is the " total volatiles, as packed."

To obtain the amount of moisture, a sample of at least 5 whole grains of not less than 20 grams will be dried in a vacuum dryer at 50° to 60° C. for two hours, cooled in a desiccator, and the resultant loss of weight determined by weighing. This loss will be considered as moisture, and the difference between this figure and that of the total volatiles by precipitation will be considered as the residual solvent; that is, this treatment will be considered as that required to convert the powder from "packed" to "dry house" condition.

The "total volatiles" or the "residual solvent" must not exceed the limits defined by the curve of volatiles, for "packed" and "dry house" conditions, respectively, which form a part of these specifications.

(*e*) *German test at 135°* C.—This test will be made on five samples in exactly the same way as for nitrocellulose, the powder being in as nearly whole grains as possible consistent with the standard weight of 2.5 grams. No sample shall turn the litmus paper completely to standard red in less than one hour and forty-five minutes, nor shall any sample explode in less than five hours.

(*f*) *Ordnance Department 115° C. test.*—Five samples will be used, each consisting of not less than 10 grains nor less than 2 whole grains. Each sample, after weighing, will be placed in a watch glass or open dish and exposed to a temperature of 115° C. ±0.5° for eight hours a day for six days. The oven is brought to the required temperature each day before inserting the samples, and at the end of the eight-hour period the samples are removed from the oven and allowed to stand over night.

At the end of the six-day period the samples are allowed to cool in a desiccator, after which they are weighed again. The total loss of weight of the samples must not exceed the limit shown by the curve forming part of these specifications.

(*g*) "*Surveillance*" *test at 65.5° C.*—The three samples required for test shall each consist of from 15 to 20 grams of powder in whole grains, the lesser weights being taken for the small-caliber powders, the greater for the large-caliber powders.

After having been exposed for twenty-four hours at 21° C., each sample is to be placed in an 8-ounce salt-mouth glass-stoppered bottle, made tight by carefully grinding the stopper. These bottles will be placed in a constant-temperature magazine at 65.5° C. in which the permissible fluctuation of temperature is ±2° C. The end of the test is the first appearance of red fumes in the bottle, and no sample shall show these fumes in less time than the limit shown by the surveillance curve.

(*h*) *Organic residue* will not be regularly determined, but if any doubt exists as to the quality of any of the materials entering into the composition of the powder a 1-gram sample will be prepared by grinding and sifting, and tested for residue as for nitrocellulose.

(*i*) *Ash.*—A sample of average slices will be tested for ash in the manner prescribed for nitrocellulose.

(*j*) *Solubility.*—A sample will be prepared by rolling and powdering slices from at least 20 powder grains and sifting through an 80-mesh sieve. A portion of about 1 gram is weighed out and the insoluble material determined, as in the test of nitrocellulose.

The sum of the per cents of volatiles, insoluble nitrocellulose, ash, and organic residue found present will be subtracted from 100 and the remainder will be considered the per cent of soluble nitrocellulose. (This test will be made without awaiting the results of the "surveillance test," provided the lot has passed all the other tests prescribed above.)

(*k*) *Ballistic tests.*—

(1) The powder must give the service velocity with a maximum pressure and a weight of charge within the limits given in the table forming part of these specifications, pages 205, 206.

(2) When the measured velocities and pressures for all rounds fired are plotted to scale, as a function of the weight of charge, the resulting curves must be reasonably smooth.

(3) For three rounds, fired under standard conditions, with the charge required to give the service velocity, the difference

MOBILE ARTILLERY.

	Weight of Projectile.	Muzzle Velocity.	Maximum Pressure.	Approximate Weight of Charge Exclusive of Igniter.		Weight of Igniter.		Weight of Lot.	Weight of Ballistic Sample.
	Lbs.	Feet per Sec.	Lbs. per Sq. In.	Lbs.	Ozs.	Lbs.	Ozs.	Lbs.	Lbs.
Guns.									
2.95-inch mountain gun	12 5 18	920 750	18,000	......	8	......	$\frac{1}{8}$	12,500	5
3-inch field gun, models of 1902, 1904, and 1905	15	1,700 1,640	33,000	1 1	10 8.5	....	$\frac{1}{4}$ *a*	25,000	15
3.2-inch field gun, model of 1885	13 5	1,685	35,000	......	......	......	......	12,500	10
3.2-inch field gun, model of 1897	13 5	1,685	35,000	......	......	......	......	12,500	10
3.6-inch field gun, model of 1891	20	1,550	35,000	1	13	......	$\frac{1}{2}$	12,500	15
3.8-inch gun, model of 1907	30	1,700	33,000	2	14	......	$1\frac{1}{4}$ *a*	25,000	25
4.7-inch gun, model of 1906	60	1,700	33,000	6	0	......	$1\frac{1}{4}$ *a*	25,000	50
5-inch siege gun, models of 1890 and 1898	45	1,830	35,000	5	0	......	$1\frac{1}{2}$	25,000	40
Howitzers.									
3-inch howitzer, model of 1907	15	900	19,000	......	8	......	$\frac{1}{8}$ *a*	12,500	10
3.8-inch howitzer, model of 1908	30	900	19,000	1	0	......	$\frac{1}{4}$ *a*	12,500	20
4.7-inch howitzer, model of 1907	60	900	20,000	2	0	......	$\frac{1}{2}$ *a*	12,500	35
4.7-inch howitzer, model of 1908	60	900	20,000	2	0	......	$\frac{1}{2}$ *a*	12,500	35
6-inch howitzer, models of 1906 and 1908	120	900	20,000	4	......	......	$\frac{1}{2}$ *a*	25,000	50
7-inch siege howitzer, models of 1890 and 1898	105	1,100	28,000	4	8	......	$1\frac{1}{2}$	25,000	35
Mortars.									
3.6-inch field mortar, model of 1890	20	690	17,000	......	8	......	1	4,000	4
7-inch siege mortar, model of 1892	125	800	20,000	4	0	......	1	15,000	35

a At one end. 110-grain percussion primer at other end of charge.

SEACOAST CANNON.

	Weight of Projectile.	Muzzle Velocity.	Maximum Pressure.	Approximate Weight of Charge, Exclusive of Igniter.		Weight of Igniter.		Number of Sections in Charge	Weight of Lot.	Weight of Ballistic Sample.
	Lbs.	Ft. per Sec.	Lbs. per Sq. In.	Lbs.	Ozs.	Lbs.	Ozs.	Lbs,	Lbs.	Lbs.
Guns.										
2.24-inch (6 pounder) gun, Driggs-Schroeder, Mark III	6	2,400	34,000	1	6	......	$a\ \frac{1}{4}$	1	12,500	10
2.24-inch (6-pounder) gun, model of 1898	6	2,400	34,000	1	6	......	$a\ \frac{1}{4}$	1	12,500	12
2.24-inch (6-pounder) gun, model of 1900	6	2,400	34,000	1	6	......	$a\ \frac{1}{4}$	1	12,500	12
3-inch (15-pounder gun, model of 1898	15	2,600	34,000	5	0	......	*a*1	1	25,000	40
3-inch (15-pounder) gun, model of 1902	15	2,600	34,000	5	0	......	*a*1	1	25,000	40
3-inch (15-pounder) gun, model of 1903	15	2,600	32,000	5	8	......	*a*1	1	25,000	50
4-inch gun, Driggs-Schroeder	33	2,300	34,000	7	8	......	*a*6	1	25,000	60
4.72-inch gun, Armstrong, 40-caliber	45	2,150	34,000	7	8	......	*a*6	1	25,000	60
4.72-inch gun, Armstrong, 45-caliber	45	2,570	34,000	11	0	......	*a*6	1	25,000	85
4.72-inch gun, Armstrong, 50-caliber	45	2,600	34,000	11	0	......	*a*6	1	25,000	85
5-inch gun, model of 1897	58	2,600	38,000	16	8	......	6	1	25,000	125
5-inch, gun model of 1900	58	2,600	30,000	22	4	......	10	1	50,000	200
6-inch gun, Armstrong	106	2,150	34,000	17	12	......	*a*6	1	50,000	150
6-inch gun, models of 1897 MI and 1908	106	2,600	38,000	29	0	......	12	1	50,000	225
6-inch gun, models of 1900, 1903, and 1905	106	2,600	38,000	31	6	1	0	1	50,000	300
8-inch gun, model of 1888 MII	316	2,200	38,000	82	0	1	8	2	100,000	660
10-inch gun, model of 1888 MII	604	2,250	38,000	158	0	4	0	2	100,000	1,200
10-inch gun, model of 1895	604	2,265	38,000	156	0	4	0	2	100,000	1,200
10-inch gun, model of 1900	604	2,250	32,000	178	0	4	0	2	100,000	1,400
12-inch gun, model of 1888 MII	1,046	2,235	37,000	275	0	7	2	4	100,000	2,100
12-inch gun, model of 1895	1,046	2,250	38,000	272	0	7	0	4	100,000	2,200
12-inch gun, model of 1900	1,046	2,250	30,000	325	0	9	0	4	100,000	2,500
14-inch gun, models of 1907 and 1907 MI	1,660	2,220	38,000	320	0	9	0	4	100,000	2,500
16-inch, model of 1895	2,400	2,250	38,000	612	0	16	8	6	100,000	5,000
Mortars										
12-inch mortar, models of 1886 and 1886-90 MI	1,046 824	910 1,050	27,500	32	0	1	4	1	50,000	500
12-inch mortar, models of 1890, 1890 MI, and 1908	1,046 824	1,050 1,300	37,000	47 57	0	1	4	1	50,000	600
Subcaliber Guns.										
1.457-inch subcaliber gun	1,057	2,000	25,000	......	$2\frac{1}{2}$	......	$b\frac{1}{8}$	1	12,500	2
2.95-inch subcaliber gun	18	750	18,000	......	7	......	$b\frac{1}{8}$	1	12,500	5

a At one end. 110-grain percussion primer at other end of charge.

b $\frac{1}{16}$ ounce only for friction firing.

between any velocity and the mean velocity for the group must not exceed ± 1 per cent of the mean velocity, and similarly, the difference in pressure must not exceed ± 5 per cent of the mean pressure. For calibers less than 3 inches the above-mentioned allowed variations will be increased 25 per cent.

(4) The pressure curve must not show any "critical" point, as evidenced by a marked and abrupt change in the law of development of pressure as a function of the weight of charge.

(5) If considered desirable, one or more rounds may be fired with weight of charge 5 per cent greater than that required to give the service velocity. Under these conditions no dangerous pressure must be shown.

Special Specifications and Tests for Smokeless Powder for Small Arms.

General Requirements.

The powder must be uniform in quality, free from dust and other foreign substances.

It must be practically smokeless. The amount of unburned powder in firing full charges must not exceed 1.25 per cent.

It must not unduly corrode or erode the barrel or corrode the cartridge case. Under this requirement the erosion of the bores of the U. S. magazine rifles, models of 1898 and 1903, after firing 5,000 rounds with powder intended for those arms must not materially exceed that exhibited by rifle barrels Nos. 21244 and 175968 of those models, respectively, each of which has been fired 5,000 rounds and will be retained at Frankford Arsenal as a present standard of reference.

It must not require an unduly strong primer for ignition.

It must not leave a hard adherent residue in the bore, especially after rapid firing.

It must not be sensitive to friction or shock.

It must not be so friable as to endanger breakage of grains in transportation incident to service.

It must not contain ingredients known to be unsuited to form a safe and reasonably stable compound.

It must admit of satisfactory machine loading, with the machines in use or that can be readily provided at the Frankford Arsenal.

It must not show a tendency to agglomeration during storage. Powder which may be found defective in this respect, and which has not already been used in the manufacture of cartridges, will be returned to the contractor at his expense, and deliveries of the powder under contract will be suspended.

Other things being equal, that powder which produces the least heating of the barrel will be preferred.

The powder for ball cartridges shall be subject to proof firing in the service arm for which intended, and in the pressure gauge for that arm used at the Frankford Arsenal, with the service cartridge case and bullet.

The weight of the powder charge is not prescribed, but will be governed by the ballistic requirements.

It is desirable that the charge shall fill the case to the shoulder. When machine loaded it must not fill the case to within 0.4-inch from the mouth.

Other powders must fulfill the general requirements for all smokeless powders and be suitable to the arm and for the purpose for which they are intended.

The methods of manufacture and the tests of raw material are essentially the same as those for cannon powders.

Test.—To test finished nitroglycerin powders for stability a sample of the powder is pulverized and after being passed though a 40-mesh sieve is dried for 48 hours at 43° C. ±3° C., and then placed in a moist atmosphere for several hours or until about 1.5 per cent of moisture is absorbed.

The test is then continued as in the potassium iodide starch paper test for nitrocellulose. The powder must give a test of not less than 40 minutes.

Cellulose prepared for nitrating.—Bleached cellulose, the material to be used, will be obtained by purifying unspun cotton wastes and thoroughly washing to remove purifying materials or salts; containing not more than 0.7 per cent extractive matter; not more than 1.25 per cent ash; of uniform character, clean and free from such lumps as will prevent uniform nitration.

POWDER FOR BALL CARTRIDGES FOR SMALL-ARMS.

The finished powder shall be a uniform ether alcohol colloid of nitrocellulose of standard quality.

Granulation.—The grain, except when otherwise permitted, shall be cylindrical in shape with one axial perforation.

The length of the dry grain must be about 2.5 times the outside diameter. No limits in the variation of web thickness are prescribed; the other requirements are expected to ensure a sufficient degree of uniformity.

Volatiles.—The powder, in the "packed" conditions. shall contain a total percentage of volatiles to be determined by the water precipitation method, not greater than 3.15 per cent.

Stability.—The powder shall be stable under any one, or both of the following tests, when conducted in the manner herein described, viz.: Potassium iodide starch paper test; German 135° C. test.

It must not show by chemical analysis or test the presence of any unauthorized ingredients, or that the nitrocellulose and other materials employed in the manufacture of the powder did not conform to the specifications.

Sampling.—After a lot of powder is blended and packed, a ten-pound sample is selected by the inspector, by taking 1 pound from each of 10 boxes and sent to Frankford Arsenal for chemical and ballistic test.

Total volatiles.—Approximately 1 gram is accurately weighed out. This sample is dissolved in 150 c.c. ether-alcohol mixture (2 to 1 by volume). When solution is complete, the nitrocellulose is precipitated by the gradual addition of a suitable amount of water. The mixture is then evaporated to dryness on a steam or water bath. When evaporation is apparently complete, the precipitate is dried for one hour at 95° to 100° C., or in a vacuum drier at 50° C., and then weighed. Correction is made for residue in water, or solvent used. The difference between weight taken and nitrocellulose found is total volatiles.

Heat test—Potassium iodide starch-paper test.—A sample is

dried forty-eight hours at 43° C. ±3°. The powder is then allowed to stand over night, or until the proper amount of moisture is obtained, and the heat test proceeded with as for nitrocellulose, except that no moisture series is taken. Not less than ten samples will be used, and each sample shall give not less than forty minutes.

German test at 135° C.—This test is made on five samples in exactly the same way as for nitrocellulose, the powder being in as nearly whole grains as possible, consistent with the standard weight of 2.5 grams. No sample shall turn the litmus paper completely red to standard in less than one hour and fifteen minutes; nor any sample explode in less than five minutes.

Organic residue.—This will not be regularly determined. If any doubt of the quality of the ingredients exists the following method will be used: A sample of 1 gram is proceeded with as in the test of nitrocellulose.

Insoluble nitrocellulose.—A portion of about 1 gram is weighed out for treatment as in the determination of insoluble nitrocellulose in nitrocellulose.

Ash.—An average sample is treated in the manner described for ash in nitrocellulose.

Soluble nitrocellulose.—The sum of the per cent of volatiles, insoluble nitrocellulose, organic residue, and ash subtracted from 100, gives the per cent of soluble nitrocellulose.

After graining, the powder is dried at a temperature not exceeding 110° F. until the amount of solvent is less than 2.75 per cent. Temperature records will be kept of the heated air, both in contact with powder and before it enters the dry houses, the thermometers in powder bins being maximum and minimum style and placed in the hottest part of the house, at least two such being used for each house. Each dry house shall be provided with a recording thermometer suitably located.

Blending and packing.—The moisture and volatiles of the powder will be determined in the dry-house condition and also in the packed condition, and the time of exposure to atmospheric condition, together with the temperature and humidity, recorded. The amount of absorbed moisture shall be as nearly a fixed quantity as possible.

The blending shall be uniformly done in lots of such size as may be prescribed.

The powder shall be packed in air-tight boxes of standard type and the contents wi h ballistic and other data stenciled thereon. Powder, or the standard ingredients of powder, shall at all times be protected from the action of direct sunlight and acid fumes.

Velocity and Pressure Tests.

The powder will be taken as received, and under the ordinary conditions governing the loading of a service powder will be tested, the powder charges being separately weighed and hand-loaded.

	.30-caliber.	.30-caliber.	.38-caliber.
	Model of 1898.	*Model of* 1906.	
Mean velocity must not be less than	1,966 ft. per sec.	2,640 ft. per sec.	775 ft. per sec.
Mean variation in velocity must not exceed	8 feet per sec	18 feet per sec.	15 ft. per sec.
Extreme variation in velocity must not exceed	30 feet per sec.	40 feet per sec.	70 ft. per sec.
Maximum pressure must not exceed	40,000 poun s	50,000 pounds	15,000 pounds

Velocities will be measured at 53 feet from the muzzle for the model of 1898 .30 caliber cartridges, 78 feet for the model of 1906 .30 caliber cartridges, and 25 feet for the caliber .38 revolver cartridges.

The velocity proof of each lot to comprise forty consecutive rounds.

The proof for pressure to comprise ten consecutive rounds fired from the pressure barrel.

Loading Test.

Fifty cartridges will be loaded by the machines in use and tested, 40 being fired for velocity and 10 for pressures. The weights of these charges will be measured before the bullets are assembled.

	.30-caliber.	.30-caliber.	.38-caliber.
	Model of 1898.	*Model of* 1906.	
Mean variation in velocity must not exceed	10 feet per sec.	12 feet per sec.	15 feet per sec.
Extreme variation in velocity must not exceed	40 feet per sec.	50 feet per sec.	70 feet per sec.
Maximum pressure must not exceed	41,000 pounds	50,500 pounds	16,000 pounds
Variation in weight of charge must not exceed	.9 grain	.9 grain	0.25 grain

VIII.

STORAGE OF EXPLOSIVES.

Magazines.—Wide variety of practice exists among the different countries in building magazines. In Austria-Hungary light wooden structures are provided for explosives, so that the debris, in case of explosion, would be projected short distances. The English laws in reference to explosives are most elaborate and rigid; the details of magazines, the character of explosives permitted for sale and the conditions of storage and transportation are carefully prescribed therein. These stringent regulations, taken in connection with those equally stringent in reference to the manufacture and tests of all explosives, make an accidental explosion of explosives in transit or storage an exceedingly rare occurrence in England.

Explosions may result from lightning or from incendiarism; to guard against such contingencies the buildings in which explosives are stored should be well protected by lightning-conductors and be made fire-proof. Some constructions have been made in which the roof and sides are of corrugated sheet iron; the roof-trusses of iron resting on brick piers; the floor of asphalt free of grit.

All doors should be double with a vestibule between; they should be strong, fire-proof, and have strong, treble-bolt locks.

According to Guttmann, the best method of protecting explosives from lightning is to build the magazine entirely of metal, extending the sides down to moist soil or connecting them well with it in several places.

A method suggested by Professor Oliver Lodge is considered efficient. This consists in covering the building completely with strong, durable iron wire netting, or running large size iron wires along all ridges and edges, with groups of wires radiating from each corner, the whole system being connected well with moist earth.

All storage-magazines should have protecting mounds or traverses of earth thrown up around them when located near other buildings or property exposed to destruction in case of explosion. When this is not possible, near-by buildings may be protected by planting thickly a deep row of trees about the magazine. As a rule, 200 yards may be regarded as a reasonably safe distance from a large storage-magazine.

Storage-magazines should not be placed within closed works if it is possible to avoid doing so.

Not more than 400 tons of black or brown gunpowder or 100 tons of nitrocellulose gunpowder should be stored in one magazine.

The English regulations prescribe that magazines for the storage of nitro-powders or high explosives shall be made of as light a form of construction as possible, compatible with sufficient strength for stability, resistance to weather, and protection against unlawful entry. The material used must not be of an inflammable nature. The temperature of magazines should be maintained at about 70° F.; if it is permitted to rise above 100° F. for any length of time the composition and stability of nitrocellulose powders may be affected; if it rises above 122° F. (50° C.), even for a few minutes, explosives stored therein should be examined for stability.

Service-magazines in coast forts are so placed as to be protected from projectiles of all kinds. The conditions as to temperature and ventilation prescribed for storage-magazines should obtain for service-magazines.

In so far as possible no iron fixtures, tools, or appliances should be used inside of a magazine.

Magazines may be heated by steam at a pressure not exceeding 15 lbs. per square inch, or by hot water; the heating pipes may be of iron, but should be placed well above the floor, not lower than 6 feet 6 inches. They need not be galvanized nor otherwise coated, nor boxed in with wood; but they should be detached and not less than 6 inches from any woodwork. They should be frequently wiped clean of all dust.

All doors and windows should be made to open outwards. They should be covered with copper sheeting. All fixtures and nails should be of copper.

Following, in a general way, the English regulations, explosives may be classified, for purposes of storage, into "Groups" and "Divisions," as follows:—

Group I. Stored in Magazines.

Explosives which must be placed *in a magazine*, each division of the group requiring a separate compartment in which "magazine conditions" must be observed, except that divisions *a* and *e* may be placed in the same compartment, and *c*, *d*, and *e* need not be under magazine conditions.

Divisions.

a. Nitrocellulose gunpowder and black and brown gunpowder, in bulk or made up in cartridges for large-caliber guns. Quick match.

b. Dry guncotton.
Dynamite.
Explosive gelatin.

c. Wet guncotton.
Picric acid and its service explosive derivatives.

d. Rapid-fire fixed ammunition for guns of 3-inch caliber and less.

e. Rapid-fire ammunition for the guns above 3-inch caliber, when the powder is in metallic cases, or in metal-lined boxes.

Group II.

Explosives which must be stored in a separate chamber of a magazine, or in a separate storeroom or building.

Divisions.

a. Percussion-caps.
Small-arm ammunition.
Priming and pyrotechnic composition; any composition in bulk containing either mercury fulminate or a chlorate.
Empty capped metallic cases.
Fuses (time, percussion, or combination).
Slow match.
Port fires.
Rockets.
Primers of all kinds (friction, percussion, or electric).

b. Mines, loaded.

c. Shells, filled and fused.
Shells, filled but not fused.

d. Detonating-caps.

All gunpowders, dry guncotton, dynamite, and explosive gelatin should always be kept *in magazines*, and magazine conditions strictly enforced.

Explosives in Group II should not be placed in the body of magazines, but in storerooms or chambers apart, and need not necessarily be under magazine conditions.

Divisions *c*, *d*, *e*, Group I, may be stored in magazines or as prescribed for Group II, whichever is most convenient.

No two divisions in either group should be placed in the same compartment or pile, except *a* and *e*, Group I, may be stored in the same magazine, and fuses and primers, Group II, may be kept in the shell-room of a service-magazine, but a box or cupboard should be provided to contain them only, and separately.

A magazine or storeroom for explosives may be divided into many compartments under the same roof for the different divisions of a group, provided they are separated by substantial brick or other walls, without openings of any kind between compartments.

Explosives of the same division may be stored in the same compartment, room, or magazine.

Nitroglycerine, dynamite, explosive gelatin, and nitrocellulose may decompose above 122° F. (50° C), and magazines containing them should *never* have a higher temperature. Nitro-powders and dry guncotton should not be exposed to a higher temperature than 104° F. (40° C) for any length of time, or repeatedly for short times.

All explosives, whether stored in magazines or in storerooms, should be kept under the following conditions:

> Lighting of fires near by should be strictly prohibited.
>
> No one should be permitted to enter rooms containing explosives stored in bulk with matches in the pockets or about the person.
>
> Oiled rags or waste, or any substance liable to spontaneous combustion, should not be kept in or near rooms containing explosives.
>
> Floors and platforms should be kept scrupulously clean.
>
> Benches, shelves, and all fittings and fixtures inside of storerooms or magazines should be kept free of grit and dust.

Magazines containing gunpowder of any kind, in bulk or in cartridges for large-caliber guns, nitroglycerine, dynamite, explosive gelatin, or dry guncotton should be kept under the following conditions, in addition to those which are given above:

> No one should be permitted to pass through the outer door of the building except those duly employed therein, or except in the presence of the officer or non-commissioned officer in whose charge the explosives are placed, and the latter should be responsible that all regulations for safety are strictly observed. To this end,

the officer or non-commissioned officer in charge should cause all persons to observe the following regulations as to clothing:

The contents of all pockets will be examined at the outer door to see that no matches or other easily combustible substances are taken within.

As soon as the outer door is entered all coats will be removed, and iron or steel articles removed from trousers' pockets. The shoes will be carefully wiped on a mat placed just inside the outer door, and magazine rubber overshoes placed on the feet of each person.

When powder is to be examined in a magazine, a paulin, carefully dusted and shaken, should be spread out on the floor, and when the work is completed the paulin should be carefully folded so as to contain within its folds all powder-dust that may have been formed; it then should be carried from the magazine and the dust shaken into water.

Door-mats should be shaken outside the outer door after each party leaves the magazine.

Packages containing explosives in Group I should not be opened in a magazine or storeroom containing other explosives of that group. They may be opened in an anteroom or outside.

Inventory lists, showing the contents of the magazine or storeroom, should be posted and kept entered to date.

Keys of magazines and storerooms containing explosives should be carefully tagged and kept in the personal possession of the officer in charge of the explosives.

When explosives are received the original packages should be carefully examined externally, the condition of the package noted to see if it has on its surfaces any nails, grit, or other objectionable substance, and, if there be any such, it will be carefully removed. If the package is broken or defective it will be set aside to be opened and have its contents examined. All marks on each separate package will be carefully entered in the receipt record book.

Shelves should be arranged in an anteroom to receive "sample bottles." On these shelves should be kept a sample of each "lot" of nitro-explosives received and in store in the magazine. These bottles should be inspected and the contents tested from time to time.

In stacking original packages they should be so placed as to exhibit the markings.

When original packages have been emptied, the markings should be scraped off before they are sent from the magazine or storeroom.

If packages are used for explosives a second time they should be carefully examined to see that all former markings are obliterated, and that they are strong and free from dust, dirt, and foreign substances of all kinds.

In the magazine and storerooms, packages should be stacked in tiers, the same divisions being kept together, and in each division each lot separate. A clear, free aisle should be left about each lot, and in each tier the bottom layer should be separated from the floor by 1-inch battens, and each layer from the one below by 1-inch battens. In each layer an inch space will be left between adjacent packages.

Filled cartridges will be stacked separately from powder in bulk, the lots being carefully separated and each lot together.

When rooms or buildings other than magazines are used for the storage of explosives they should be thoroughly repaired, washed, dried, swept, and cleared of all movable articles before the explosives are introduced.

If there be no anteroom or vestibule in connection with a room used for the storage of explosives one should be improvised. If it is not practicable to observe the strict regulations prescribed for permanent magazines, it is possible always to require that no matches or other easily combustible substance should be taken within the building or room, and that the feet should be carefully wiped inside the outer door.

Ventilation of Magazines.

It is very important that magazines containing gunpowder should be carefully ventilated. If powder be stored in damp magazines in cases not hermetically sealed, the powder absorbs moisture and its ballistic value is thereby reduced.

With smokeless powders the temperature of the magazine has also a special influence on the muzzle-velocity. It has been found by trial that powders tested in summer and used in target-practice in winter give velocities lower than the test velocity, and those tested in winter and used in summer give higher velocities. Corrections allowing for difference of temperature of powder in firing have been ascertained and tabulated.

Powders should be tested ballistically at a standard temperature, say 70° F., and the temperatures of service-magazines should be such as to permit the powder to be delivered to the guns at as near this temperature as possible. If this is not done a temperature correction must be introduced in applying range tables.

The humidity and temperature of the air in magazines are, therefore, a matter that must be carefully watched.

It is especially important with all nitro-explosives that there should be free circulation of air, so that in case any incipient decomposition should occur, at any spot in any package, the fumes would be directly carried off, thereby preventing an accumulation of pressure and temperature, and also favoring detection of the decomposition by the odor of the escaping gases.

The air inside of magazines should be kept always above its dew-point to avoid condensation. The problem is, therefore, to keep the air circulating and to maintain it at a temperature above its dew-point. Three methods are practised to accomplish this:

1. The air of magazines may be kept above the dew-point by providing that it pass over heated steam- or hot-water pipes, using a fan or natural circulation.

2. Air-shafts with revolving hoods, like those of ships, may be arranged to face the wind and conduct a large volume of air through all rooms and galleries. It is found that a shaft about 20 inches in diameter with a well-flared hood will work efficiently in wind above 5 miles per hour. With little wind and on damp days the shafts are closed. The principle applied in this type is that the volume of air passing through must be sufficient to give its temperature to the surfaces of the rooms and galleries.

3. Some magazines are not provided with circulating air, but the air is renewed as often as possible by opening all doors and windows to the outside air whenever the conditions of temperature and dew-point are such as to make the air let in a drying air.

That is, it is necessary to establish a proper dew-point inside by heat or otherwise, and to cause a constant circulation of air by blowers, or to make use intermittingly of the natural weather conditions as they may warrant.

The regulation of the air within a magazine by natural ventilation is effected by means of thermometers inside of magazines and wet- and dry-bulb hygrometers outside. The wet- and dry-bulb hygrometers are permanently placed outside the magazine, protected from the direct and reflected rays of the sun, and from wind and rain. Magazines should be arranged with a window, and the inside thermometers should be placed at this window, so that the inside temperature may be read without opening the magazine. Before installing the inside thermometer and the outside hygrometer, the former and the dry-bulb thermometer of the latter should be compared as to their readings under the same conditions. If a difference of reading is noted, this should be entered as a correction on both instruments and applied in all computations.

The scale of the dry-bulb thermometer of the hygrometer will give the temperature of the outside air; the reading of the wet-bulb thermometer will always be below that of the dry bulb, and the amount of this difference is the argument with which the humidity tables are entered, as explained below.

In using the wet- and dry-bulb hygrometer care must be exercised to have the well of the wet-bulb thermometer always supplied with clean, pure water, and to see that the cloth leading to the wet bulb is wet before taking any reading.

Readings should be taken in the morning and in the afternoon. These readings and the readings of the inside thermometers should be entered in a record book.

The dampness of magazines results from two causes:

1. The *condensation* of moisture from the air of the magazine on the walls, ceiling, floors, and all surfaces in the magazine. Outside air at a given temperature and relative humidity admitted to a magazine at a lower temperature may, by simply having its temperature lowered, become supersaturated and deposit moisture by condensation.

2. *Percolation* of water through the ceiling and walls often causes dampness. This is sometimes seen in magazines, especially when Rosendale cement has been used in the construction, and when sheet lead or asphaltum has not been placed over the ceilings. Such water, running into the magazine, collects in small pools and tends to keep the air constantly saturated.

After magazines have been opened the greatest care should be exercised to see that they are closed tightly as soon as the conditions favorable to opening cease to exist, or before this limit is reached.

Subject to the above conditions, magazines should be opened as often and for as long a time as possible, and every means used to get a good circulation of air.

Two tables, A and B, are provided for guidance of the person in charge of the magazine. Copies of these tables should be attached to boards hung up in each magazine. Table A gives the weight of water-vapor per cubic foot of air for each degree from 13° to 100° F., when the reading of the wet bulb is from 0 to 14 degrees lower than that of the dry bulb. The table is not carried below one grain of water-vapor per cubic foot of air, as this is a condition seldom met with, and no harm would be done in ventilating a magazine at any temperature likely to occur with air so dry as this. Table B gives the temperature which must be shown by the inside thermometer corresponding to the weight of water-vapor per cubic foot, before the magazine should be opened for ventilation. The table gives two columns of temperature: column I gives the temperature for the magazine at or above which ventilation would be advantageous, namely, that at which the water-vapor that is in the air outside would cause a degree of humidity of 70 per cent or less inside; column II gives the low limit of temperature for the magazine below which it should never be opened for ventilation, as its degree of humidity would become 85 per cent or more, and if it is necessary to open the doors for any purpose, they should be closed again as quickly as possible.

Method of Reading the Tables.—To work these tables the readings of the wet- and dry-bulb thermometers are taken, and from Table A the weight of water-vapor per cubic foot of air is ascertained. The temperature is then taken from Table B, which is opposite that weight in the first column.

Application of Tables.— Should the thermometer in the magazine read at or above the temperature taken from column I, Table B, the magazine may safely and advantageously be opened for ventilation. If this condition is not fulfilled for a month, the first opportunity should be taken for ventilating the magazine when the thermometer in it reads between the temperatures taken from columns I and II for the weight of water-vapor per cubic foot of air at the time; but the temperature taken from column II is the minimum for the thermometer in the magazine for any ventilation to be attempted.

Length of Time to be Opened.—It must be borne in mind that conditions favorable for ventilation may not last long, especially when the temperature inside the magazine is below that outside,

TABLE A.

WEIGHT (IN GRAINS) OF WATER-VAPOR IN ONE CUBIC FOOT OF AIR FOR USE WITH TABLE B, WHEN THE READING OF THE WET BULB IS BELOW THAT OF THE DRY BULB, AS FOLLOWS:

Dry Bulb (Fahrenheit).	Difference Between Dry and Wet Bulbs.														
	0°	1°	2°	3°	4°	5°	6°	7°	8°	9°	10°	11°	12°	13°	14°
110°	18.9	19.8	18.0	17.2	16.3	15.5	14.8	14.1	13.4	12.7	12.1	11.5	10.9	10.4	9.9
99	19.3	18.4	17.5	16.7	15.9	15.1	14.4	13.7	13.0	12.3	11.7	11.1	10.5	10.0	9.5
98	18.7	17.8	17.0	16.2	15.4	14.6	13.9	13.2	12.6	12.0	11.4	10.8	10.2	9.7	9.2
97	18.2	17.3	16.5	15.7	14.9	14.2	13.5	12.8	12.2	11.6	11.0	10.4	9.9	9.4	8.9
96	17.7	16.8	16.0	15.3	14.5	13.8	13.1	12.4	11.8	11.2	10.7	10.1	9.6	9.1	8.6
95	17.2	16.3	15.5	14.8	14.1	13.4	12.7	12.1	11.5	10.9	10.3	9.8	9.3	8.8	8.3
94	16.7	15.9	15.1	14.4	13.7	13.0	12.3	11.7	11.1	10.5	10.0	9.5	9.0	8.5	8.0
93	16.2	15.4	14.7	14.0	13.3	12.6	11.9	11.3	10.7	10.2	9.6	9.1	8.7	8.2	7.8
92	15.7	14.9	14.2	13.5	12.9	12.2	11.6	11.0	10.4	9.9	9.3	8.8	8.3	7.9	7.5
91	15.3	14.5	14.8	13.1	12.5	11.8	11.2	10.6	10.1	9.5	9.0	8.5	8.1	7.7	7.2
90	14.8	14.1	13.4	12.7	12.1	11.4	10.8	10.3	9.7	9.2	8.7	8.3	7.8	7.4	7.0
89	14.4	13.7	13.0	12.3	11.7	11.1	10.5	10.0	9.4	8.9	3.4	8.0	7.5	7.1	6.7
88	14.0	13.3	12.6	12.0	11.4	10.8	10.2	9.6	9.1	8.6	8.1	7.7	7.3	6.9	6.5
87	13.6	12.9	12.2	11.6	11.0	10.4	9.8	9.3	8.8	8.3	7.9	7.4	7.0	6.6	6.3
86	13.2	12.5	11.8	11.2	10.8	10.1	9.5	9.0	8.5	8.1	7.6	7.2	6.8	6.4	6.1
85	12.8	12.1	11.5	10.9	10.3	9.7	9.2	8.7	8.3	7.8	7.4	7.0	6.6	6.2	5.9
84	12.4	11.7	11.1	10.5	10.0	9.4	8.9	8.5	8.0	7.5	7.1	6.7	6.3	6.0	5.6
83	12.0	11.4	10.8	10.2	9.7	9.1	8.6	8.2	7.7	7.3	6.9	6.5	6.1	5.8	5.4
82	11.7	11.1	10.5	9.9	9.4	8.9	8.4	7.9	7.5	7.1	6.7	6.3	5.9	5.6	5.2
81	11.3	10.7	10.1	9.5	9.1	8.6	8.1	7.6	7.2	6.8	6.4	6.0	5.7	5.4	5.1

TABLE A—*Continued.*

Dry Bulb (Fahrenheit).	Difference Between Dry and Wet Bulbs.														
	0°	1°	2°	3°	4°	5°	6°	7°	8°	9°	10°	11°	12°	13°	14°
80°	11.0	10.4	9.8	9.3	8.8	8.3	7.8	7.4	7.0	6.6	6.2	5.8	5.5	5.2	4.9
79	10.6	10.1	9.5	9.0	8.5	8.0	7.6	7.2	6.8	6.4	6.0	5.6	5.3	5.0	4.7
78	10.3	9.8	9.2	8.7	8.2	7.8	7.3	6.9	6.5	6.2	5.8	5.5	5.1	4.8	4.5
77	10.0	9.5	8.9	8.4	8.0	7.5	7.1	6.7	6.3	5.9	5.6	5.3	4.9	4.6	4.3
76	9.7	9.2	8.6	8.2	7.7	7.2	6.8	6.4	6.1	5.7	5.4	5.1	4.8	4.5	4.2
75	9.4	8.9	8.4	7.9	7.4	7.0	6.6	6.2	5.8	5.5	5.2	4.9	4.6	4.3	4.0
74	9.1	8.6	8.1	7.6	7.2	6.8	6.4	6.0	5.6	5.3	5.0	4.7	4.4	4.1	4.3
73	8.8	8.3	7.8	7.4	7.0	6.6	6.2	5.8	5.4	5.1	4.8	4.5	4.2	4.0	3.7
72	8.5	8.0	7.6	7.1	6.7	6.3	5.9	5.6	5.3	5.0	4.7	4.4	4.1	3.8	3.6
71	8.3	7.8	7.3	6.9	6.5	6.1	5.7	5.4	5.1	4.7	4.4	4.2	3.9	3.7	3.4
70	8.0	7.5	7.1	6.7	6.3	5.9	5.5	5.2	4.9	4.6	4.3	4.0	3.8	3.5	3.3
69	7.8	7.3	6.9	6.5	6.1	5.7	5.3	5.0	4.7	4.4	4.1	3.9	3.6	3.4	3.2
68	7.5	7.1	6.6	6.2	5.8	5.5	5.2	4.8	4.5	4.2	4.0	3.7	3.5	3.2	3.0
67	7.3	6.8	6.4	6.0	5.6	5.3	5.0	4.7	4.4	4.1	3.8	3.6	3.3	3.1	2.9
66	7.1	6.6	6.2	5.8	5.5	5.1	4.8	4.5	4.2	3.9	3.7	3.4	3.2	3.0	2.8
65	6.8	6.4	6.0	5.6	5.3	4.9	4.6	4.3	4.0	3.8	3.5	3.3	3.1	2.9	2.7
64	6.6	6.2	5.8	5.4	5.1	4.8	4.5	4.2	3.9	3.6	3.4	3.2	3.0	2.8	2.6
63	6.4	6.0	5.6	5.2	4.9	4.6	4.3	4.0	3.7	3.5	3.3	3.0	2.8	2.6	2.4
62	6.2	5.8	5.4	5.1	4.7	4.4	4.1	3.9	3.6	3.4	3.1	2.9	2.7	2.5	2.4
61	6.0	5.6	5.2	4.9	4.6	4.3	4.0	3.7	3.5	3.2	3.0	2.8	2.6	2.4	2.3

60	5.8	5.4	5.1	4.7	4.4	4.1	3.8	3.6	3.3	3.1	2.9	2.7	2.5	2.3	2.2
59	5.6	5.2	4.9	4.6	4.3	4.0	3.7	3.5	3.2	3.0	2.8	2.6	2.4	2.2	2.1
58	5.4	5.0	4.7	4.4	4.1	3.8	3.6	3.3	3.1	2.9	2.7	2.5	2.3	2.1	2.0
57	5.2	4.8	4.5	4.2	3.9	3.7	3.4	3.2	3.0	2.7	2.5	2.4	2.2	2.0	1.9
56	5.0	4.7	4.4	4.1	3.8	3.5	3.3	3.1	2.8	2.6	2.4	2.3	2.1	1.9	1.8
55	4.8	4.5	4.2	3.9	3.7	3.4	3.1	2.9	2.7	2.5	2.3	2.1	2.0	1.8	1.7
54	4.7	4.4	4.1	3.8	3.5	3.3	3.0	2.8	2.6	2.4	2.2	2.1	1.9	1.7	1.6
53	4.5	4.2	3.9	3.6	3.4	3.1	2.9	2.7	2.5	2.3	2.1	2.0	1.8	1.7	1.5
52	4.4	4.1	3.8	3.5	3.3	3.0	2.8	2.6	2.4	2.2	2.0	1.9	1.7	1.6	1.4
51	4.2	3.9	3.6	3.4	3.1	2.9	2.7	2.5	2.3	2.1	1.9	1.8	1.6	1.5	1.4
50	4.1	3.8	3.5	3.3	3.0	2.8	2.6	2.4	2.2	2.0	1.8	1.7	1.5	1.4	1.3
49	4.0	3.7	3.4	3.1	2.9	2.7	2.5	2.3	2.1	1.9	1.8	1.6	1.5	1.3	1.2
48	3.8	3.5	3.3	3.0	2.8	2.6	2.4	2.2	2.0	1.8	1.7	1.5	1.4	1.3	1.2
47	3.7	3.4	3.2	2.9	2.7	2.5	2.3	2.1	1.9	1.7	1.6	1.5	1.3	1.2	1.1
46	3.6	3.3	3.1	2.8	2.6	2.4	2.2	2.0	1.8	1.7	1.5	1.4	1.3	1.1	1.0
45	3.4	3.1	2.9	2.7	2.5	2.3	2.1	1.9	1.7	1.6	1.4	1.3	1.2	1.1	1.0
44	3.3	3.0	2.8	2.6	2.4	2.2	2.0	1.8	1.7	1.5	1.4	1.2	1.1	1.0	
43	3.2	2.9	2.7	2.5	2.3	2.1	1.9	1.7	1.6	1.4	1.3	1.2	1.1	1.0	
42	3.1	2.8	2.6	2.4	2.2	2.0	1.8	1.7	1.5	1.4	1.2	1.1	1.0		
41	3.0	2.7	2.5	2.3	2.1	1.9	1.7	1.6	1.4	1.3	1.2	1.0			
40	2.9	2.6	2.4	2.2	2.0	1.8	1.6	1.5	1.4	1.2	1.1	1.0			
39	2.8	2.5	2.3	2.1	1.9	1.7	1.6	1.4	1.3	1.2	1.0				
38	2.7	2.4	2.2	2.0	1.8	1.7	1.5	1.3	1.2	1.1	1.0				
37	2.6	2.3	2.1	1.9	1.7	1.6	1.4	1.3	1.1	1.0					
36	2.5	2.2	2.0	1.8	1.7	1.5	1.3	1.2	1.1	1.0					
35	2.4	2.1	1.9	1.7	1.6	1.4	1.2	1.1	1.0						
34	2.3	2.1	1.8	1.6	1.4	1.2	1.1	1.0							
33	2.2	2.0	1.7	1.5	1.3	1.1	1.0								
32	2.1	1.9	1.6	1.4	1.2	1.0									
31	2.1	1.8	1.5	1.3	1.1	1.0									

TABLE A—*Continued.*

Dry Bulb (Fahrenheit).	Difference Between Dry and Wet Bulbs.														
	0°	1°	2°	3°	4°	5°	6°	7°	8°	9°	10°	11°	12°	13°	14°
30°	2.0	1.7	1.4	1.1	1.0										
29	1.9	1.6	1.3	1.0											
28	1.8	1.5	1.2												
27	1.7	1.4	1.1												
26	1.7	1.3	1.0												
25	1.6	1.2													
24	1.5	1.1													
23	1.5	1.1													
22	1.4	1.0													
21	1.4	1.0													
20	1.3														
19	1.3														
18	1.2														
17	1.1														
16	1.1														
15	1.1														
14	1.0														
13	1.0														

TABLE B.

SHOWING TEMPERATURE AT WHICH MAGAZINES MAY BE OPENED FOR VENTILATION, ACCORDING TO THE MOISTURE IN THE OUTSIDE AIR ASCERTAINED FROM TABLE A.

Weight of Water-vapor in One Cubic Foot of Air (Outside) Ascertained from Table A.	Temperature of Magazine When It May be Opened.		Weight of Water-vapor in One Cubic Foot of Air (Outside) Ascertained from Table A.	Temperature of Magazine When it May be Opened.	
	I.—Minimum for Good Ventilation.	II.—Limit below which Ventilation is Injurious.		I.—Minimum for Good Ventilation.	II.—Limit below which Ventilation is Injurious.
Grains.[1]	Degrees F.	Degrees F.	Grains.[1]	Degrees F.	Degrees F.
17.0	107	100	5.2	68	62
16.5	106	99	5.0	67	61
16.0	105	98	4.9	66	60
15.5	104	97	4.7	65	59
15.0	103	96	4.6	64	58
14.6	102	95	4.4	63	57
14.2	101	94	4.3	62	56
13.8	100	93	4.1	61	55
13.5	99	92	4.0	60	54
13.1	98	91	3.8	59	53
12.7	97	90	3.7	58	52
12.3	96	89	3.5	57	51
12.0	95	88	3.4	56	50
11.6	94	87	3.3	55	49
11.2	93	86	3.2	54	48
10.9	92	85	3.1	53	47
10.5	91	84	3.0	52	46
10.2	90	83	2.9	51	45
9.9	89	82	2.8	49	44
9.6	88	81	2.7	48	43
9.3	87	80	2.6	47	42
9.0	86	79	2.5	46	41
8.8	85	78	2.4	45	40
8.5	84	77	2.3	44	39
8.2	83	76	2.2	43	38
8.0	82	75	2.1	41	37
7.7	80	74	2.0	40	36
7.4	79	73	1.9	39	35
7.2	78	72	1.8	38	34
7.0	77	71	1.7	37	33
6.8	76	70	1.6	35	32
6.6	75	69	1.5	33	31
6.3	74	68	1.4	31	30
6.1	73	67	1.3	29	28
5.9	72	66	1.2	27	26
5.7	71	65	1.1	25	24
5.6	70	64	1.0	23	21
5.4	69	63			

[1] When the number of grains of water-vapor per cubic foot of air is not found exactly in the column, the nearest higher figure should be taken.

as the latter will soon fall after entering the magazine when the doors are opened, and the relative humidity of the outside air which has entered the magazine be increased. Under these circumstances about five minutes should be long enough for ventilating a small magazine; but when the temperature inside is above that outside the magazine and other conditions are fulfilled, there is no limit to the time during which ventilation may be continued, provided outside conditions remain favorable.

Lighting.

Magazines of permanent seacoast works are lighted, as a rule, by electricity. When lighted by lamps, or when it is necessary to take a lamp into a magazine or a room containing explosives, only some authorized type should be used.

Great care must be exercised in protecting electric lamps from being broken, and the insulation of all parts of electric circuits within the magazine, or the room containing explosives should be of the most approved form.

It has been ascertained by experiment that the incandescent filament of the electric light will fire gunpowder dust if the globe be broken in an atmosphere containing such dust in suspension. It is considered necessary, therefore, to have all incandescent lamps protected by a strong outer glass globe and this latter by a strong, copper-wire cage; the outer glass globe should have an inlet and outlet tube admitting a circulation of air; the capacity of the globe and ventilating pipes should be such as to keep the temperature inside the outer globe not greater than 140° F.

In the case of very dusty and dangerous localities, the outer globe may be arranged to contain water instead of air, and a circulation of water provided, the lamp being immersed therein. In all cases where complete globes are used, one side should be painted to prevent the focussing of heat rays.

Lamps should be attached in such a manner as to make it

impossible to be broken by a fall; for this purpose a light wire cage is placed immediately about the lamp globe. No wire carrying a current should be used to *support* a lamp, or be otherwise subjected to a mechanical stress.

Lead wires should be inclosed in metal tubing up to the lamps, and the lamp wires should be soldered to the leads. No mere contact-joints should exist in the leads within or near the magazine. Each lamp should be provided with a fuse cut-out outside the magazine, so placed as to be readily inspected. The fuses should consist of tin wire about 0.036 inch in diameter, additional wires in parallel being used if necessary.

Each lamp should be supplied with a double-throw switch outside the magazine, by means of which the circuit may be completely broken. Before attempting to repair or replace a lamp, this switch should be thrown off for that lamp.

An efficient leakage-detector and lightning-arrester should be placed in each magazine-lighting system.

The difference of potential between any parts of the circuit within magazines should not be greater than 110 volts.

The system should be thoroughly tested from time to time in all its parts.

Special Storage Regulations for High Explosives.

High explosives in storage should have blue litmus strips placed in each package. These packages should be examined once a month, the litmus strip replaced, and the boxes turned over. Methyl violet paper has been substituted for blue litmus paper by recent orders. The methyl violet paper should not turn white in 30 days.

The floor under packages containing nitroglycerine explosives should be covered with clean sawdust, to absorb any nitroglycerine that might exude. This sawdust should be renewed from time to time, the old sawdust being burned in the open air.

In case a floor, or package, becomes coated or stained with free nitroglycerine, the latter should be decomposed by washing the floor or package with a solution of flowers of sulphur in carbonate of sodium. This soda-sulphur solution should be kept on hand wherever nitroglycerine in any form is stored.

Dynamite should be stored so that the sticks are horizontal; the tendency of dynamite to exude nitroglycerine is greater if the sticks stand on end.

It is important that dynamite-cartridges be kept dry. If exposed to a moist atmosphere, there is a tendency of the water condensed from the air on all exposed surfaces to displace the nitroglycerine.

A little sodium carbonate is usually placed in dynamite. Moisture often causes this to leave to some extent the body of the cartridge and to appear as a white efflorescence on the outside of the wrapper. If the dynamite is not otherwise changed, particularly if blue litmus is not reddened and there is no leaking of nitroglycerine, the efflorescence does not in itself indicate deterioration. It does suggest, however, that an examination of the dynamite should be made with a view to determining its condition as to the other defects named.

Guncotton is always stored in a saturated condition, containing from 30 to 35 per cent of water. In this condition it is practically non-explosive. If not stored in hermetically sealed cases, guncotton should be examined monthly and resaturated.

Dry guncotton is required as a primer in detonating wet guncotton. Dry guncotton primers should be stored apart from wet guncotton. The disks may be kept dry by immersing in melted paraffin. If dry primers so prepared are not on hand, wet disks should be dried out at temperature not above 110° F.

Liquid nitroglycerine is very rarely kept in storage. If it becomes necessary to store it, it should be stored in earthen crocks only, and should be kept covered with water. These crocks should be placed on supports of wood, near the floor, and over a trough containing sawdust or other absorbent. Like dynamite and guncotton, it should be examined monthly with blue litmus for evidences of acidity.

All buildings and rooms containing these explosives should

have a free circulation of air and should be under other magazine conditions.

Examination of Smokeless Powder in Magazines.

A sample of each accepted lot of powder is kept at the works of the manufacturers, where it is observed from time to time and tested. A part of each sample should be kept exposed at about 104° F. (40° C.), under conditions resembling as near as possible those which obtain in storage-magazines. This part of the sample should be carefully examined from time to time, and subjected to the stability test once every three months for the period of one year and thereafter, as long as any of the lot is in the service, once every six months. A small part of the original sample should be kept permanently in a glass bottle, in a suitable place, where it can be under observation. Another sample of each lot of the powder should be placed in a glass-stoppered bottle, with a piece of moistened litmus paper suspended just clear of the powder. This should be kept in position for six hours, moistening the litmus paper from time to time, noting whether the litmus paper reddens and to what extent, and being careful not to confuse the pink color due to the ordinary bleaching of litmus with the reddening due to free acid. In order to determine what the *acid color* for a given piece of litmus should be, a piece of the paper should be dipped in vinegar and the true acid color will result. If this color develops in the bottle, it is due to escaping nitro fumes.

Care should be taken to prevent the direct rays of the sun from falling upon powder or powder-boxes.

External Examination.

In making superficial examinations of smokeless powder a small scoopful should be taken into a good light, where a change of color may be most readily detected. Decomposing powder becomes lighter in color all over or in spots, showing a decidedly yellow tinge, and, when the decomposition is well established, the grains become in a measure soft, yielding to the pressure of the thumb nail. If nitro fumes are given off, the inside of the box, tank, or bag would probably show a yellowish appearance, and an acrid, pungent odor of nitric-peroxide gas would be present. Close observation is necessary to detect these signs in the case of incipient decomposition, but, if discovered at any time, such powder should at once be subjected to the stability heat-test.

In case the heat test gives evidence of diminished stability, all powder of the lot involved should be immediately segregated at a distance from other powders in a place where, if combustion should ensue, no harm could be done. The place of storage should be cool and dry. If the powder should give evidence of advanced decomposition, indicated by unmistakable odor of nitrous fumes; a very low methyl violet paper and surveillance test, whitening one-tenth normal methyl violet paper in less than 20 days; grains friable and crumbling easily or in a mushy condition, it should be removed at once and destroyed either by burning in the open air or immersing in water.

Samples of each lot of powder received at a magazine should be kept in glass-stoppered bottles and so placed in the magazine that they can be regularly and carefully examined twice a week.

The present practice of the U. S. Ordnance Department is to issue, in the cartridge-storage case, with each fifth charge of a lot, a small bag containing 8 ounces of the same lot as the charge, and this is enough for two 4-ounce observation samples. The cases containing such samples have the words "Observation Sample" stamped in red on the linen tag attached to the case, outside. When one or both 4-ounce samples are removed suitable note should be made on the tag. Charges or sections, packed in cases with samples, should be expended last.

When no such samples are available for separate loading or fixed ammunition, samples must be taken by opening a propelling charge or a fixed round, and when this is done, it is necessary to put in the place of the amount removed a like quantity of a suitable powder of recent manufacture, in order that the muzzle velocity may remain normal. Powder for this purpose has generally been issued in 5-pound containers, and where no such powder is on hand for the particular model of cannon in question, more must be obtained by requisition. No charge or round should have a second sample taken from it, and therefore suitable marks should be put on the container whenever samples are taken.

When a shipment of powder is received at a storage-magazine, each box or package which shows signs of rough handling and liability that its hermetical sealing has been destroyed should be opened and a superficial examination made of its contents to ascertain if it is in normal condition.

Fixed ammunition received for storage should have a few rounds taken apart for superficial examination.

Heat- and litmus-tests should be made in each case where superficial indications of incipient decomposition are observed, and unless the powder meets both of these tests it should not be placed in the magazine.

In preparing fixed ammunition, care must be exercised to see that the inside of the case is free from grease or any other foreign substance, and that the base of the projectile is perfectly clean.

The temperature and hygroscopic conditions of magazines should be constantly watched. Maximum and minimum thermometers should be placed one in the hottest part of the magazine and the other in the coolest. The temperatures should be taken daily and noted in the Magazine Record Book

Magazines should be inspected each day and the fact noted in the Record Book over the signature of the person who makes the inspection.

At these inspections the general condition of the magazine and its contents should be examined and noted in the Record Book. If the condition of the magazine is such as to indicate that everything is in a satisfactory state the word "Normal" should be entered. If otherwise, the particular defects noted should be spread upon the Record, and the matter reported at once to the proper officer.

No loose powder should be permitted in any building, except such as is actually being used in preparing cartridges.

Large quantities of powder should not be permitted in cartridge-filling rooms; only just enough to supply the immediate need.

As rapidly as cartridges are filled and prepared for use, they should be removed from the filling-rooms and placed in storage.

Neatness and cleanliness should be insisted upon at all times; no foreign substances, such as oakum, waste, rags, paper, paint-pots, -brushes, etc., should be allowed in any building assigned for the storage or preparation of cartridges.

If it should at any time become necessary to dry smokeless powder, it should be done out of the direct rays of the sun.

Smokeless powder should not be stored in magazines wherein the temperatue runs at any season above 95° F., or which ever reaches 104° F. If the temperature tends to rise so high artificial cooling must be resorted to.

If the odor of ether is noticeably strong in any magazine, such magazine should be blown out with portable fans or otherwise ventilated.

A naked light should never, under any circumstances, be taken into a room containing any quantity of powder.

The following tests and examinations should be made of smokeless powders kept in service-magazines at posts: [1]

Daily.—A sample from each lot of smokeless powder in the magazine is to be kept in a glass-stoppered bottle [2] in a

---

[1] These tests do not apply when powder is stored in soldered metallic cases.

[2] The style of bottle desired is that known as "salt-mouth" bottles and of a capacity of about two pounds; they should be filled about two-thirds full.

conspicuous place, and frequently examined in a good light as to its external appearance.

Fortnightly.—The powder in one or more boxes or bags of each lot to be examined externally for evidences of incipient decomposition.

Monthly.—The sample in the index-bottles will be subjected monthly to a moist litmus-paper test for 30 minutes.

Quarterly.—A sample from each lot in the magazine to be subjected to the potassium-iodide-starch-test for 40 minutes once a quarter, and also to a six-hour litmus-test.

In case a pungent odor is detected it should be investigated.

The following regulations, with regard to the care and preservation of smokeless powders in store, are prescribed by the Ordnance Department, U. S. Army:

All lots of smokeless powder will, as far as practicable, be shipped from the manufacturers to one of the powder depots; except, under unusual circumstances, issues to posts will be made only from such depots.

In issuing smokeless powder from the depots the oldest lots in store will be issued first, unless instructions to the contrary be given.

All powders stored at the powder depots shall be tested as follows:

1. By the usual stability tests at the Ordnance Laboratory. For this purpose an 8-ounce sample from each lot of powder in store will be sent to the laboratory for test.

These tests of powder shall be made each six months after delivery. The samples will be selected as follows: From lots for the 10-inch and 12-inch B. L. rifles not more than one grain shall be taken from a box; from lots for guns of other calibers 5 per cent of the boxes shall be opened and a proportionate part taken from each.

2. A litmus paper test will be made every three months for six hours from a sample taken from one or more boxes of each lot. The sample is placed in a clean glass-stoppered

bottle, and a piece of litmus paper moistened with water (distilled, if practicable) is suspended just clear of the powder.[1]

3. In each magazine samples of each lot stored therein should be placed in glass-stoppered bottles and examined semi-weekly. The appearance of yellowish or brownish-red fumes gradually assuming a red color as the quantity increases is a sign of deterioration. The fumes have a disagreeable, sharp, acrid odor similar to that of nitric acid, and are very irritating to the eyes and nose.

Should there be any indication of fumes the bottle should be opened and two pieces of litmus paper moistened with water (distilled water, if possible) quickly inserted, one in contact with the powder and one hanging from the stopper. If there are any fumes being evolved, the litmus paper should be reddened in a few hours. The moist paper will gradually dry out; if any doubts exist as to its reddening, the paper should be again moistened and replaced. The papers should be exposed in the bottles or boxes for at least six hours.

4. Small samples of each lot should be kept in glass bottles, either in the offices or in some suitable place for purposes of daily observation. These bottles should not be exposed to the direct rays of the sun, nor in any place where they would be liable to be overheated.

Recently methyl violet paper has been substituted for litmus paper in the storage tests of powder at fortifications. This paper placed in a test bottle will turn white in less than 30 days if the powder be bad. The test consists simply in placing a standard methyl violet paper in a test bottle with a sample of powder the first of each month, noting the color at the end of the month and renewing the paper.

This test is the only powder-test of a chemical nature required at posts.

The Methyl-violet test paper used at posts is what is called "tenth-normal," that is, it is of one-tenth the sensitiveness of

[1] The caution mentioned on page 231 as to the true acid color should be kept in mind.

normal methyl-violet paper, as used in the laboratory. As received fresh at posts, it should have the violet color shown at the top of the color scale in the Ordnance Dept. pamphlet cited. Old paper or paper showing soiling or discoloration should not be used. The paper may be handled with clean dry hands, but the less it is handled, the better. It is not affected by diffused light, but should not be exposed to direct sunlight. Its test-value depends upon its property of gradually losing its violet color in the presence of oxides of nitrogen given off by decomposing powder. The time of test is the number of days required for it to become entirely white, no trace of the violet, pink, or yellow colors, through which it passes in the change from original color, remaining.

To make the test, a piece of fresh 1/10 normal methyl violet paper is marked with date in *lead pencil* and inserted *dry*, in the glass-stoppered bottle containing the 4-ounce sample of powder, in such a way that the tight closure of the bottle is not interfered with and that the date may be read without opening the bottle. Wedging the paper between stopper and bottle prevents good closure and should be avoided, as should unnecessary opening of the bottle; as absorption of moisture and loss of volatiles due to exposure to the atmosphere tend to affect the powder, while escape of any nitrous fumes that may have formed would tend to show an unduly long test of the powder-sample.

Only one test-paper should be in a bottle at one time, and it should be replaced by a fresh one at the end of each month. Examine about the 10th, 20th and end of each month for loss of color, making this examination without opening bottle.

Powders of new manufacture will give, at ordinary temperatures, a test of two months or more. A test of one month is considered as indicating a satisfactory degree of stability. If a sample gives a test of less than one month, the test should be repeated upon a fresh sample taken from the same lot. If a test is less than 20 days, the powder of lot represented should in the absence of special instructions, be segregated from other lots. Results of all tests completed in 30 days or less should be reported, using blank form provided for the purpose, and test papers, with the dates of commencement and completion, and the lot-numbers written thereon, should accompany the report.

Fresh samples should be taken about January first and July first, observing all the precautions laid down in the pamphlet.

MINIMUM DISTANCES THAT SHOULD SEPARATE STORAGE MAGAZINES FROM INHABITED BUILDINGS.

RECOMMENDED BY COL. B. W. DUNN, CHIEF INSPECTOR, BUREAU FOR THE SAFE TRANSPORTATION OF EXPLOSIVES.

Pounds of Explosives.	Minimum safe distance of barricaded [1] storage magazines from inhabited buildings. Feet.
50	120
100	180
200	260
300	320
400	360
500	400
600	430
700	460
800	490
900	510
1,000	530
1,500	600
2,000	650
3,000	710
4,000	750
5,000	780
6,000	805

[1] "Barricaded," as here used, signifies that the building containing explosives is screened from other buildings or from railways by either natural or artificial barriers. Where such barriers do not exist, the distances shown should be at least doubled.

Pounds of Explosives.	Minimum safe distances of barricaded [1] storage magazines from inhabited buildings. Feet.
7,000	830
8,000	850
9,000	870
10,000	890
20,000	1055
30,000	1205
40,000	1340
50,000	1460
60,000	1565
70,000	1655
80,000	1730
90,000	1790
100,000	1835
200,000	2095
300,000	2335
400,000	2555
500,000	2755
600,000	2935
700,000	3095
800,000	3235
900,000	3355
1,000,000	3455

[1] See footnote on preceding page.

IX.

HANDLING HIGH EXPLOSIVES.

WHILE the explosives herein treated have enormous potential energy stored up in them, they are perfectly safe unless a definite act be taken to let loose this energy.

If they are so handled that no particle of any given mass is brought to a certain definite temperature by application of heat, friction, or shock, they are as safe as any other solids or liquids. The solid nitro-explosives are at least no more dangerous than the old black gunpowder. The precautions to be kept in mind have been pointed out as the several explosives have been taken up in succession. Some of the more important of these may, perhaps, with advantage be collected and repeated here.

Summary of Precautions of a General Nature to be Observed in Handling Explosives.

Avoid bringing any matches or other easily combustible substances near an explosive.

Avoid the use of hard, rigid tools, implements, or apparatus in connection with explosives. A particle of explosive pinched between two hard surfaces, and subjected to a blow or to sliding friction, is apt to explode. The minutest particle caught in this way and exploded has the power to initiate the explosion of a large mass. Copper is the only metal that should be used about explosives.

Use only the quantity of explosive necessary for the work in hand, and keep the main supplies far removed from the point of explosion, and well protected from all possible exposure to fire or shock, or to handling by unauthorized persons.

Keep explosives and means of exploding them apart until it is desired to arrange a charge for explosion.

Explosives and primers, fuses or caps, should never be transported or stored together.

Nitroglycerine, dynamite, dry guncotton, and explosive gelatin, if transported, should be protected against violent shock by preparing a soft, elastic bed of hay, straw, excelsior, or similar substance in the cart, wagon, or car. Rough pavements and roads should be avoided in so far as practicable.

Never prepare a dynamite or explosive gelatin primer cartridge near other dynamite or explosive gelatin.

Never try to thaw nitroglycerine or a nitroglycerine derivative over a naked flame or on heated metal. Use always a closed vessel in a water-bath.

In case a charge at any time misses fire, *do not be in haste to investigate the cause.* Wait at least ten minutes, and, then, when satisfied that no explosion is to take place, remove the tamping, cut the lead-wires of the fuse, and prepare another primer. Open up the charge as little as possible and not near the old primer.

In using an electric current for firing, the wires should not be connected to the source of electricity until the circuit is otherwise complete, the primer in place, and charge all ready for firing. One man should be detailed to see that the firing ends of the wires are not tampered with while the charge is being arranged.

Before firing a charge, warning should be given to all persons connected with the firing, and a lookout stationed to warn off all friends.

Precautions to be Observed in Charging Torpedoes and Shell with High Explosives.

The work should be done in light frame buildings apart from other buildings. The floor must be swept frequently, and the sweepings burned at a distance.

The temperature of the loading-room should not be above 90° F. nor below 50° F.

No acids or primers should be allowed near explosives in bulk. Magazine conditions will be strictly enforced, both as to persons engaged in the work and to the surroundings.

In connecting together parts of material by screwing, as in fusing shell and arranging the tropedo fuse, great care must be exercised that no particle of explosive is caught in the screw-threads.

Shell loaded with picric acid or its derivatives should not have screw-threads coated with white or red lead.

Great care must be taken that particles of explosive are not dropped on the floor.

A torpedo loaded with dynamite should be kept carefully protected from the sun's rays. The direct rays of the sun would soon heat the interior to a high degree, and the sensitiveness of all high explosives increases rapidly with the temperature. Loaded torpedoes should, therefore, be kept in the shade, and, if necessary, covered with paulins.

Safety Precautions in Preparing to Fire Demolition Charges.

1. In testing fuses or detonators never attach a wire to either lead, unless the fuse or detonator is safely inclosed or at a safe distance.

2. Always hold a cap or primer pointing from you.

3. Be careful not to bend, strike hard, or heat a cap or primer.

4. Do not place caps or primers near strong acids.

5. Be careful not to allow any strain to be put on the leads of a primer in making up a charge or in connecting up the circuit.

6. Any one who connects a wire to the lead of a primer is responsible for his own safety. He should not make the connection unless he *knows that the circuit is broken* between him and the source of electricity. To increase safety, the outer ends of the circuit should be put in charge of some person, with instructions to keep the leads apart.

7. All persons except those directly engaged in the work should withdraw to a safe distance or take cover while the charge is being made up and the circuit prepared.

8. The exploding-machine, electric battery or other firing apparatus, should not be brought to the firing-point until all preparations for firing have been made. The *last* thing before firing is to connect the leads with the source of electricity.

9. Place the exploding apparatus or machine as near the charge as safety permits. Before using, test the machine by seeing if it will redden, by heating, a small piece of platinum wire, or if it will explode a spare primer, or take the throw of a galvanometer, or the shock of the current between ends of short leads attached.

10. If a charge is to be fired by using a firing-key, examine carefully to see that there is a *real and sufficient break* when the key is "off," and that there are no loose wires or other means near to form a circuit except through the key. In firing, connect one terminal of the firing-key with the positive pole of the firing-battery, and, lastly, connect with the battery's negative pole.

11. *Immediately after firing, disconnect both leads and place them in charge of some responsible persons, as explained in 6.*

12. In testing circuits and primers, not more than 1/20 ampere should flow through any primer.

13. For certainty of ignition, a single large charge should have two or more primers connected up in parallel, thus:

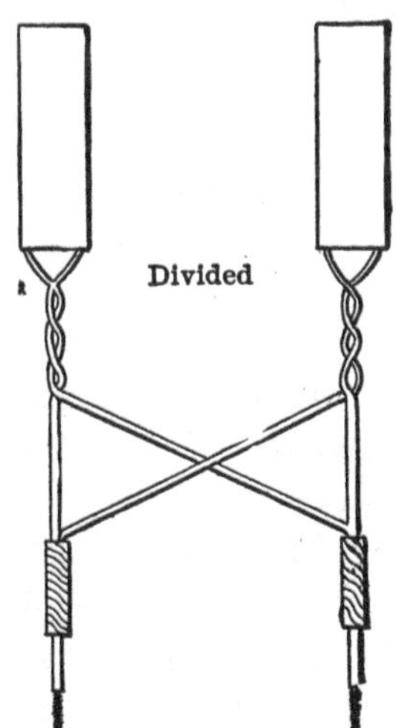

FIG. 1.

14. Always use the same kind of primers in the same circuit.

The utmost care must be always exercised in handling all kinds of explosives and in their preparation for firing. The tendency of those charged with the duty of handling explosives is to become careless and indifferent, and to neglect those precautions and that carefulness which should always be observed in connection therewith. Only the constant, utmost watchfulness will avoid accidents. No relaxation of these precautions or of the rules and regulations governing magazine duties should be permitted.

Preparing a Charge for Firing.

In arranging a charge for firing, the primer-cartridge of dynamite or the primer-disk of guncotton is placed as near as possible in the middle of the charge, and the mass of explosives packed tightly around it.

The charge may be ignited by a time-train fuse, or by an electric primer or cap.

If a time-train is used, its normal rate of burning in open air must be ascertained by trial. A single-tape time-train fuse will burn at the rate of about 1 foot in 18 seconds, a double-tape fuse, 1 foot in about 20 seconds, a triple-tape fuse, 1 foot in about 25 seconds.

The time-fuse is cut to the desired length, placed in the open end of the cap, and the latter pinched down tightly on it, as shown in Fig. 2.

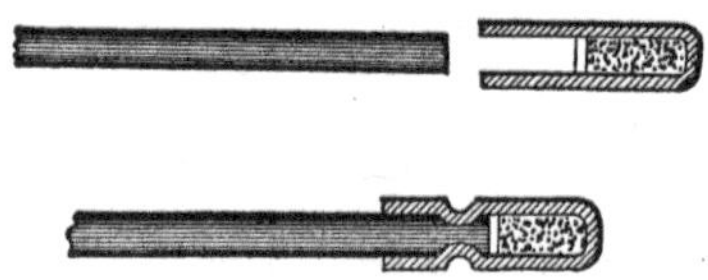

FIG. 2.

If the fuse is to be used under water, the cap must be well coated with paraffin, tar, or shellac, so as to make the joint water-tight.

The cap is next inserted in the cartridge. In doing this,[1] open that end which has the longest paper-folds. Punch a hole in the center of the end of the cartridge with a round-pointed stick, making the hole slightly larger than the cap. Insert the cap (about two-thirds of its length) until it is almost but not quite covered by the explosive. Bring the paper of the cartridge close around the fuse-train and tie tightly with a strong

[1] The description contemplates a dynamite stick-cartridge.

string. The primer-cartridge thus made will appear in longitudinal section, as shown in the following figure.

The charge having been arranged with the primer-cartridge as near as possible in the center, the train is led off in the direction of cover, its free end is ignited, and the operator quickly withdraws.

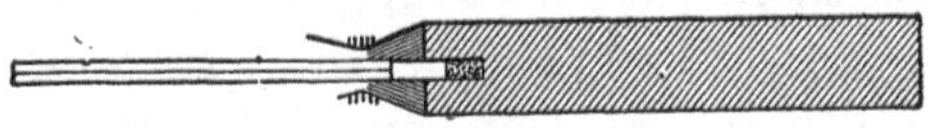

FIG. 3.—Primer-cartridge arranged with time-train fuse.

In firing by electricity, an electric primer is used. A primer-cartridge is prepared as follows: The paper is unfolded at one end of the cartridge, an opening is made in the center of the end with a pointed round stick, a little larger than the primer-cap. The cap is inserted until the upper end is nearly but not quite flush with the upper surface of the explosive in the cartridge. The lead-wires are then bent sharp over the end of the cartridge and along its side to the opposite end, leaving the free ends of the wires at that end. In passing along the cartridge, two half-hitches should be taken around the cartridge with the lead-wires, one near the end in which the cap is placed, to prevent the latter from being disturbed; the other near the opposite end. When completed, the primer-cartridge should appear as in Fig. 4.

FIG. 4.—Primer-cartridge arranged for electric firing.

To allow for this arrangement, and to allow also for ample free ends, the lead-wires should be at least 6 feet long.

This primer-cartridge should be placed at the center of the charge and the components of the charge packed tightly about it, the free lead-wires passing out through the charge in the direction of the point from which it is to be fired.

In the case of guncotton, a dry block is taken for the primer-block. The primer is placed in the hole of the block and packed in tightly with scraped dry guncotton taken from the corners of the block. The leads are then bent over and around the block, making a close-fitting half-hitch. If it is to be fired under water the whole should be dipped in melted paraffin.

In jointing wires, strip off the insulation for about two inches, leaving the end of the insulation conical, like the wood part of a pointed lead-pencil, and clean the wire carefully with the back of a knife, or other suitable tool, until a smooth, even, bright metallic surface is obtained, being careful not to nick or roughen the surface of the bared wire if possible. Cross the wires at right angles, as shown in Fig. 5. Then bend each wire around the

FIG. 5.

FIG. 6.

other spirally in the direction of the pointed insulation of the other wire, keeping the turns of the spiral close together, as shown in Figs. 6 and 7. Three or four turns should be made, *pressing the turns tightly down on the standing part of the other wire*, using pincers, preferably, to make the turns regular and tightly pressed on the other wire. Cut off the spare ends and pinch the cut ends close down, as shown in Fig. 8.

FIG. 7.

FIG. 8.

In jointing stranded wires, each strand should be separately cleaned, and each strand wrapped around the standing part of the other wire, as explained above for a solid wire.

A three-way joint is made by first making a simple joint, as explained above, and then opening the wires at the first crossing sufficiently to insert the bared end of the third wire, as shown in Fig. 9. This third wire is wrapped closely down

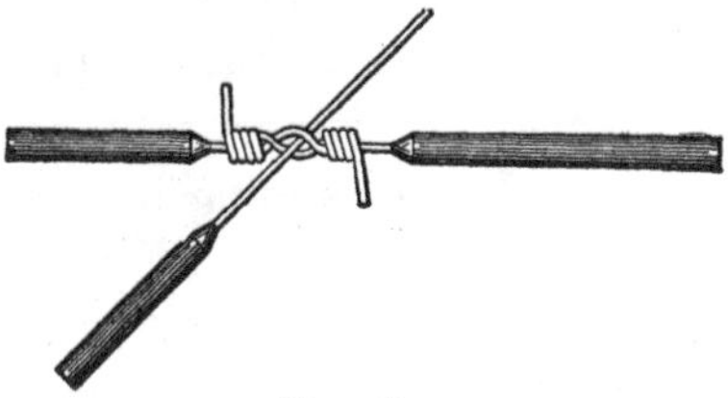

FIG. 9.

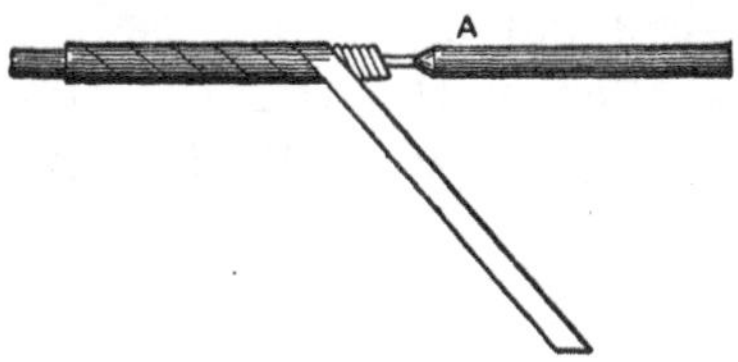

FIG. 10.

on the turns of the first wires. Other wires may be connected in, in the same manner.

Important joints should be soldered if time allows. To solder a joint, first wash the joint with zinc chloride, heat the soldering-iron until it will readily melt the solder. Rub one face of the iron with a coarse file, then rub over a little sal ammoniac; or dip it quickly in a solution of sal ammoniac, then rub the solder on this cleaned face of the iron and apply to the joint. The solder should be hot enough to run freely into the spaces between the wires. The joint is then washed clean with carbonate of soda or other alkaline solution. Instead of zinc chloride, a solution of resin in spirits of wine may be used.

Great care should be taken to keep the bare hands off the scraped wires, and to keep the latter free from all grease.

All joints, whether soldered or not, should be insulated. This is accomplished by the usual insulating rubber tape. Begin well down on the wire insulation and wrap spirally well over on the insulation of the other side of the joint; letting each turn overlap the previous one one-half, ending in a half-hitch (see Fig. 10).

If the joint is to lie under the water, each turn of the insulation-wrapping should be carefully smeared with india-rubber solution before the next turn is laid over it. In unsoldered joints, the india-rubber solution should not be placed over tape lying next to and immediately over the twisted wires. Care must be taken to notice that the tape *adheres* to the rubber solution as it is laid down, and especially to the insulation of the wires on each side. To insure this, the insulation of the wires and the tape to be laid down should be cleaned off with a little naphtha, and the insulation smeared with rubber solution.

A good water-tight joint may be made by slipping a piece of rubber tubing on the wire before the jointing, then, after the jointing, slipping it over the joint and binding it on each side tightly down on the wire insulation with strong twine or with pliable wire.

If neither tape nor tubing is available, a fairly good insulated joint, suitable for use in damp places, may be made by slitting longitudinally the insulation of a spare piece of wire, detaching it carefully from the wire, cutting this piece in two across, and then applying the two sections over the joint and binding down tightly with twine or fine wire.

A joint should be made in that part of the circuit least liable to moving or bending. If necessary, the joints should be fixed in position by weights or stakes or staples.

Before a circuit is connected up for firing, the joints should be tested for continuity. The complete circuit should finally be tested by a weak current.

The service-exploder is known as the Laflin & Rand Magneto-electric Machine, or the Laflin & Rand Exploder.

The internal arrangement (see Figs. 11, 12, and 14) consists of a Siemens armature, *B*, which revolves between soft-iron prolongations of the cores of an electromagnet, *A*.

The electricity is generated by forcing the armature to revolve in the field of the magnet and is transformed by a commutator, *F*, from an alternating to a continuous current. The circuit passes from the commutator-springs into the adjacent ends of the windings of the magnet. The back-strap ends of the windings of the two halves of this magnet are extended to the terminals, or binding posts, *G*, for the connecting wires;

and thence to a brass spring, *D*, and collar, *E*, where, by platinum points, they are joined together, thus completing an interior short circuit as a shunt. The magnet is wrapped with 1.76 ohms of cotton-insulated copper wire, No. 18, B. W. G., and the armature with 0.92 ohms of No. 21 of the same. The novelty of the machine lies in the mode of giving rotation to the Siemens armature, and of switching into the firing circuit the powerful induced current. Both objects are accomplished by the firing-bar, which consists of a square brass

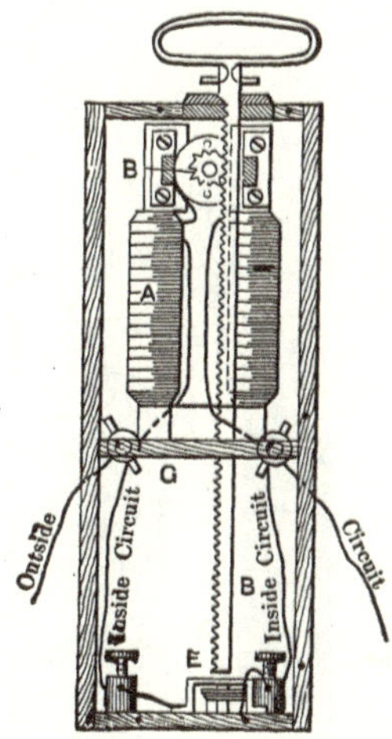

FIG. 11.—End View.

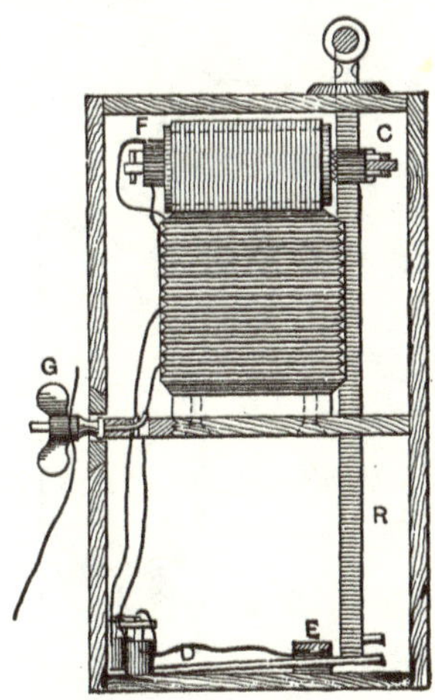

FIG. 12.—Side View.

rod, 14 by ½ by ½ inches, fitted with a wooden handle at one end, the other end passing down into the box. One side of the bar is provided with teeth which engage in a loose pinion, *C*, fitted over the prolongation of the armature spindle. A clutch holds the pinion to the spindle when the rod is descending, but leaves it free when the latter is raised, thus restricting the revolutions of the armature to one direction only. When the firing-bar reaches its lowest position, it strikes the brass spring which forms part of the interior circuit; and, if in rapid motion, the shock breaks the circuit and thus shunts the current into the firing circuit.

In passing from the top to the bottom of the box, the rod causes seven and one-half complete revolutions of the armature; and, if the movement be the result of a sudden and downward pressure, this is enough to develop a powerful electrical current.

This form of exploder is very compact and strong, and not liable to get out of order except through very rough usage.

The machine may become temporarily deranged through two causes:

1st. Dust or some foreign substance may find its way between the platinum contact-points between D and E, Fig. 11.

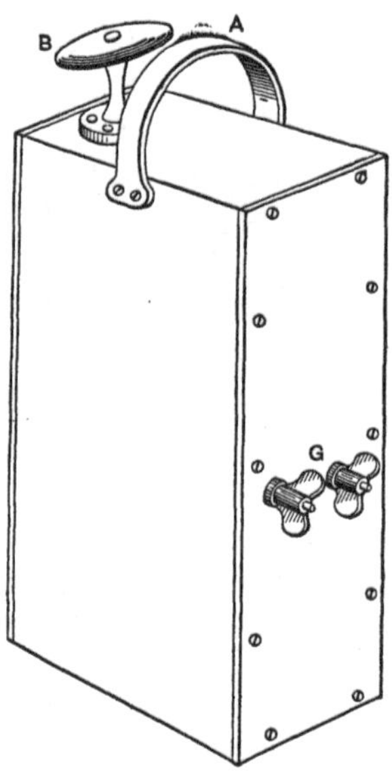

Fig. 13.

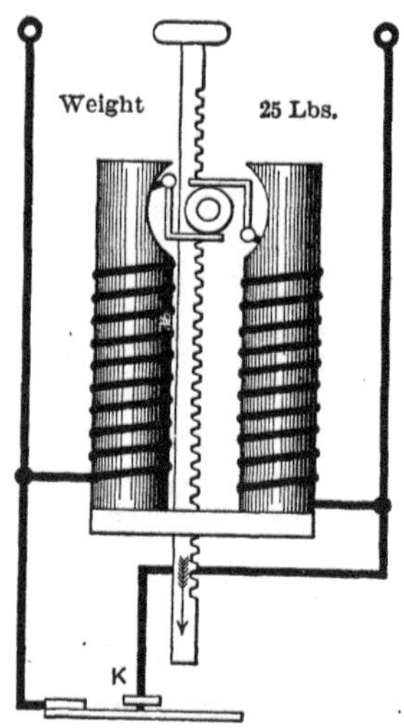

Fig. 14.

By removing the screws that hold it in place, the rear of the case may be removed and the trouble remedied by using a piece of fine emery-cloth.

2d. Trouble may arise from the surface of the commutator becoming tarnished. In order to cleanse it, remove the rear of the case as before, and also the small pin near the lower end of the firing-bar, and then withdraw the firing-bar from the case.

The works of the machine, with the shelf upon which they rest, are next partially removed from the case, and the springs which press upon the commutator, and the yoke which holds in place the spindle upon which the commutator revolves, are disconnected. The commutator may then be cleaned with a piece of fine emery-cloth.

Proper attention to these details and careful preparation of the wires and fuses save a vast deal of trouble, and cannot be too strongly insisted upon when success is absolutely necessary and time is to be saved.

To use the exploder, note that safety precautions have been taken by all persons; clean the lead ends; attach cleaned ends to the binding-posts (*G*, Fig. 13) of the exploder; raise the firing-bar[1] (*B*, Fig. 13) to its full height; force the firing-bar down with firm, rapid, uniform stroke, keeping the bar vertical.

In some recent forms of this exploder, there are three binding-posts for firing a larger number of primers than can be fired by two. The third post is connected at a central point of the group of fuses; the current goes out on this central line and divides over the two return routes. The resistance is thus lowered, so that a sufficient current is developed to fire the primers in each return route.

---

[1] The firing-bar should be kept down at all times, except in the act of firing.

X.

DEMOLITIONS.

DEMOLITIONS may be divided into two kinds: (1) deliberate and (2) hasty.

In the case of deliberate demolitions, time is not an important factor in the preliminary arrangements, and economy of means and material may be given due consideration.

In hasty demolitions, the saving of time is the controlling consideration. Tamping, and other means of economizing the quantity of explosive required for a given demolition, must often be neglected, and hence hasty demolitions require relative larger quantities of explosives than deliberate demolitions. Hasty demolitions only are considered in these notes.

When the demolition requires *mass effect,* a progressive explosive like gunpowder is to be preferred to a high explosive. If a local shattering effect is desired, the latter is to be preferred.

With gunpowder, tamping is essential if a good effect is to be had. Tamping is not so important with dynamite, guncotton, and other high explosives. The full effect of dynamite is obtained when the tamping is equal in thickness to the thickness of the mass to be destroyed; with gunpowder, the tamping should be 1½ to 2 times thicker.

Demolitions may be "moderate," in which the fragments remain at or near the point of explosion; or "violent," in which the fragments are scattered and thrown to some distance.

In destroying masonry revetment walls, the charge should be placed on the back of the wall on a level with the foot of it, and along the length of the wall to be demolished. For this purpose a gallery must be driven through the revetment and extended right and left behind it. The charge should be sufficient to destroy the wall, and should be covered in the gallery through the revetment with earth $1\frac{1}{2}$ times the thickness of the wall. If the wall have buttresses, there should be an additional charge and tamping opposite these points. The foot of the wall may be reached by a shaft from above, instead of a gallery through it. The lateral galleries should be run the same, however.

The resistance of ordinary masonry may be taken at $1\frac{1}{2}$ times that of a similar thickness of earth. A tamping of earth over the charge double the thickness of the wall should be sufficient.

Buildings.

Large buildings with substantial masonry walls should have the charges laid at intervals all along the ground at the foot of the outside walls. A ditch dug parallel to the line of charges will furnish earth for tamping.

If the charges be let a short distance into the wall, the charge may be smaller and the tamping reduced.

It would be better to place the charges inside, but, as a rule, the interior arrangements, floors, etc., interfere, and it is difficult to get sufficient earth for tamping.

When there is difficulty in getting earth for tamping it may be necessary to *blast* the walls down.

Blasting is effected by relatively small charges of explosives placed in holes of small diameter called "bore-holes." It is resorted to only where hard, rigid material is to be removed, such as rock, masonry, etc. The charge must be put in the form to fit the bore-holes. The stick form of dynamite is a convenient one to charge bore-holes.

The positions of bore-holes with respect to the mass to be demolished are important.

The direction of maximum effect is at right angles to the bore-hole opposite the center of the charge. The charge should be so placed that the "burden" of the charge is on this line. This line of the "burden" of the charge is the "line of resistance," abbreviated L.R. It is the longest line from the charge at right angles to the bore-hole in the direction the explosive effect must be carried.

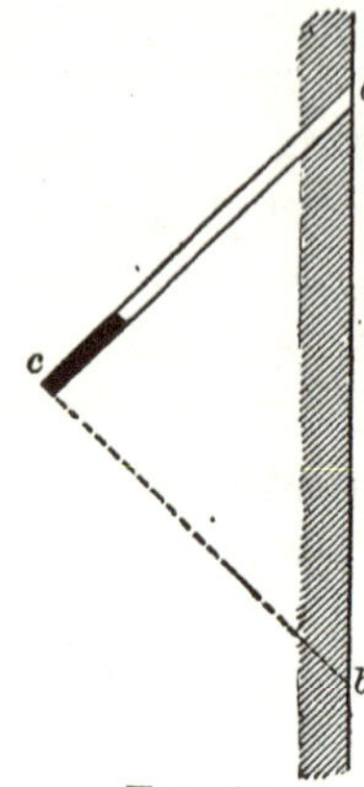

FIG. 15.

The angle of the bore-holes should be less with the face of the mass, the harder and more tenacious the latter.

When there are two free surfaces the bore-hole should be run parallel to the longest free side, as illustrated in Fig. 16: *acb* = probable crater.

If the mass be vertical and have an undercut, as in Fig. 17, the bore-hole should be driven at least beyond the angle at *d*. The depth of the bore-hole should be at least ¾ A.D. If the side *AD* is not parallel to the bore-hole *ac*, then L.R. is the *longest* perpendicular to the charge. In all cases the size of

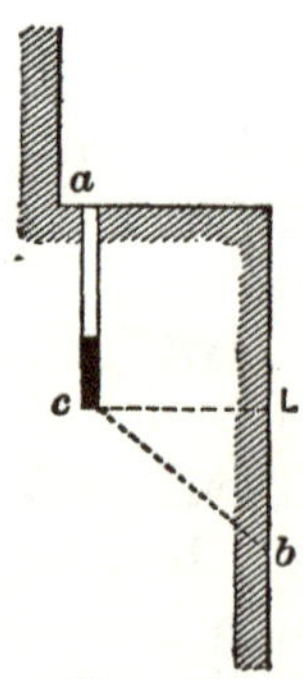

FIG. 16.

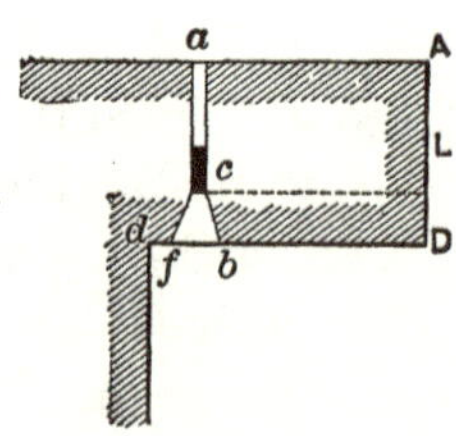

FIG. 17.

the charge must be adjusted to this longest perpendicular. If this is not done, a small crater like *fcb* might be made, leaving the rest of the mass undisturbed.

A vertical-face undercut without a top surface should be arranged as in Fig. 18, the bore-hole being parallel to the undercut face.

When several bore-holes are placed in series the distance between them should be equal to $1\frac{1}{2}$ L.R. when fired separately, and equal to 2 L.R. when fired simultaneously.

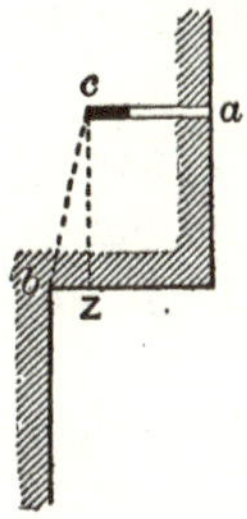

FIG. 18.

In charging a bore-hole, as many sticks of dynamite or other explosive as may be required, according to the computation for the charge, are placed in the hole, pressed firmly with a wooden drift until the sticks are in close contact with each other and with the sides of the bore-hole. The primer-cartridge is placed in last. A paper or cloth wad is placed over this, and the whole is tamped with sand or other material.

The weight of charge in ounces may be computed by the following formula:

Let C = total charge in ounces.

c = charge per foot run of bore-holes in ounces, i.e.,

$$= \frac{C}{\text{length of bore-hole in feet}}.$$

L.R. = line of resistance in feet.

k = coefficient of resistance of the mass to be blasted.

B = length of bore-hole in feet.

Then

$$C = k(\text{L.R.})^2,$$

$$k = \frac{C}{(\text{L.R.})^2},$$

$$c = \frac{C}{B}.$$

k is determined by experiment for the material to be blasted. When not known, and there is not time to determine it, it may be taken as 0.2.

For blasting purposes dynamite or explosive gelatin is, as a rule, more convenient than gunpowder or guncotton.

Buildings can, as a rule, be demolished more economically and readily by blasting charges placed *in* the walls than by charges placed along the bottoms of walls and covered with earth.

Charges of black gunpowder will be effective in demolishing walls when placed at the middle of the wall, provided the charge is in compact form, and the diameter of the bore-hole is greater in inches than the wall is thick in feet.

In boring into walls, the holes should slant downward toward the middle of the wall at an angle of 45°. The middle of the wall will be reached when the bore-hole is $1\frac{4}{10}$ L.L.R.[1] The hole must then be lengthened so as to contain one-half the charge, and to bring the center of the charge at the middle of the wall.

The amount of explosive may be reduced by cutting away portions of the wall, leaving only piers to be demolished.

If the strength of the wall varies from point to point by buttresses or other construction, the charge must be increased at such points. Bore-holes may be driven as follows:

Single.—Slant downwards at 45°, alternating on opposite sides of the wall.

V-shaped.—Same, but directly opposite each other, meeting at the middle of the wall.

X-shaped.—Same, but crossing at middle of wall.

The table on page 253 gives the charges of black powder required for demolitions when placed twice the line of least resistance apart.

If the holes have to be made with a diameter in inches less than $\frac{2}{3}$ L.L.R., *V* or *X* holes may be used with diminished intervals, or two parallel holes may be cut side by side and the partition between them cut away.

---

[1] L.L.R = line of least resistance, it is that line drawn outward from the charge along which the resistance is smallest. It is always expressed in feet.

Diameter of Hole in Inches.	Charge of Powder in Pounds.	Depth to which each Hole is to be Bored in Feet.	Kind of Hole.	Length of Hole Occupied by Powder in Feet.	Remarks. Charges to be Fired Simultaneously.
2 L.L.R.	$\frac{1}{3}$ (L.L.R.)^3	$1\frac{1}{2}$ L.L.R.	Single	$\frac{1}{4}$ L.L.R.	This is the best size of hole.
$1\frac{1}{3}$ "	$\frac{4}{10}$ (L.L.R.)^3	$1\frac{3}{4}$ "	"	$\frac{2}{3}$ "	
"	$\frac{1}{2}$ (L.L.R.)^3	$2\frac{1}{5}$ "	"	$1\frac{1}{2}$ "	
"	$\frac{1}{3}$ (L.L.R.)^3	$1\frac{4}{10}$ "	V	$\frac{1}{2}$ "	Half the charge in each hole; overlap *slightly*.
$\frac{2}{3}$ "	$\frac{1}{3}$ (L.L.R.)^3	2 "	X	$1\frac{1}{4}$ "	Half the charge in each hole; overlap equally, forming ×. L. L. R. always expressed in feet.

Bridges.

The destruction of bridges is an important division of demolitions. Usually the time available for preparation is brief; traffic over the bridge cannot be interrupted during the preparation; and, finally, the destruction must be accomplished suddenly when the proper time has arrived, and the demolition must be certain and complete.

The proper way to destroy a masonry bridge of a single arch is to demolish one or both haunches.

A bridge having piers should have the charges placed at the bottom of the piers, and several charges should be placed rather than one large one, since the risk of failure of a single charge should not be run; several charges should be placed at intervals apart equal to 2 L.L.R.

The arch of a bridge offers greater resistance to destruction than a plane surface. The charge should always be placed on the haunch and so that *cb* is the L.L.R.; its resistance being less than *ca*, or any other line out from *c* to any surface.

In order to insure these relations, *ca* or any other line should be equal at least to 3 *cb*. The distance between charges across the width of the bridge should not be greater than 2 *cb*.

If the bridge is to be destroyed with a single charge, the L.L.R (*cb*) should be made equal to at least ¼ the width of the bridge. Except with very narrow bridges, it would be better to use multiple charges.

A single charge placed at the crown is not advisable, for the reason that it may simply blow out the crown, as indicated

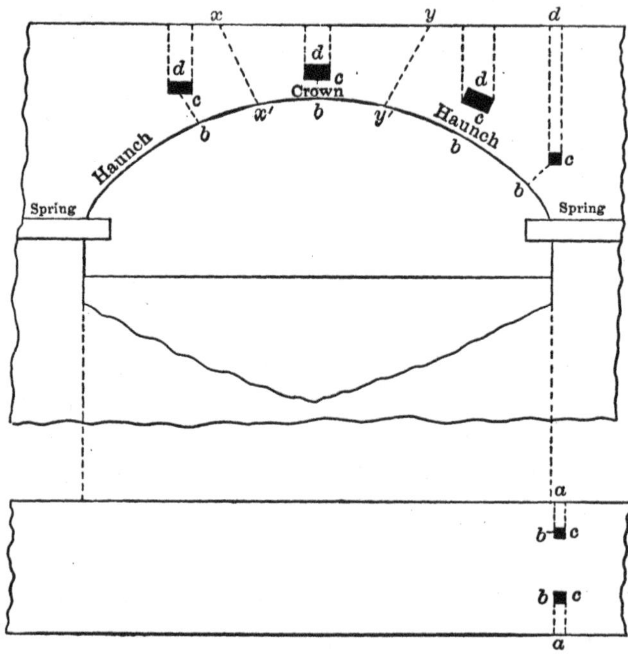

FIG. 19.

by the lines *xx′* and *yy′*, making the repair of the bridge a comparatively simple matter. It might be that this, in some special case, would be desired; then an overcharge should be distributed across the bridge along the crown, midway between the roadway and the surface of the crown.

When there is not sufficient time to place charges to destroy the haunches, several rows of charges should be placed over the arch, as shown at *ddd*. The distances between these charges across the width should not be greater than 2 L.L.R. The

L.L.R. should be regulated by the depth of the stones forming the arch. It should not, as a rule, be less than $1\frac{1}{2}$ feet nor more than 5 feet: if less than the former, the charges would be too small; if greater than the latter, too large.

Another method of arranging the charge is to place it in a trough suspended below the arch. This answers better for high explosives than for gunpowder.

The following empirical formula is given by Captain H. Schaw, R.E., for determining the charge of powder required to demolish a strongly built masonry arched bridge, when the charge is well tamped and placed over the haunch, at a depth below the roadway equal to twice the distance through to the surface of the arch: $C = \frac{2}{3}(\text{L.L.R.})^2 \times B$.

If on the arch: $C = \frac{2}{3}(\text{L.L.R.})^2 \times B$ {Placed in a shallow trench along crown on the keystone, with excavated material placed over it.}

In which C is the total charge of powder in pounds required for the charge in a single mass, or in line across the bridge; L.L.R is the line of least resistance; B is the breadth of the bridge in feet.

When a bridge is wide, the charges may be placed, without stopping traffic, by sinking a shaft in the middle of the roadway and placing a board cover over the shaft. When the bridge is narrow, the charges may be placed by running galleries from the side walls. If the mining be difficult and the time limited, it may be necessary to resort to overcharged mines.

Wooden bridges may be destroyed by explosives, cutting through the important ties or struts of the middle section, or by burning or cutting or sawing through the important members.

Iron-girder Bridges.

These bridges should, as a rule, be destroyed by demolishing the girders, their members or parts, rather than by blowing up the piers, unless there be ample time and it is desired to effect the greatest damage possible.

Girders may be solid and continuous, as in the simple I-beam girder, or they may be in the form of a built-up truss.

Where there is a continuous truss across several spans, the shore spans should be cut near the first pier, thus:

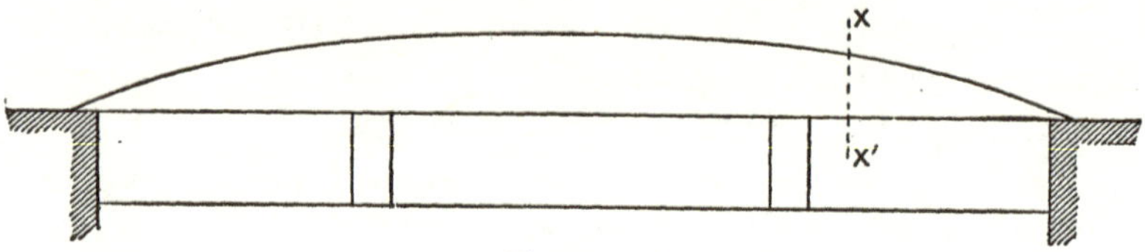

Fig. 20.

Cut at *XX′*. If the spans are large, usually it will be sufficient to cut one span.

When the girder is not continuous, but rests separately as a single span:

(*a*) If it consist of a single span of uniform cross-section throughout, as is usually the case with small bridges, cut near both ends, thus:

Fig. 21.

(*b*) If it consist of a truss, or strengthened beam, cut at a point near each support just before the first strengthening or thickening of the parts begins, thus:

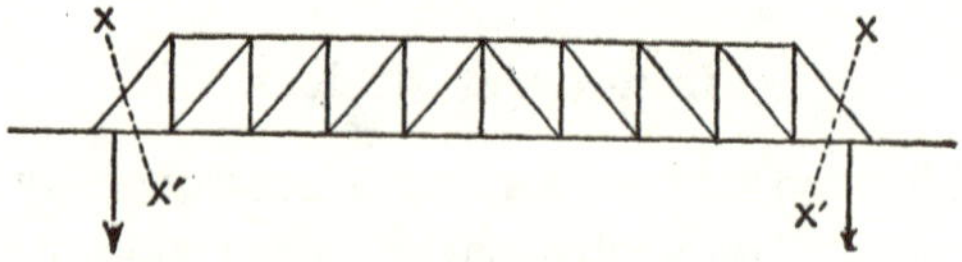

Fig. 22.

Specific rules cannot be laid down for cutting each separate type of truss, but there are certain general rules, such as those just given, which may be taken as a guide.

To insure complete destruction, the cut should be made through the entire truss. When there is not available sufficient explosive for cutting through a whole truss, the upper and lower chords should be cut. If there is not enough for both chords, cut the *tension*-chord of the panel rather than the compression-chord.

With a solid I-beam girder, the explosive should be placed on both top and lower flange and against the web between.

Curved girders, whether solid or built up open, should be cut completely through on both haunches, if possible.

Suspension bridges should be cut through each cable, either at the middle of the cables or near the anchors; the former for large bridges and the latter for small ones.

Large iron-truss bridges on stone piers may be most effectually destroyed by blasting the piers, but this should be attempted only when there is ample time. Small girder-bridges may be pried by levers off their piers or abutments, if no explosives be at hand.

Iron-truss bridges may be destroyed also by fires built against the important struts or ties; when red-hot, the heated members will give way and the structure will collapse.

Suspension bridges may be destroyed by uncovering and destroying the anchorages of the supporting wires, by destroying the supporting pier below the saddle, or by cutting through the wires at the middle.

In blasting stone piers, charges should be at 2 L.L.R. intervals apart in the middle of the pier, computing the size of the charge by the following formula:

$$C = \frac{1}{3}(\text{L.L.R.})^3.$$

Iron Plates.

To cut iron plates, the charge must extend along the entire line to be cut. The weight of charge in pounds may be computed approximately from the following formulas:

For wrought iron or soft steel: $C = \frac{3}{2}Bt^2$.
For cast iron: $C = \frac{3}{4}Bt^2$.
B = length to be cut in feet.
t = thickness of plate in inches.

Laminated plates should be treated as solid. Care should be taken that the contact of the charge with the plate is close throughout.

Subaqueous Demolitions.

The most common subaqueous demolitions are the blowing-up of sunken hulks, cutting down piles, and removing rocks from channels.

Hulks are broken up by exploding large single charges inside of the hulk. For this purpose, it is necessary for divers to go down into the hulk to place the charge.

Guncotton is a convenient explosive for under-water demolition, as its explosive force is not diminished by being wet. It is only necessary to arrange in the charge a primer of dry guncotton.

If dynamite or powder is used, it is necessary to inclose the charge in a water-tight case. Various common articles may be found to answer for a case, such as beer-barrels, iron sewer-, gas-, or water-pipes, lead pipes, rubber tubing, fire-hose, etc.

Explosive gelatin is unaffected by water, and, like guncotton, may be detonated if a primer of the dry explosive be used.

Single piles may be cut by using an encircling charge, in the form of tubing or hose, or by a single charge held in place at the proper height. The single charge may be fastened to a long beam, and the latter used to press the charge against the pile.

If a row of piles is to be cut down, the same principle may be applied. Fasten an extended charge to a heavy plank; attach the latter to two or more beams or scantling; lower until the ends of the beam bite into the bottom; lash the upper ends of the beams to the top of the piling, pressing the charge tightly up against the piles.

The charges for subaqueous demolitions may be considered as "tamped" charges, and the weight of charges computed for piles by the same formulas as given for hard-wood trees and stockades.

Masonry Tunnels.

Either the crown of the arch or the side-walls may be attacked. To prepare crowns of arches for demolition shafts may be sunk from above or galleries run from the ends, or openings made through the wall or arch and galleries run laterally from these. The side-walls may be prepared for demolition by opening holes through the wall, and running galleries laterally, or running galleries from the ends behind the walls, as explained for masonry revetment walls.

If time is limited, the charges may be placed along the foot of each wall and tamped.

If it is desired to break-in several yards in length of the tunnel, "over-charge" charges should be placed some distance along the arch or walls behind them, reckoning the resistance equal to two or three times the thickness of earth.

The part of a tunnel selected for destruction should be, if possible, some distance from either end. Ventilating shafts may easily be destroyed, and some tunnels thereby rendered unserviceable. If the subsoil is plastic, or contains water under pressure, great damage may be done by opening a hole through the foundation.

Stockades or Barriers.

The charge should be placed along the bottom and tamped; a single row of charges of dynamite or other explosive will usually be sufficient. The strength and character of the barrier must be considered.

An ordinary stockade or barrier-gate will be broken in by the equivalent of 40 to 100 lbs. of black powder fastened near the lock. Larger and stronger fort-gates should be attacked with the equivalent of 200 lbs. of powder placed along its bottom.

Demolition of Railroads.

The destruction of railroads may be divided into three classes of operations:

1. Those looking to the rendering of a particular portion of the line unserviceable for a limited time.
2. Those looking to the total destruction of the railroad, its works and rolling-stock.
3. Hasty demolitions having in view the production of the maximum amount of damage at some point or section in a limited time.

In classes 1 and 3, it is necessary to know the time limit. A reconnaissance should precede each, so that the precise nature of the work to be done may be ascertained and the necessary tools, material, and men may be determined.

The railroad may be within the enemy's line and be in use by him, or it may be within our own lines and its destruction made advisable, in order to prevent its use by the enemy at a subsequent time. In the latter case, all rolling-stock and movable property should be collected at a safe interior point.

Buildings, storehouses, workshops, etc., need not be destroyed. The machines may be rendered useless and engines disabled, but buildings should not be destroyed; water-supplies especially should be subjected only to injury that may be repaired later. The demolitions should include lighting and signal appliances, switches, bridges, tunnels, embankments, cuts, etc.

Apart from the removal and destruction of particular

pieces of property, the simplest and quickest method is to destroy the rails by explosives. Two sticks of dynamite or one block of guncotton, fastened by wire or cord close to the web of a steel or iron rail and detonated in that position, will completely destroy that portion of the rail. A string of cartridges may be applied in this manner, one charge to each rail, placed in series and exploded at the same time, thus destroying a great length of track instantaneously.

Land-mines.

The nomenclature and essential data connected with the use of explosives in land-mines are here briefly given:

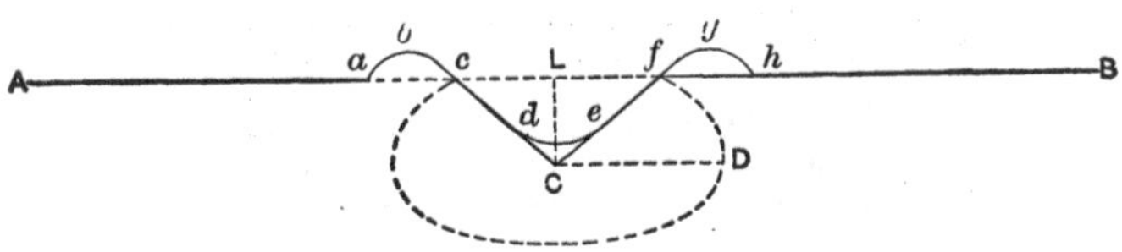

FIG. 23.

Let *AB* represent the original surface of the ground; *C*, the position of the center of the charge; *CL*, the line of least resistance.

After explosion, the crater will take the form *cdef*, with the crest, *abc-fgh*, about it.

The line *cf* is the diameter of the crater; *Lf* is the radius of the crater.

The radius of explosion is *DC*, the distance through the earth to which the effects of the explosion extend. When a crater is formed, the horizontal radius of explosion is greater than the vertical radius; when there is no crater, these two radii are equal. In the former case the volume included in the effects of rupture is a spheroid; in the latter case it is a sphere.

When the radius of explosion is greater than the line of least resistance, the mine is an "overcharged mine"; when less, an "undercharged mine"; when equal, a "common mine."

The following formulas give the charge of black powder, or equivalent, required to form these mine craters:

For overcharged mine: $C=\frac{k}{10}[\text{L.L.R.}+0.9(r-\text{L.L.R.})]^3$.

For undercharged mine: $C=\frac{k}{10}[\text{L.L.R.}-0.9(\text{L.L.R.}-r)]^3$.

For common mine: $C=\frac{k}{10}(\text{L.L.R.})^3$.

C = charge in pounds.

L.L.R. = line of least resistance in feet.

r = radius of the crater in feet.

k = a constant depending on the nature of the soil. It may be given the following values:

For very light earth	0.8
" common earth	1.
" hard sand	1.25
" earth and stones	1.45
" clay	1.55
" inferior brickwork	1.65
" rock and good brickwork	2.25
" best brickwork and masonry	2.50

Arrangement of Charges.

The charge may be applied either concentrated in one mass, or extended in a long line. In case the object to be demolished is a piece of rectangular shape and small in dimensions, like a beam, or round like a tree or pile or mast, a modification of the latter form of charge may be used by encircling the beam or tree with the extended charge. A piece of rubber hose is a convenient means of holding the explosive.

In case a piece of hose is not available, an encircling charge may readily be arranged by distributing the explosive on a piece of canvas, or other strong cloth of suitable length and width, and the cloth rolled over so as to form a long cylinder; this should be overwrapped spirally with strong twine and lashed snugly about the object to be destroyed.

In all cases, all parts of the charge should be brought into the closest possible touch with each other, and the whole charge with the surface of the object to be demolished.

For breaching or cutting through a plane surface of any kind, the charge may be attached to a plank, the parts being lashed tightly to the plank and in close contact with each other.

The whole plank may then be applied to the surface of the object to be demolished.

Such objects as trees and wooden beams may be cut conveniently by charges placed in auger-holes bored into them. The auger sould be about two inches across its bit. The hole should be bored along a diameter of the tree, or perpendicular to the axis of the beam. If one hole will contain the charge, only one should be bored; if one hole is not sufficient, others should be bored, meeting at the center, or parallel to the first. The centers of charges should be at middle of the tree or beam in each hole.

TABLE OF RELATIVE STRENGTHS OF VARIOUS HIGH EXPLOSIVES.

Name of Explosive.	Order of Strength.
Explosive gelatin	106.17
Nitroglycerine	100.00
Guncotton	83.12
Dynamite, No. 1	81.31
Rackarock	61.71
Melinite and other picric-acid explosives	50.82
Black gunpowder in small grains	28.13

A stick of dynamite weighs about 6.73 ozs. (190 grms.).
A disk of guncotton weighs about 10.63 oz. (300 grms.).
A stick of explosive gelatin weighs about 1.42 oz. (40 grms.).

ENERGY OF COMMERCIAL EXPLOSIVES.

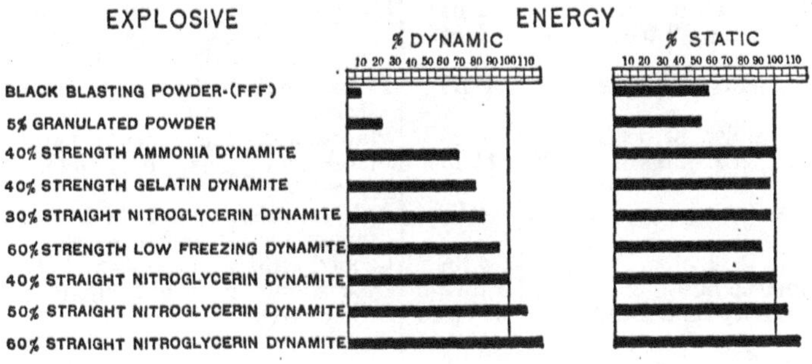

Dynamic energy as represented by the average of Trauzl lead block, small lead block, and rate of detonation tests.

Static energy as represented by the average of ballistic pendulum and pressure gauge tests.

SUMMARY OF CHARGES FOR HASTY DEMOLITIONS.

(USING DYNAMITE OR GUNCOTTON CHARGES.)

B = length of breach to be made in feet.
T = thickness of object to which charge is applied in feet.
t = thickness in inches of iron plate.

These charges are for untamped conditions; if tamped, they may be reduced one-half.

When prepared in great haste in the presence of the enemy, increase the charges one-half.

Object	Lbs.	Remarks.
Hard-wood trees, round.	$5T^3$	Also piles, masts, etc., encircling charge.
Hard-wood beam, rectangular	$3BT^2$	B = longer side of cross-section, encircling charge.
Hard-wood stockade or barrier.	$3BT^2$	B = length of breach; T = maximum thickness of stockade; single charge.
Earth and wood stockade or barrier.	4 per foot.	This is for breastworks 2 to 3 feet thick, made of earth rammed between planks or railway sleepers.
Iron-rail stockade or barrier. .	7 per foot.	This made of iron rails touching each other, placed in ground on end.
Hard-wood tree, round. . . . . . (Soft-wood objects require only one-half of the charge required by the same object in hardwood.)	$\frac{3}{8}T^2$	T = smallest diameter of tree; auger-hole charge. Hole bored radially, so that center of charge shall be at center of the tree.
Brick and masonry revetments.	$\frac{1}{4}BT^2$	Charge placed behind revetment against its back surface; for scarp-walls of forts and surfaces of tunnels.
Heavy gates.	50 lbs.	Gates of forts, armories, etc.
Iron plates, wrought or steel.	$\frac{3}{2}Bt^2$	t = thickness in inches. Laminated plates same as solid.
Detached masonry or brick wall, over 2 feet thick.	$\frac{1}{2}BT^2$	If over 2 feet thick.
Detached masonry or brick wall less than 2 feet thick.	2 per foot.	Charge calculated by last formula would be too heavy, and simply blow a hole through the wall.
Masonry piers of bridges. . . .	$\frac{2}{3}BT^2$	Placed against the pier in close contact.
Masonry arches of bridges. . .	$\frac{3}{4}BT^2$	Placed along the crown of haunches.
Field- or siege-guns or R. F. guns.	$1\frac{1}{2}$ lbs.	Placed on the chase near the muzzle.
Large seacoast guns.	4 lbs.	In bore tamped from the breech and muzzle with sand or earth.
Steel rails.	$\frac{2}{3}$ lbs.	Lashed tightly to the web of the rail.
Inflammable buildings or materials may be ignited by	1 disk of dry guncotton.	Disk should be simply ignited, not detonated.

Explosive gelatin would require charges 20 per cent less than those above.
Gunpowder would require charges 4 times greater than those above.

APPENDIX I.

LABORATORY EXPERIMENTS AND NOTES.

APPENDIX I.

LABORATORY EXPERIMENTS.

The following simple experiments illustrate the chemical principles set forth in Principles of Chemistry, Part I:

EXPERIMENT No. 1.

To illustrate the formation of a metallic oxide, and the influence of temperature in the action of chemical affinity (paragraphs 31 and 116).

Apparatus and Materials:

1. Blowpipe.
2. Small piece of charcoal, about three inches long.
3. Gas- or lamp-flame.
4. Forceps.
5. Small piece of iron.
6. Small piece of copper.
7. Small piece of zinc.
8. Small quantity of mercury.

Preparation: Make a small depression near one end of the charcoal. Scrape clean the surface of charcoal in this depression and the surface adjacent thereto before using the blowpipe.

Procedure:

(*a*) Take a small piece of iron, brighten it with a file or emery-paper, place it in the depression in the charcoal and bring to bear on it the outer point of the blowpipe-flame. The bright surface of the iron becomes dull, due to the combination of the oxygen of the air with the iron under the influence of the heat of the flame, and the formation thereon of a film of black iron oxide.

(*b*) Repeat (*a*), using a piece of copper; its oxide is also black.

(*c*) Repeat (*a*), using a piece of zinc; note the coating of zinc oxide on the surface of the charcoal near the depression, which is yellow when hot and white when cold.

(*d*) Repeat (*a*), using a small globule of metallic mercury; note the coating of mercury oxide on the charcoal, which is red.

Experiment No. 2.

To illustrate the formation of metallic hydroxides (paragraphs 58 to 63).[1]

Apparatus and Materials:

1. Small piece of metallic sodium.
2. A porcelain surface.
3. Distilled water.
4. A small glass tube for use as a dropper.
5. A small quantity of fat (unslaked) lime.
6. A porcelain bowl.
7. Solution of zinc chloride.
8. Solution of potassium hydroxide.
9. Two small beakers.

Procedure:

(*a*) The formation of the hydroxides of the alkaline metals (K, Na, Li, Cs, Rb). Cut a thin slice of metallic sodium and place it on the porcelain surface. Add water carefully with a dropper. Hydrogen is liberated from the water. A slight explosion may occur. A crusty grayish residue of sodium hydroxide is left on the porcelain surface. The reaction is as follows:

$$Na + H_2O = NaHO + H.$$

(*b*) The formation of the hydroxides of the alkaline-earth metals (Ca, Ba, Sr, Mg). Place a piece of fat (unslaked) lime, about the size of a bean, in the porcelain bowl. Add water until the lime is half covered. The process of "slaking" will take place, the fat lime swelling and crumbling up and finally reducing to a fatty, pasty mass with evolution of considerable heat. The resultant pasty mass is calcium hydroxide. The reaction is as follows:

$$\underset{\text{Fat-lime}}{CaO} + \underset{\text{Water}}{H_2O} = Ca(HO)_2.$$

[1] See also Experiments Nos. 22, 23, and 24.

(c) The formation of the hydroxides of metals other than the alkaline and alkaline-earth metals. Take a small quantity of the solution of potassium hydroxide in one of the beakers, and a small quantity of the solution of zinc chloride in the other beaker. Pour one solution into the other. The mixed solution now has a milky-white opaque appearance. This is caused by the production of the insoluble zinc hydroxide. This reaction also illustrates the principle of insolubility (paragraph 116). The reaction is written as follows:

$$\underset{\text{Solution of zinc chloride}}{ZnCl_2} + \underset{\text{Solution of potassium hydroxide}}{2KHO} = \underset{\text{Solution of potassium chloride}}{2KCl} + \underset{\text{Solid precipitate of zinc hydroxide}}{Zn(HO)_2}.$$

Experiment No. 3.

To illustrate the formation of non-metallic oxides (paragraph 31).

Apparatus and Materials:

1. Small quantity of calcium carbonate (marble or chalk).
2. Small quantity of hydrochloric acid.
3. Small quantity of roll sulphur.
4. Porcelain dish.
5. Small quantity of alcohol.
6. Small glass funnel.
7. Small piece of filter-paper, colored blue by having been dipped in solution of indigo.
8. Nitric acid.
9. Small piece of tin.
10. About 3 feet of rubber tubing to fit funnel above.

Procedure:

(a) *Carbon dioxide.* Drop a small quantity of hydrochloric acid on the calcium carbonate. Effervescence will occur due to the escaping carbon dioxide. A piece of moistened blue litmus held in the escaping gas will be turned red, this being a test of the acidity of the escaping gas. A lighted match held in the gas is extinguished, exhibiting the power of carbon dioxide to extinguish flame. If the escaping gas is collected under the glass funnel, the rubber tube be attached to the neck of the funnel, and the gas conducted into some clear lime-water, the latter will become turbid, due to the formation of the precipitate of insoluble calcium carbonate. The reactions are as follows:

$$1.\ CaCO_3 + 2HCl = CaCl_2 + H_2O + \underset{\text{Gas}}{CO_2}.$$

$$2.\ CO_2 + Ca(HO)_2 = \underset{\substack{\text{Solid}\\\text{precipitate}}}{CaCO_3} + H_2O.$$

(*b*) *Sulphur dioxide.* Take a piece of roll sulphur about the size of a bean, place it in the porcelain dish, pour a little alcohol in the dish, and ignite the latter. The sulphur will soon be ignited by the burning alcohol, and will burn with a blue flame, giving off an exceedingly pungent odor, due to the gas, sulphur dioxide, which has been formed. This gas has the property of extinguishing flame, and gives the acid test with moistened blue litmus. It also has the property of bleaching, as may be illustrated by moistening the blue filter-paper and placing it in the neck of the glass funnel while the latter is held over the burning sulphur. The reaction is as follows:

$$S + O_2 \text{ (oxygen of the air)} = SO_2.$$

(*c*) *Nitrogen dioxide.* Take a small piece of tin, about $\frac{3}{16}''$ square, place it in the porcelain dish and pour on it some nitric acid. Nitrogen tetroxide (N_2O_4) will be evolved as a gas; if the reaction does not readily take place, dilute the acid with water. The gas, in coming off, gives rise to reddish fumes. The odor is very pungent. The reaction is as follows:

$$HNO_3 + 2H_2O + Sn = H_2SnO_3 + NO_2 + H_3.$$

Experiment No. 4.

To illustrate the direct combination of an acid oxide and a basic oxide or basic hydroxide (paragraph 32).

Apparatus and Materials:

1. Small piece of lime.
2. Shallow porcelain dish.
3. Distilled water.
4. Filter-paper.
5. Small glass funnel.
6. Short piece of rubber tubing.
7. Small beaker.
8. Woulfe bottle.
9. Calcium carbonate (marble, chalk).
10. Hydrochloric acid.

Procedure:

(*a*) *Acid oxide and basic oxide.* Water for this purpose may be considered an acid oxide, being the combination of a non-metal with oxygen, and lime the basic oxide. Place a small quantity of lime in the porcelain dish. Cover it half with distilled water. The phenomenon of "slaking" described in (*b*), Experiment No. 2, will take place. The experiment and reaction are in all respects the same as in that experiment.

(*b*) *Acid oxide and basic hydroxide.* Take carbon dioxide as the acid dioxide, and calcium hydroxide as the basic hydroxide. Generate the carbon dioxide as follows: Place a small quantity in the Woulfe bottle. Attach the rubber tubing to one neck. Pour hydrochloric acid in through the other neck, then close the latter with a rubber stopper. Carbon-dioxide gas will be generated in the bottle and pass out through the rubber tubing. Conduct this into a beaker filled with lime-water (water containing calcium hydroxide—slaked lime—in solution). The clear lime-water will become turbid as soon as the carbon dioxide enters, due to the formation of insoluble calcium carbonate. The reaction is

$$Ca(HO)_2 + CO_2 = H_2O + CaCO_3.$$

Experiment No. 5.

To illustrate the formation of an oxyacid (paragraph 45).

Apparatus and Materials:

1. Apparatus and materials required for (*b*), Experiment 3.
2. Apparatus and materials required for (*b*), Experiment 4.
3. Glass funnel.
4. Iron-ring support for funnel.
5. Rubber tubing attached to neck of funnel.
6. Beaker.
7. Distilled water.

Procedure:

(*a*) Generate SO_2 as in (*b*), Experiment No. 3. Support funnel with tubing attached over burning sulphur. Conduct SO_2 through tubing into distilled water in beaker. The water and SO_2 unite, forming sulphurous acid. The reaction is

$$SO_2 + H_2O = SO_3H_2.$$

(*b*) Generate CO_2 as in (*b*), Experiment No. 4. Conduct through tubing into distilled water. A certain quantity of CO_2 will remain in the water, this quantity depending on the pressure. The resulting liquid is carbonated water. It is sometimes called carbonic acid.

Experiment No. 6.

To illustrate the formation of a hydracid (paragraph 50).

Apparatus and Materials:

1. Solution of common salt in a beaker.
2. Sulphuric acid.

Procedure:

Add sulphuric acid to solution of common salt. Hydrochloric acid will escape as a gas. The reaction is

$$2NaCl + H_2SO_4 = Na_2SO_4 + 2HCl.$$

Experiment No. 7.

To illustrate the property of an acid to exchange its hydrogen for a metal (paragraph 52).

Apparatus and Materials:

1. Metallic zinc.
2. Silver nitrate solution.
3. Hydrochloric acid.
4. Beaker.

Procedure:

(*a*) Place a small quantity of HCl in the beaker. Drop in small pieces of zinc until effervescence ceases. The HCl will have been changed to ZnCl. The gas escaping is hydrogen (H). The reaction is

$$2HCl + Zn = ZnCl_2 + H_2.$$

(*b*) Place a small quantity of HCl in the beaker. Add silver nitrate. The clear HCl will turn white with insoluble silver chloride formed. The liquid remaining is nitric acid. The reaction is

$$AgNO_3 + HCl = HNO_3 + AgCl.$$

The silver has displaced the hydrogen in the acid, and the hydrogen has been united with NO_3, forming nitric acid.

Experiment No. 8.

To illustrate the formation of an *ous* acid (paragraph 47).

Same as (*a*), Experiment No. 5.

Experiment No. 9.

To illustrate the formation of an *ic* acid (paragraph 47).

Apparatus and Materials:

1. Potassium chlorate.
2. Manganese dioxide.
3. Ignition-tube.
4. Rubber tube.
5. Beaker.
6. Sulphurous acid from Experiment No. 8.

Procedure:

Mix the $KClO_3$ and the MnO_2 and place in ignition-tube. Attach rubber tube to side neck of tube. Place cork lightly in top of tube. Apply heat gently. Oxygen will be generated and pass out through rubber tube. Test for O by holding a match that has been lighted and extinguished, but still has a spark. The latter will glow brightly and reignite the match in the O. Conduct this oxygen into sulphur*ous* acid made as in Experiment No. 8. The oxygen will combine and produce H_2SO_4. The reaction is

$$O + H_2SO_3 = H_2SO_4 \text{ (sulphuric acid).}$$

Experiment No. 10.

To illustrate the formation of an *ite* salt (paragraph 48).

Apparatus and Materials:

1. Material and apparatus for making SO_2 (*a*, Experiment No. 5).[1]
2. Sodium carbonate in solution in water.

---

[1] Instead of generating SO_2 by burning sulphur, it may be obtained from sulphuric acid as follows: Arrange a Woulfe bottle with rubber tube on one neck. Place some copper filings on the bottom of the bottle. Add H_2SO_4 through the other neck until the filings are well covered. Close the latter neck of the bottle with a rubber cork. Heat gradually and carefully. Bubbles of SO_2 will soon rise and pass out through the rubber tube. The reaction is

$$Cu + 2H_2SO_4 = CuSO_4 + 2H_2O + SO_2.$$

Procedure:

(*a*) Place a small quantity of the solution of sodium carbonate in a small beaker. Pass the gas SO_2 into the solution of sodium carbonate. Sodium sulphite will be formed. The reaction is

$$Na_2CO_3 + H_2O + 2SO_2 = \underset{\text{Acid sodium sulphite}}{2NaHSO_3} + CO_2.$$

(*b*) If more Na_2CO_3 be added to the solution of acid sodium sulphite, the normal sodium sulphite will be formed. The reaction is

$$2NaHSO_3 + Na_2CO_3 = \underset{\text{Normal sodium sulphite}}{2Na_2SO_3} + H_2O + CO_2.$$

(*c*) If the normal sodium-sulphite solution be mixed with a solution of a non-alkali metallic salt, the insoluble sulphite of the latter metal will be precipitated. The reaction is

$$Na_2SO_3 + 2AgNO_3 = 2NaNO_3 + Ag_2SO_3.$$

(*d*) Silver sulphite may be formed directly by passing the gas SO_2 into a solution of silver nitrate, the reaction is

$$SO_2 + 2AgNO_3 + 2H_2O = Ag_2SO_3 + 2HNO_3.$$

Experiment No. 11.

To illustrate the formation of an *ate* salt (paragraph 49).

Apparatus and Materials:

1. Small quantity of H_2SO_4 in a test-tube.
2. Zinc filings.
3. Beaker containing a little alcohol.

Procedure:

Drop zinc filings into the test-tube containing H_2SO_4 until bubbles cease to rise (heat gently if necessary). The H_2SO_4 has been changed to $ZnSO_4$. The bubbles escaping are hydrogen. The reaction is

$$H_2SO_4 + Zn = ZnSO_4 + H_2.$$

Since the sulphate of zinc is insoluble in alcohol, it will be precipitated as a solid if poured into the beaker containing alcohol.

EXPERIMENT No. 12.

To illustrate a synthetical reaction (paragraph 27).

Apparatus and Materials:

1. Piece of charcoal.
2. Piece of sulphur.

Procedure:

(*a*) Hold charcoal in flame of lamp or gas. It will glow and waste away, illustrating "combustion." The carbon of which charcoal is constituted combines with the oxygen of the air, forming CO_2. The reaction is

$$C+O_2+\text{heat}=CO_2.$$

(*b*) Same as (*b*), Experiment No. 3.
(*c*) Same as (*a*), (*b*), (*c*), and (*d*), Experiment No. 1.

EXPERIMENT No. 13.

To illustrate an analytical reaction (paragraph 27).

Apparatus and Materials:

1. Small quantity of $CaCO_3$.
2. Ignition-tube with rubber tubing attached to side neck.
3. Beaker containing lime-water.

Procedure:

Pulverize $CaCO_3$, and fill the ignition-tube nearly half-full. Close top of tube. Heat gradually until gas passes out through rubber tube. This is CO_2. If this gas be passed into the lime-water, the latter will become turbid from the reformation of the insoluble $CaCO_3$. The reactions are

$$1.\ CaCO_3+\text{heat} = CaO+CO_2.$$
$$2.\ CO_2+Ca(HO)_2 = CaCO_3+H_2O.$$

All nitrates and all carbonates and sulphates, except those of the alkalies, are decomposed by heat, illustrating analytical reactions.

EXPERIMENT No. 14.

To illustrate a metathetical reaction (paragraph 27).

Apparatus and Materials:

1. Solution of silver nitrate.
2. Dropper.
3. Solution of common salt in a test-tube.
4. Zinc filings in a test-tube.
5. Hydrochloric acid.

Procedure:

(*a*) Drop HCl on the zinc filings. The H is displaced, passing off as a gas and leaving $ZnCl_2$. The reaction is

$$2HCl + Zn = ZnCl_2 + H_2.$$

(*b*) Drop $AgNO_3$ into the solution of common salt. The solution will be filled with the white curdy precipitate of silver chloride (principle of insolubility). The reaction is

$$NaCl + AgNO_3 = AgCl + NaNO_3.$$

EXPERIMENT No. 15.

To illustrate the influence of temperature on the action of chemical affinity.

See Experiments Nos. 1 and 12.

EXPERIMENT No. 16.

To illustrate the influence of the liquid state on the action of chemical affinity (paragraph 116).

Apparatus and Materials:

1. Solid iron sulphate.
2. Solid barium chloride in a porcelain dish.
3. Solution of iron sulphate in a test-tube.
4. Solution of barium chloride.

Procedure:

Mix the solids in the porcelain dish. There will be no chemical action, however finely the substances be pulverized and mixed. Mix the solutions of the same substances by dropping a little of the barium chloride in the test-tube containing iron sulphate. Instantly a reaction takes place, barium sulphate being formed as a white precipitate (principle of insolubility). The reaction is

$$FeSO_4 + BaCl_2 = BaSO_4 + FeCl_2.$$

Experiment No. 17.

To illustrate the influence of insolubility in producing reaction (paragraph 116).

(*a*) See last part of last experiment; also (*b*), Experiment No. 14.

(*b*) *Apparatus and Materials:*

1. Woulfe bottle with rubber tube attached to side neck.
2. Small quantity of iron sulphide, FeS.[1]
3. H_2SO_4. (dilute).
4. Solution of lead nitrate in a test-tube.
5. Bottle of distilled water.

Procedure:

Place a small quantity of powdered FeS in Woulfe bottle. Add dilute H_2SO_4. Heat gently. H_2S, sulphydric acid, is formed and passes off as a gas through rubber tube. Collect in bottle of distilled water until water will absorb no more; this is sulphydric-acid solution. Drop a little of the sulphydric in the lead nitrate; instantly the black insoluble lead sulphide is formed as a black precipitate. The reaction is

$$H_2S \text{ (solution)} + Pb(NO_3)_2 = PbS + 2HNO_3.$$

Experiment No. 18.

To illustrate the influence of volatility in producing reactions (paragraph 116).

Apparatus and Materials:

1. Powdered $CaCO_3$.
2. Powdered NH_4Cl.
3. A large test-tube.

Procedure:

Mix one part of $CaCO_3$ with two parts of NH_4Cl. Place mixture in the test-tube. Heat gently. Since the substances contain between them the constituents of the volatile salt, ammonium

[1] FeS may be produced by mixing and heating together iron filings and powdered sulphur in a strong porcelain or earthen dish or in a crucible.

carbonate, we find the principle of volatility operating and this salt formed and passes off as a gas; it may be condensed and collected as a solid by conducting it into a cooled receptacle. The reaction is

$$CaCO_3 + 2NH_4Cl + \text{heat} = (NH_4)_2CO_3 + CaCl_2.$$

EXPERIMENT No. 19.

To illustrate the influence of the gaseous envelope (paragraph 116).

Apparatus and Materials:

1. Iron filings.
2. Small still or other apparatus for generating steam.
3. Rubber tube attached to still.
4. Glass tube attached to other end of rubber tube.
5. Woulfe bottle.

Procedure:

(*a*) Set up the still with some water in it over source of heat. Place some iron filings in the glass tube, and connect latter with rubber tube. The bright iron filings will become oxidized to Fe_3O_4, the black oxide of iron, and hydrogen gas will pass off out of the free end of the glass tube. That is, iron is oxidized in an atmosphere of water-vapor.

(*b*) Substitute for the still the Woulfe bottle, with some HCl in the bottle. Drop in some zinc filings, generating H. Leave the Fe_3O_4 in the glass tube. The H will now pass out through the rubber tube over the Fe_3O_4 in the glass tube. Apply heat under the glass tube.[1] The H will combine with the O of the Fe_3O_4, passing off as H_2O vapor, leaving Fe behind, thus reversing the reaction in (*a*).

EXPERIMENT No. 20.

To illustrate catalytic action; that is, when a reaction appears to take place more readily, due simply to the presence of some substance, the latter undergoing no apparent change.

Apparatus and Materials:

1. $KClO_3$.
2. MnO_2.
3. Woulfe bottle with rubber tube attached.

Procedure:

(*a*) Place some $KClO_3$ in Woulfe bottle and apply heat. Note the degree of heat required to decompose the $KClO_3$.

[1] Care must be taken to allow the H to drive off all the O in flask and tube before applying heat, otherwise there may be an explosion

(b) Place a mixture of $KClO_3$ and about one-fifth its weight of MnO_2 in same bottle and apply heat. Note how much more readily the O passes off at a comparatively low temperature. The reaction is

$$KClO_3 + MnO_2 + heat = KCl + MnO_2 + O_3.$$

This is the usual method of producing oxygen gas. Test O with match having a spark; its glow will be greatly increased when O is coming off.

Experiment No. 21.

To illustrate the principle of "disposing affinity"; that is, a chemical reaction that is due to the presence of a third substance and the latter is decomposed.

Apparatus and Materials:

1. NaCl.
2. H_2SO_4.
3. MnO_2.
4. Woulfe bottle with rubber tube attached.

Procedure:

Place a mixture of NaCl and MnO_2 in the Woulfe bottle and add H_2SO_4. Chlorine gas is given off, passing out through the rubber tube. If the experiment is tried without MnO_2, HCl is produced instead of Cl. This is the usual method of producing chlorine gas. The reaction is

$$2NaCl + MnO_2 + 2H_2SO_4 = Na_2SO_4 + MnSO_4 + 2H_2O + Cl_2.$$

If the MnO_2 is not present, the reaction is

$$NaCl + H_2SO_4 = HCl + NaHSO_4.$$

Experiment No. 22.

To produce the alkalies (paragraph 58).[1]

(a) *Hydroxide of Potassium.*

Apparatus and Materials:

1. Solution of potassium carbonate.
2. Clear filtered solution of slaked lime.
3. Glass funnel.
4. Small beaker.
5. Filter-papers.
6. Test-tube.

[1] See also Experiment No. 2.

Procedure:

Place a small quantity of the solution of potassium carbonate in a test-tube. Bring it to a boil over the flame. Add small quantity of lime-water. Calcium carbonate is precipitated as a white finely divided precipitate. Arrange a glass funnel and filter-paper over small beaker. Pour the clouded liquid on the filter-paper. The clear liquid that passes through is a solution of potassium hydroxide. It may be obtained in the solid form by evaporation.

(b) *Hydroxide of Sodium.*

Hydroxide of sodium is produced in the same manner, using sodium carbonate instead of potassium carbonate.

(c) *Hydroxide of Ammonium.*

Apparatus and Materials:

1. Ammonia-gas, manufactured as explained in Experiment No. 29.
2. Distilled water.

Procedure:

Pass the ammonia-gas through a rubber tube into the distilled water. The water will absorb the gas, and the resulting liquid is ammonium hydroxide (ammonia-water). The reaction is

$$NH_3 \text{ (ammonia)} + H_2O = NH_4HO.$$

This substance exists only in the state of solution.

EXPERIMENT No. 23.

To produce the alkaline earths (paragraph 62).[1]

(a) *Calcium Hydroxide.*

Apparatus and Materials:

1. Small portion of unslaked lime.
2. Distilled water.
3. Small porcelain bowl.

Procedure:

Place lime in the bowl and about half cover it with water. The process of slaking will proceed, the fat lime swelling, crumbling, and forming a white paste, which is the hydroxide of calcium. The reaction is

$$CaO + H_2O = Ca(HO)_2.$$

[1] See also Experiment No. 2.

If sufficient water be added to the hydroxide, it will be dissolved therein, forming the solution of calcium hydroxide or lime-water.

(b) *Barium Hydroxide.*

Apparatus and Materials:

1. Solution of barium nitrate.
2. Solution of sodium hydroxide.
3. Arrangements for filtering.

Procedure:

Add the barium nitrate to the sodium hydroxide. and filter the resulting turbid liquid. The filtrate is a solution of barium hydroxide; the solid may be collected by evaporation.

Experiment No. 24.

To produce the hydroxides of other metals (paragraph 63).[1]

(a) *Zinc Hydroxide.*

Apparatus and Materials:

1. Solution of sodium hydroxide.
2. Solution of zinc chloride.
3. Test-tube.
4 Filtering arrangements.

Procedure:

Place a small portion of sodium hydroxide in test-tube and add a small quantity of zinc chloride. Zinc hydroxide will be formed as a white gelatinous precipitate. The reaction is

$$2NaHO + ZnCl_2 = 2NaCl + Zn(HO)_2.$$

(b) *Iron Hydroxide.*

Method of procedure same as just explained, substituting iron chloride for zinc chloride. Iron hydroxide forms a white precipitate.[2]

Experiment No. 25.

To produce oxygen.[3]

Apparatus and Materials:

1. Potassium chlorate.
2. Manganese dioxide.
3. Test-tube with rubber tube attached.

Procedure:

Heat a mixture of potassium chlorate with about one-fourth, by weight, of manganese dioxide in an ordinary test-tube. Oxygen

[1] See also Experiment No. 2.

[2] Changing to green from the production of ferroso-ferric oxide and ultimately brown by passing to the ferric hydroxide.

[3] See also Experiment No. 20.

will be given off. Test for oxygen with a match with a spark at end. It will glow and ignite in the atmosphere of oxygen immediately above the test-tube. The reaction is

$$2KClO_3+MnO_2+heat=2KCl+O_6+MnO_2.$$

Experiment No. 26.

To produce hydrogen.

Apparatus and Materials:

1. Metallic zinc filings.
2. Hyrochloric acid.
3. Shallow porcelain dish.

Procedure:

To a small quantity of zinc filings placed on a porcelain dish add hydrochloric acid. Hydrogen is evolved rapidly as a gas. It will burn or explode on the application of a lighted match. The reaction is

$$Zn+2HCl=ZnCl_2+H_2.$$

Experiment No. 27.

To produce chlorine.

Apparatus and Materials:

1. Manganese dioxide.
2. Hydrochloric acid.
3. Test-tube.

Procedure:

Add hydrochloric acid to small quantity of manganese dioxide placed in test-tube. Chlorine is given off as a greenish-yellow gas having a very pungent odor. It has an acid action on blue litmus paper. It has bleaching properties, and will bleach filter-paper that has been stained in indigo solution. The reaction is

$$MnO_2+4HCl=MnCl_2+2H_2O+Cl_2.$$

Experiment No. 28.

To produce carbonic-acid gas (carbon dioxide).

Apparatus and Materials:

1. Calcium carbonate.
2. Hydrochloric acid.
3. Small beaker.

Procedure:

Add hydrochloric acid to a small quantity of powdered calcium carbonate in a small beaker. Carbon dioxide will be given off rapidly as a gas. The reaction is

$$CaCO_3+2HCl=CaCl_2+H_2O+CO_2.$$

The gas may be detected by its taste and smell. Flame of lighted match introduced in the beaker is extinguished. Gas has acid action on blue litmus paper.

Experiment No. 29.

To produce ammonia-gas.

Apparatus and Materials:

1. Powdered ammonium chloride.
2. Powdered unslaked lime.
3. Small porcelain dish.

Procedure:

Intimately mix small quantity of the two substances in the porcelain dish and apply heat. The odor of ammonia-gas (NH_3) is soon detected. Moistened red litmus paper is turned blue if held in this gas, showing its alkaline action. The reaction is

$$2(NH_4)Cl+CaO=CaCl_2+H_2O+NH_3.$$

Experiment No. 30.

To produce hydrogen sulphide.

Apparatus and Materials:

1. Iron filings.
2. Roll sulphur.
3. Sulphuric acid.
4. Porcelain dish.

Procedure:

Mix a small quantity of iron filings with powdered roll sulphur in a porcelain dish and heat the same. Chemical combination takes place between the iron and the sulphur, forming iron sulphide (FeS). Add to this sulphuric acid, and gas is evolved which is hydrogen sulphide. It may be detected by its characteristic odor, which is that of decomposing flesh. The reaction is

$$FeS+H_2SO_4=FeSO_4+H_2S.$$

Experiment No. 31.

To produce nitric acid.

Apparatus and Materials:

1. A few crystals of potassium nitrate.
2. Small quantity of sulphuric acid.
3. Test-tube.

Procedure:

Place a few crystals of potassium nitrate in the test-tube. On addition of sulphuric acid, strong odor of nitric-acid vapor (HNO_3) will be detected. It gives acid reaction to blue litmus paper. The reaction is

$$KNO_3 + H_2SO_4 = KHSO_4 + HNO_3.$$

Experiment No. 32.

To produce hydrochloric acid.

Apparatus and Materials:

1. Sodium-chloride solution.
2. Sulphuric acid.
3. Test-tube.

Procedure:

Place a small quantity of sodium chloride in a test-tube. On addition of sulphuric acid, the vapor of hydrochloric acid will be given off (HCl), which may be detected by its strong pungent odor. Gives acid reaction to litmus paper. The reaction is

$$NaCl + H_2SO_4 = NaHSO_4 + HCl.$$

Experiment No. 33.

To test any solution for a soluble chloride.

Apparatus and Materials:

1. A solution containing an unknown soluble chloride.
2. Small quantity of silver nitrate.
3. Test-tube.

Procedure:

Place a small quantity of the supposed chloride solution in the test-tube; add a drop of silver nitrate: if there be a chloride present in the solution, the insoluble silver chloride will be formed as a white curdy precipitate which turns dark in the sunlight, and is soluble in ammonia-water.

Experiment No. 34.

To test a solution for the presence of a soluble sulphate.

Apparatus and Materials:

1. A solution containing an unknown soluble sulphate.
2. A solution of barium chloride.

Procedure:

Place in the test-tube a small quantity of the solution supposed to contain the sulphate; add a few drops of barium chloride solution: if a sulphate be present, the barium sulphate will be formed as a white finely divided heavy precipitate.

Experiment No. 35.

To test a solution for the presence of a soluble hydroxide.

Apparatus and Materials:

1. A solution containing an unknown hydroxide (hydroxides of the alkaline earths are soluble).
2. Small portion of zinc chloride in solution.
3. Test-tube.

Procedure:

Place in the test-tube a small quantity of the supposed hydroxide solution; add a small quantity of zinc chloride: if an hy droxide is present in the solution, zinc hydroxide will be formed as a white precipitate.

Experiment No. 36.

To test a solution for the presence of a soluble carbonate. The carbonates of the alkalies are soluble.

Material and Apparatus:

1. A solution containing a soluble carbonate.
2. Calcium chloride.
3. Test-tube.

Procedure:

Place a small quantity of the supposed soluble carbonate in test-tube; add a small quantity of calcium chloride. Insoluble calcium carbonate will be formed as a white precipitate.

Experiment No. 37.

To test a solution for the presence of a soluble calcium salt.

Apparatus and Materials:

1. Small quantity of a solution containing the soluble calcium salt.
2. Small quantity of solution of ammonium carbonate.
3. Test-tube.

Procedure:

Place a small quantity of the supposed soluble calcium salt in test-tube; add small quantity of ammonium carbonate in solution. Calcium carbonate will be produced as a white precipitate.

Experiment No. 38.

To test a solution for the presence of a soluble nitrate.

All nitrates are soluble.

Apparatus and Materials:

1. Small quantity of solution of any nitrate.
2. Small quantity of solution of ferrous sulphate.
3. Small quantity of concentrated sulphuric acid.
4. Test-tube.
5. Copper filings.

Procedure:

(*a*) Place a small quantity of the acid in the tube. Mix a small quantity of the ferrous sulphate and the supposed nitrate solution in a test-tube; add, carefully, a few drops of the latter, allowing it to run down the side of the tube: if a nitrate is present, a reddish-brown or purple layer will be formed at the junction of the sulphuric acid and the other liquid.

(*b*) Introduce, in test-tube containing a supposed nitrate solution, a few copper filings, and add a few drops of concentrated sulphuric acid; apply heat carefully until solution boils freely: dark reddish-brown pungent fumes of nitrogen peroxide (NO_2) will be evolved if a nitrate is present.

Experiment No. 39.

Test for solution containing a soluble iron salt.

Apparatus and Materials:

1. A solution containing a soluble iron salt.
2. A solution of ammonium sulphide.
3. Test-tube.

Procedure:

Place in the test-tube a small quantity of the supposed solution of iron salt; add a small quantity of the ammonium sulphide: if iron be present, the insoluble ferrous sulphide will be precipitated, first having a bluish color, which turns quickly to black.

Experiment No. 40.

To make an acetone colloid.

Apparatus and Materials:

1. Small quantity of acetone.
2. Small quantity of guncotton (cotton that has been dipped and allowed to steep for a few minutes in a mixture of nitric and sulphuric acid and afterwards cleansed by thorough washing in water).[1]
3. Small beaker.
4. Small porcelain dish.

Procedure:

Dissolve a piece of the guncotton about the size of a lima bean in about 55 c.c. of pure acetone; dissolve in the beaker; decant the solution into the shallow porcelain dish; evaporate to dryness over a water bath, being careful to evaporate only to dryness and to avoid burning or igniting. A thin film of transparent colloid will be left on the porcelain dish. Note the difference in rate of burning by igniting first a small piece of raw nitrocellulose and then a piece of the dry colloid.

Experiment No. 41.

To make an ether-alcohol colloid.

Apparatus and Materials:

1. Small quantity of nitrocellulose, containing about 12.5 per cent of N.[1]
2. Small quantity of ether and alcohol, in the proportion of 60 grams of ether to 20 grams of alcohol.
3. Small beaker.
4. Small porcelain dish.

Procedure:

Dissolve a piece of the nitrocellulose about the size of a lima bean in a portion of the ether-alcohol solution placed in the beaker. Allow the nitrocellulose to become thoroughly dissolved. Decant the solution to shallow porcelain dish and evaporate carefully to dryness over a water-bath, avoiding igniting. A thin film of colloid is left on the dish. Compare the rate of burning of the colloid with that of unchanged nitrocellulose.

---

[1] See p. 140 et seq.

MATERIALS AND APPARATUS FOR LABORATORY DESK.

FOR USE IN CONNECTION WITH THE FOREGOING EXPERIMENTS.

Solution of sodium hydroxide
Solution of potassium hydroxide
Solution of calcium chloride
Solution of barium chloride
Solution of copper sulphate
Solution of silver nitrate
Solution of ammonium sulphide
Hydrochloric acid
Nitric acid
Commercial sulphuric acid
Concentrated sulphuric acid
Pure ether
Pure alcohol
Pure acetone
A solution of ammonium carbonate and crystals
A solution of ammonium chloride and crystals
Calcium carbonate
Calcium oxide (fat lime)
Metallic iron, filings and turnings
Metallic copper, turnings
Metallic zinc, strips
Metallic mercury
Metallic sodium
Zinc chloride, solution
Roll sulphur
Solution of indigo
Metallic tin
Sodium chloride
Manganese dioxide
Iron sulphate
Barium chloride
Solution of lead nitrate
Iron sulphide
Potassium chlorate
Potassium carbonate
Sodium carbonate
Ammonia-water

Barium nitrate, solution
Sodium nitrate, solution
Iron chloride
Potassium nitrate
Silver nitrate
Ammonium sulphide
Nitrocellulose
A piece of charcoal about 3″ long and 1″ square cross-section
Platinum foil $1\frac{1}{2}'' \times 1''$
Platinum wire 3″ long
1 test-tube rack
1 test-tube cleaner
1 test-tube stand
1 glass funnel
1 shallow porcelain dish
1 porcelain crucible
1 porcelain mortar and pestle
1 Woulfe bottle
6 test-tubes, assorted
3 beakers, assorted
1 iron tripod
2 watch-crystals
1 blowpipe
1 pair of tongs
1 pair of forceps
1 spatula
1 glass dropper
1 glass rod
1 test-tube holder
1 asbestos pad
1 water-bottle for distilled water
Filter-papers
Litmus papers
Source of heat: gas or lamp
Rubber tubing
Iron ring-support
Ignition-tube
Small still

LABORATORY NOTES.

Throw all solid waste materials in the earthen crocks provided at each desk, and *not in the sinks.*

In rinsing apparatus containing acids, allow the water to run for a moment to dilute the acids and thereby protect the pipes.

When through with a source of heat, extinguish it.

Always keep the reagent-bottles in their proper places, with labels to the front.

In using a liquid reagent, grasp the stopper first between the little finger and palm of the hand, then grasp the bottle between the thumb and other fingers of the same hand, the label of the bottle being against the palm of the hand. Pour out slowly and carefully the smallest amount of reagent possible for the reaction and, at the last, touch the lip of the bottle against the edge of the vessel, so that the last drop will not run down the sides of the bottle. *Replace the stopper and put back the bottle at once.* Neither bottle nor stopper should ever be put on the table.

Dry reagents and the more unusual wet reagents should be kept on a separate stand for general use.

All glass and porcelain articles should be cleansed immediately after using, and in no case left or put away dirty.

In performing experiments which give rise to pungent or offensive fumes, such as NO_2, SH_2, etc., go to the hood and perform the experiment there.

On leaving the laboratory, be careful to label distinctly any solution or substance which is to be further examined or used, and mark the slip with the word "preserve" and your name.

Leave the desk in order so that the attendant may dust it and clean it.

If a solution has to be put aside even for a few minutes, *label* it over your initials.

Laboratory notes may be entered either in rough form on

a pad to be entered later in the note-book, or directly in the note-book. The latter is the better method. Time is too valuable to spend it in *copying*.

Lecture-notes in abbreviated form must, of course, first be taken down in rough and then *expanded* into the note-book, but a distinction should be drawn between *mere copying* and *expansion of abbreviated notes*.

When a glass stopper sticks tightly, heat the neck gently and gradually, keeping the stopper entirely out of the flame. Then press the stopper gently from side to side. While heating the neck, turn it round and round in the flame.

Test-tubes are little cylinders of thin glass, closed at one end, in which most tests and liquid reactions are conducted. They vary in size from 4 to 8 inches long and from ½ to ¾ inch in diameter. They should not be so large in diameter that the open end may not be closed by the thumb. They may be used for heating liquids in a flame, holding either in the bare fingers, or, if too hot, in a test-tube holder.

Two precautions must always be observed in heating test-tubes and all glass vessels.

1. The outside should be wiped perfectly dry just before placing in the flame.
2. The tube should be brought gradually into the flame and moved in and out and rolled between the finger and thumb, so that the heating shall be *gradual and uniform*.

The reactions which take place in test-tubes, and the boiling of liquids therein, often cause portions of the liquid to be ejected. To guard against accident from this cause, the operator should never hold the mouth of the tube toward himself or another person near him.

Test-tubes are cleaned by a *test-tube cleaner*, consisting of a bunch of bristles caught between twisted wires and a small piece of sponge held at the end, or a round end of bristles.

Test-tubes are kept in *racks*, a set of holes being provided for tubes in use, and a set of *draining-pegs* for those not in use.

These racks usually contain a dozen tubes. The tubes should be thoroughly washed before placing on the pegs.

Flasks are bottle-shaped glass vessels having a neck and globe; the latter may have a round or flat bottom. They are used for boiling liquids in, and are often placed in iron ring supports over the source of heat. The same rules as to heating and cleaning apply to these as to test-tubes. In arranging flasks for experiment, *be careful to allow sufficiently large exit for gases generated*—an explosion of a flask is liable otherwise.

Beakers are thin glass, open, tumbler-shaped vessels with a flare edge and, often, a small spout. They are used chiefly to receive filtered liquids, or for reactions on a larger scale than in test-tubes.

Glass funnels should be thin and light and have the throat cut off obliquely. Their sides incline at 60°, which angle permits a filter-paper folded twice to fit exactly. They are used for transferring liquids from one vessel to another, and for holding filter-papers. Agate-iron, iron, and porcelain funnels are also furnished for rougher work.

Some funnels are arranged with corrugations or cut channels specially to accelerate filtering.

Filtering-papers are used to separate the precipitates from the liquids in which they were formed; the latter, after separation, is often called the *filtrate*. A good filter-paper should be porous enough to filter rapidly and yet sufficiently close in texture to retain the finest powder. The paper should be strong enough to bear when wet the pressure of the liquid poured on it. Good filter-paper should be free from all salts and as near pure cellulose as is possible; when burned, it should leave a very small proportion of ash. White paper is more likely to fulfill these conditions than the colored varieties.

Filter-paper comes in sheets, but cut filter-papers are supplied as a rule. The separate papers are in circular form.

Small papers and funnels should be used in experiments. A paper about three inches in diameter is the most convenient size, except for reactions involving large quantities of materials.

A filter is prepared for placing in a funnel as follows:

1. Fold across on one diameter.
2. Fold each end of semicircle back on 45° radius.
3. Fold each of the 45° folds in its middle.
4. Open out between the folds.

Or a second method is as follows:

1. Fold across a diameter as before.
2. Fold across the semicircle on the 90° radius.
3. Open out 3 layers on one side and 1 on the other.

The first method is the better, as it gives quicker filtration.

Filter-papers are placed in funnels so as to fit closely to the sides, and after they are in place they are wetted down with distilled water, using a wash-bottle for this purpose. The rate of filtering may be increased by using larger filter-papers or by lengthening the throat of the funnel and letting it dip down into the filtrate.

Strong acid or alkaline solutions should be filtered through asbestos wool placed in the throat of the funnel.

A filter-paper of less than 2 inches in diameter may be placed directly in the mouth of a test-tube, and those between 2 and 3 inches may be placed in a funnel and the funnel placed directly in the mouth of the test-tube without other support.

When, however, a large quantity of liquid is to be filtered, larger papers are necessary and larger funnels; these latter are supported in *stands* or *rings* independently of the vessel arranged to receive the filtrate. A beaker or a porcelain dish may be arranged to receive the filtrate. Care should be taken that the lowest point of the throat of the funnel touches the side or edge of the vessel, in order that the liquid passing through may not fall in drops, but run quietly down the side without splashing.

Porcelain *evaporating-dishes* of various sizes are used. These dishes will bear the heat of a lamp- or gas-flame without cracking. The best are the "Berlin" dishes glazed on both sides. With these dishes a solution may be evaporated to dryness, or even to ignition over the open flame of a lamp- or gas-burner. It is well, however, to support the dishes in such cases on a piece of iron wire gauze; otherwise the dish may be supported on a small wire triangle.

Porcelain crucibles are made of very thin porcelain and may be subjected to even higher heat than the dishes. They are made with covers. They are supported over the flame by small wire triangles.

Both porcelain dishes and crucibles should be brought gradually to the full heat.

Two kinds of lamps are used—the common *spirit-lamp,* and the circular-wick lamp, also known as the *Berzelius lamp.* The former is used for ordinary heating of test-tubes, etc.; the latter when a higher temperature is required and a larger flame, especially for water- and sand-baths, for evaporation, and ignition of residues. Spirit-lamps, when not in use, should be covered over to prevent evaporation.

Supports.—Several forms of supports are used in heating:

1. The *iron tripod,* consisting of a ring to which three legs are attached. The flask, dish, or crucible is supported on this ring and the lamp is placed below. The proper height is given by wooden blocks, either blocking up the tripod or the lamp.
2. The *iron-rod support* consists of an iron rod attached to a heavy cast-iron base. Several rings of different diameters are secured to the rod by binding-screws, and may be adjusted vertically and laterally, like the stand of the Berzelius lamp.
3. *Iron-wire gauze*—a piece about 6 inches square.
4. *Iron-wire triangle*—three pieces of iron wire formed into an equilateral triangle, with the wires twisted together at the vertices for a distance of an inch or two.

A water-bath[1] consists of a copper vessel with a set of covers of different diameters. It is used to evaporate at moderate heat, or to dry precipitates or other substances which must be kept below a certain temperature. This temperature is fixed by the boiling-point of the liquid placed in the bath. If, for

[1] Other liquids than water may be used. The bath takes its name, in any case, from the liquid used. By taking liquids of different boiling-points different constant temperatures may be had.

example, an aqueous solution is to be evaporated without ebullition, it must not rise above the boiling-point of water, nor permitted quite to reach that point. To accomplish this, fill the bath two-thirds full of water, place on it those particular cover rings that will permit the greater part of the dish containing the solution to be below the cover but not in the water of the bath. Support the bath on either the tripod or ring support and apply the heat. The dish holding the solution is thus heated by an atmosphere of steam, and the temperature will not exceed 212° F. The water in the bath must never be allowed to boil away. There are several modifications of the water-bath.

If a gradual and uniform temperature higher than the water-bath be desired, this may be accomplished by the *sand-bath*. This consists simply of a shallow dish or pan in which sand is placed, and the body to be heated is placed in a dish on this sand. The thickness of the sand layer regulates the temperature for a given flame.

The blowpipe is used to oxidize and deoxidize samples and to give a high degree of heat. Deoxidization is often called "reduction."

In using the blowpipe, the air should be forced from the lungs into the mouth-cavity, distending the cheeks, and the air then forced through the blowpipe by the muscles of the cheeks. A steady uniform pressure may thus be maintained.

For oxidization purposes the sample should be held just beyond the tip of the outer luminous flame; for reducing purposes it should be held at the tip of the inner blue flame. The hottest part of the blowpipe flame is between the luminous and blue flame; for melting metals, and when a high degree of heat is desired, the sample should be held at this point.

Specimens may be supported and held before the blowpipe either on charcoal, on platinum-foil, or on a platinum loop.

(*a*) On charcoal: Take a piece of charcoal about 3″×1″×1″. Near the end of one of the longer faces cut with knife or scraper a small depression about $\frac{1}{3}$″ diameter and $\frac{1}{4}$″ deep. Place the sample in this depression.

Hold the charcoal between the thumb and forefinger of the left hand, slanting at about 30° downward, the sample being at the lower end. Present the sample to the blowpipe in this position.

(*b*) On platinum-foil: Take a piece of platinum-foil about $1\frac{1}{2}''$ by $1''$. Clean its surface with moist sand. If wrinkled, rub out the wrinkles on the bottom of the agate mortar, using the agate pestle. Bend over one corner slightly. Take hold of this corner with the forceps. Place the sample on the foil. Present to the flame, holding the forceps in the left hand.

(*c*) Platinum wire loop: Fuse a fine platinum wire to the end of a glass rod. Straighten out all kinks in the wire by making a single loop over a round lead-pencil or other similar article, and pulling the pencil along the wire without turning. Make a small circular loop at the end of the wire about $\frac{2}{10}''$ in diameter. Heat the loop to red heat, and wipe after cooling with clean filter-paper. Prepare the sample with proper fluxes, place it on the loop and present to the blowpipe.

Never heat any metal or any substance from which a metal can be reduced on platinum, as the latter forms alloys with other metals, which alloys have a lower fusing-point than platinum and injure its properties otherwise. The alkaline sulphides and hydroxides also act on platinum. It is dissolved by aqua regia and chlorine-water.

Wash-bottle.—This is a large bottle of distilled water for general use in carrying out experiments. It is used particularly for diluting specimens in test-tubes, for wetting down filter-papers so they will adhere closely to the sides of funnels, for washing down precipitates from the sides of vessels, and for washing precipitates. Two tubes enter the bottle through a rubber cork. One is straight and projects about 4″ above the cork, and the other at a point about 1″ above the cork is bent sharply downward at an angle of about 45°, and terminates at about 4″ from the bend in a pointed aperture. The first tube stops inside of the bottle above the surface of the water, the bent tube extends inside the bottle well down to near its bottom.

The water is poured out through the straight tube, holding the bent tube uppermost.

By blowing down the straight tube, using some little force in the act, the water is forced up through the bent tube and out at the pointed aperture.

Glass Tubing.—Various sizes of glass tubing are used; the larger sizes for joining parts of apparatus, in connection with rubber-tubing; the smaller sizes for exits through corks from bottles and large test-tubes. A piece of small-caliber glass tube is used as a dropper. The tube, when used for this purpose, must be perfectly clean. It is inserted in the reagent-bottle, the reagent rises in the tube, the end of the finger is placed over the top and the tube then withdrawn, bringing with it the small quantity of reagent held in the tube.

Ordinary glass tubing may be cut in the simplest way by placing it lengthwise in a V trough, the point to be cut resting just beyond the trough; passing a diamond around at the point with one hand, holding the tube tightly with the other, then grasping the tube firmly with both hands on either side of the cut and near it, break the tube at the cut by turning the hands evenly, upward and outward, using the necessary force.

The sharp corners of the ends of glass tubing may be rounded by holding in the Bunsen or alcohol flame. This should always be done before attempting to insert a tube in corks or in rubber tubing, as the tube inserts much more easily if the corners are rounded. Care should be exercised not to change the size of the orifice. It will be sufficient to bring the very outer edges to a good red heat and rub a second heated rod gently over these edges.

Very thin glass tubing, which cannot be cut as described above, may be cut by filing a slight cut at the point, then apply gradually a hot point progressively around the tube, starting at the file-cut. It may be necessary sometimes to chill the tube at the file-cut by placing it in cold water or ice for a minute or so, and then wiping dry, before applying the heated point.

Glass tubes are bent by heating them over a flame until plastic, then bent carefully with force applied very slowly; only the heat necessary should be used.

To close a glass tube, heat the end until plastic, press together opposite points of circumference until they meet, make weld complete, then shape.

To form a bulb in a glass tube, heat the tube in the point at which it is desired to have the bulb until the glass is plastic at that point. Blow through the tube, using sufficient force to cause the plastic glass to expand to the size desired.

To make an opening in the side of a glass tube, heat the tube at the point until the glass there is plastic. Perforate the side with a pointed rod, open the perforation to the size desired, round off and smooth the edges.

Rubber tubing of various sizes is used to connect the glass and metal parts of apparatus. There is a great advantage in this means of connection by reason of the pliability of the tubing, the air-tight joints that are made, and the fact that alkalies and dilute acids do not act on rubber.

The cork-borer consists of a nest of metal tubes of various sizes, with one end bevelled to a cutting circular edge. It is used to bore holes through rubber and cork stoppers for glass tubes.

In *putting a glass tube through a bored stopper,* see that the edges of the tube have been rounded by heating, grasp the tube firmly, close to the stopper, press in easily and directly along the axis of the tube with a screw motion. Wet the tube with alcohol or with soap-suds, if it moves with great difficulty. Avoid lateral pressure. Do not hold the body of a funnel in forcing the neck through a stopper nor a bent tube at the bend.

Rubber stoppers are used when absolutely air-tight closing of bottles is important. They may be perforated for glass tubes by a brass cork-borer; the latter should be moistened with alcohol to facilitate the process. They have a further advantage over cork stoppers by reason of the non-action of alkálies and weak acids.

Sheet rubber is used to make tight joints between glass tubes of different sizes, or between the neck of a bottle or a flask and a large glass tube entering it.

Cork stoppers should be softened by rolling or squeezing before using. There is difficulty in finding perfectly round corks; eccentric parts may be removed by using a fine flat file. The size of corks may be reduced somewhat by squeezing or filing or both.

Double-neck bottles are convenient for generating gases; one neck being used for the reagent, and the other, with glass tube and rubber tubing attached, for transferring the gas generated.

There are four kinds of *mortars* in common use: (1) an *iron mortar*, for heavy material requiring great strength to pulverize; (2) *porcelain mortars*, for ordinary solid reagents; (3) *agate mortars*, for minerals and reagents having high degree of hardness; (4) *diamond mortar*, consisting of small steel cylinder, anvil, and piston, in which very hard and tough materials are pulverized or broken before using the agate mortar.

Spatulas are thin, knife-like blades made of steel, horn, or porcelain. They are used in handling solid reagents and samples.

Watch-glasses are used in pairs, with a suitable metal clasp to hold them tightly together, in holding samples for weighing, drying, and for preserving them safely from loss or change during experimentation.

The clothing should be covered by overalls or aprons during laboratory work. In case strong acid gets on the clothing or skin, it should be neutralized at once with ammonia-water or other strong base, or washed for some time in running water.

APPENDIX II.

REGULATIONS FOR THE TRANSPORTATION OF EXPLOSIVES BY FREIGHT AND EXPRESS.

CONTENTS.

REGULATIONS FOR THE TRANSPORTATION OF EXLOSIVES BY FREIGHT.

Prescribed under act of March 4, 1909. Originally formulated and published January 15, 1910, effective 90 days thereafter. Revision formulated and published July 2, 1914, effective October 1, 1914, and superseding the regulations published January 1, 1912.

GENERAL NOTICE.

1401. (*a*) As the use of explosives is essential to various business activities throughout the country, it is the duty of the interstate railroad carriers to accept and transport such explosives under these regulations. When local conditions make the acceptance, transportation, or delivery unusually hazardous, local restrictions, as provided by paragraph 1429 herein, may be imposed.

(*b*) It is also the duty of carriers and shippers to make these regulations effective and to thoroughly instruct their employees in relation thereto.

1402. The Bureau for the Safe Transportation of Explosives and other Dangerous Articles, hereinafter called Bureau of Explosives, organized by the railways under the auspices of the American Railway Association, is an efficient bureau in charge of an expert chief inspector. This bureau will make inspections and conduct investigations, and will confer with manufacturers and shippers with a view to determining what specifications and regulations will within reasonable limits afford the highest degree of safety in packing and preparing these dangerous articles for shipment and in transporting the same. The Commission will seek to avail itself of the expert knowledge thus developed, and in formulating amendments to these regulations or specifications supplemental thereto, while not bound thereby, will give due weight to such expert opinions.

1403. The Commission will make further provision as occasion may require for new explosives not included in or covered by the following regulations.

1404. These regulations apply to all shipments of explosives as defined herein, including carrier's material and supplies.

GENERAL RULES.

1421. Unless specifically authorized by these regulations, explosives must not be packed in the same outside packages with each other or with other articles.

1422. Explosives, when offered for shipment by rail, must be in proper condition for transportation and must be packed, marked, loaded, stayed, and handled while in transit in accordance with these regulations. All packages of less-than-carload shipments must also be plainly marked on the outer covering or boxing (outside package) with the name and address of the consignee.

1423. Except on through bills of lading to a foreign destination, shipments of dangerous explosives as named in paragraph 1661 must not be accepted when consigned " order notify."

1424. **Empty boxes or kegs** previously used for high explosives are dangerous and must not be again used for shipments of any character. Empty boxes which have been used for the shipment of other explosives than high explosives must have the old marks thoroughly removed before being used for the shipment of other articles. **Empty metal kegs** which have been used for the shipment of black powder which was not contained in an interior package must not be again used for shipment of any explosive, and if used for the shipment of other articles must have the old marks thoroughly removed or obliterated.

1425. (*a*) Explosives except such as are forbidden (see par. 1501) may be offered for transportation to railroads engaged in interstate commerce, provided the following regulations are complied with, and provided their method of manufacture, packing, and storage, so far as it affects safe transportation, is open to inspection by a duly authorized representative of the initial carrier or of the Bureau of Explosives. Shipments of explosives that do not comply with these regulations must not be received.

(*b*) Shipments offered by the War and Navy Departments of the United States Government may be packed, including limitations of weight, as required by their regulations.

1426. When practicable at any point, **regular and separate days** should be assigned for receiving from shippers less-than-carload lots of dangerous explosives named in paragraph 1661.

1427. To enable the carrier to provide proper cars at stations where less-than-carload shipments of the dangerous explosives named in paragraph 1661 are accepted for loading by the carrier, the shipper must give to the carrier not less than **24 hours' notice** of his intention to offer such shipments, and state their destinations. When a regular day to receive all explosive shipments offered at such a station has been appointed this notice may be waived by the carrier, but the explosive shipments must be delivered on such days in time to permit proper inspection, billing, and loading on that day.

1428. Before any shipment of dangerous explosives as named in paragraph 1661 destined to a point beyond the lines of the initial carrier is accepted from a shipper, the initial carrier must ascertain that the shipment can go forward via the route designated.

1429. All carriers must report to the chief inspector of the Bureau of Explosives for compilation and publication, full information as to **restrictions** which may be imposed as provided in paragraph 1401 (*a*) against the acceptance, delivery, or transportation of explosives over any portion of their lines.

1430. (*a*) **Forbidden explosives,** as defined in paragraph 1501, and explosives condemned by the Bureau of Explosives (except properly repacked samples for laboratory examination), must not be offered for shipment.

(*b*) Samples of any **new explosives must be examined** and approved as safe for transportation by the Bureau of Explosives before shipments (except samples for this examination not exceeding 5 pounds net in weight) can be accepted. For this purpose a **new explosive** is defined to be the product of a new factory or an explosive of an essentially new composition made by any factory.

1431. **Leaking or damaged packages** of explosives must not be offered for shipment. Should any package of high explosives when offered for shipment show excessive dampness or be mouldy or show outward signs of any oily stain or other indication that absorption of the liquid part of the explosive is not perfect, or that the amount of the liquid part is greater than the absorbent can carry, it must be refused in every instance. The shipper must substantiate any claim that a stain is due to accidental contact with grease, oil, or similar substance. In case of doubt, the package must be rejected.

1432. **Condemned or leaking dynamite** must not be repacked and offered for shipment unless the repacking is done by a competent person in the presence of, or with the written consent of the inspector, or with the written authority of the chief inspector of the Bureau of Explosives.

1433. Carriers must forward shipments of explosives promptly and within 48 hours after acceptance at originating point or receipt at transfer station or at interchange point, and consignees must remove such shipments from the carrier's property within 48 hours after notice of arrival at destinations, Sundays and holidays not included.

1434. (*a*) Serious **violations** of these regulations (such as defective packing, improper staying, rough treatment of car, broken packages, etc.) and **accidents** or **explosions** occurring in connection with the transportation or storage on carrier's property of explosives must be reported by the carrier to the chief inspector, Bureau of Explosives, 30 Vesey Street, New York City.

All violations must be corrected before forwarding the explosives.

(*b*) Consignees should report promptly to the chief inspector, Bureau of Explosives, all instances of improper staying and broken or defective packages of explosives in shipments received by them.

SECTION 1.

INFORMATION AND DEFINITIONS.

Grouping.

1500. For transportation purposes, explosives are divided into the following groups:

1. **Forbidden explosives.**
2. **Acceptable explosives.**

Group 1.—Forbidden Explosives.

1501. The following are forbidden explosives:

(*a*) **Liquid nitroglycerin.**

(*b*) **Dynamite** containing over 60 per cent of nitroglycerin (except gelatin dynamite).

(*c*) **Dynamite** having an unsatisfactory absorbent, or one that permits leakage of nitroglycerin under any conditions liable to exist during transportation or storage.

(*d*) **Nitrocellulose** in a dry and uncompressed condition, in quantity greater than 10 pounds, net weight in one exterior package.

(*e*) **Fulminate of mercury in bulk** in a dry condition, and fulminates of all other metals in any condition, except as a component of manufactured articles whose transportation is not forbidden herein.

(*f*) **Fireworks** that combine an explosive and a detonator or blasting cap.

(*g*) **Explosive compositions** including **fireworks** that ignite spontaneously, or undergo marked decomposition when subjected for 48 consecutive hours to a temperature of 75° C. (167° F.).

(*h*) **Firecrackers** whose dimensions exceed 5 inches in length or three-quarters of an inch in diameter or whose explosive charges exceed 45 grains each in weight.

(*i*) **Toy torpedoes** exceeding 1½ inches in diameter, or toy caps containing more than an average of thirty-five hundredths of a grain of explosive composition per cap.

(*j*) **Fireworks** that can be exploded en masse by a blasting cap placed in one of the units or by the impact of a rifle bullet or otherwise.

Such articles may be shipped when packed, marked, and certified in accordance with these regulations and offered for shipment as high explosives.

(*k*) **Fireworks** containing a match tip or head, or similar igniting point or surface, unless each individual tip, head, or similar igniting point or surface is entirely covered and securely protected from accidental contact or friction with any other surface.

(*l*) **Fireworks or explosives** containing an ammonium salt and a chlorate.

(*m*) **New explosives** (except samples for laboratory examinations) until approved for transportation by the Bureau of Explosives.

(*n*) **Explosives properly condemned** by the Bureau of Explosives (except properly repacked samples for laboratory examinations). Parties who are dissatisfied with the decision of the Bureau may appeal to the Commission, and if it is a question of chemical composition the Commission will arrange to have a test made by a disinterested Government laboratory.

Group 2.—Acceptable Explosives.

1502. **Low explosives and black powder** are general names that may be used to describe all explosives having a composition similar to that of ordinary black powder, such as carbonaceous material, sulphur, and a nitrate of sodium or potassium, and that cannot be detonated by a commercial blasting cap. Examples are, rifle, sporting, cannon, and blasting powders.

1503. **High explosives** are all explosives more powerful than ordinary black powder, except smokeless powders and fulminates. Their distinguishing characteristic is their susceptibility to detonation by a blasting cap. Examples of high explosives are the dynamites, picric acid,[1] picrates, chlorate powders, nitrate of ammonia powders, dry trinitrotoluol, dry nitrocellulose (gun cotton,) and fireworks that can be exploded en masse.

1504. **Smokeless powders** are those explosives from which there is little or no smoke when fired. They include smokeless powder for can-

[1] Picric acid for medicinal purposes, and not exceeding 4 ounces in one outside package, may be shipped without other restrictions when in securely closed glass bottles, properly cushioned to prevent breakage.

non and smokeless powder for small arms. **Smokeless powder for cannon** used in the United States at the present time consists of a nitrocellulose colloid, and is comparatively safe to handle and transport. **Smokeless powders for small arms** may consist of nitrocellulose or nitrocellulose combined with nitroglycerin. Detonable picrate or chlorate mixtures are classed as high explosives.

1505. **Fulminates** are fulminates of mercury or of other metals in bulk form, that is, not made up into percussion caps, detonators, blasting caps, toy torpedoes, or exploders.

1506. **Small-arms ammunition** (such as is used in sporting or fowling pieces, or in rifle or pistol practice, etc.), consists usually of a paper or metallic shell, the primer, and the powder charge, with or without shot or bullet, the materials necessary for one firing being all in one piece.

1507. **Ammunition for cannon** embraces all fixed or separate-loading ammunition too heavy for use in small arms. When the component parts are packed in separate outside packages such packages may be shipped as smokeless powder for cannon, explosive projectiles, empty (including solid and sand loaded) projectiles, primers, or fuses. Igniters composed of black powder may be attached to packages in shipments of smokeless powder for cannon.

1508. **Explosive projectiles** are loaded shells for use in cannon. They are not liable to be exploded except by fire of considerable intensity, and the flying fragments would then be very dangerous.

1509. **Blasting caps** contain from 5 to 50 grains of dry fulminate of mercury, or other substance similar to or in combination with fulminate of mercury, packed in a thin copper shell and fired by a slow-burning safety fuse. When a small " bridge " of fine wire is embedded in a suitable priming material and arranged to fire the fulminate by heating the bridge by means of an electric current, the cap is called an "electric blasting cap." They cause explosions of a high order, or "detonations." This means the instantaneous conversion of the entire explosive into gas instead of the gradual conversion known as "combustion." Dynamite "detonates" and smokeless powder for cannon "burns."

1510. **Detonating fuses** are used in the military service to detonate the high explosive bursting charges of projectiles or torpedoes. In addition to a powerful detonator they may contain several ounces of a high explosive, such as picric acid or dry nitrocellulose, all assembled in a heavy steel envelope.

1511. **Primers, percussion and time fuses** are devices used to ignite the black powder bursting charges of projectiles, or the powder charges of ammunition. For small-arms ammunition the primers are usually called "small-arms primers" or "percussion caps." **Tracer fuses** consist of a device which is attached to a projectile and contains a slow-burning composition to show the flight of projectiles at night.

1512. **Safety fuse** consists ordinarily of a core of granular black powder, which is surrounded by yarn, tape, pitch, rubber, etc.

Cordeau detonant is a fuse containing trinitrotoluol, assembled in a drawn lead tube.

Safety squibs are small paper tubes containing a small quantity of black powder, one end of each tube being twisted and generally tipped with sulphur.

1513. **Fireworks** include everything that is designed and manufactured primarily for the purpose of producing a visible or an audible pyrotechnic effect by combustion or by explosion. They consist of common fireworks and special fireworks. (See par. 1501 (*j*).)

1514. **Common fireworks** include all that depend principally upon nitrates to support combustion and not upon chlorates; that contain no phosphorus and no high explosive sensitive to shock and friction; that produce their effect through color display rather than by loud noises. If noise is the principal object, the units must be small and of such nature and manufacture that they will explode separately and harmlessly, if at all, when one unit is ignited in a packing case. They must not be designed for ignition by shock or friction. Examples are Chinese firecrackers, Roman candles, pinwheels, colored fires, rockets, serpents, railway fuses, flash powders, etc.

1515. **Special fireworks** include all that contain any quantity of phosphorus, a fulminate, or other high explosive sensitive to shock or friction, or that contain units of such size that the explosion of one while being handled would produce a serious injury, or that require a special applicance or tool, mortar, holder, etc., for their safe use, or that are designed for ignition by shock or friction. Examples are giant firecrackers, bombs, and salutes (not high explosives), toy torpedoes and caps, ammunition pellets fired in a special holder, railway torpedoes, etc.

SECTION 2.

PACKING, MARKING, AND CERTIFYING ACCEPTABLE EXPLOSIVES.

1531. (*a*) The construction of **shipping containers** purchased hereafter and used in shipping explosives must conform to specifications approved by the Interstate Commerce Commission that apply, and excepting shipments offered by the United States Government, each container must be stamped, labeled, or marked "Complies with I. C.

C. Spec'n. No. ——" or equivalent marking, as stated in the specification.

(*b*) In addition to standing the tests prescribed, the design and construction of packages must be such as to prevent the occurrence in individual packages of defects that permit leakage of their contents under the ordinary conditions incident to transportation. The results of experience, gained by an examination of broken or leaking packages on arrival at destination, must be reported to and recorded by the Bureau of Explosives, to the end that further use of any particular kind of package shown by experience to be inefficient may be prohibited by the Commission.

(*c*) Pending approval and promulgation by the Commission of specifications for types of shipping containers other than those for which specifications are published herein, containers may be used which after investigation made by the Bureau of Explosives or by other competent testing laboratory in the presence of a representative of the Bureau of Explosives, are shown to possess the general efficiency and the protection against leakage of contents afforded by the standard types of corresponding capacity described in the specifications published herein, provided they are labeled or marked to show compliance with this requirement.

Samples of Explosives Together.

1532. **Samples of explosives** (except blasting caps) in separate interior containers may be packed in the same outside package of gross weight not exceeding 50 pounds, provided the weight of any one sample does not exceed 8 ounces, and provided the interior packages are so cushioned and protected as to insure their transportation without rupture or leakage of contents. The package must be marked and described with the name of the most dangerous explosive included among the samples, such as "HIGH EXPLOSIVE," "BLACK POWDER," etc.

Low Explosives and Black Powder.

1533. **Packing.**—Packages containing less than 12½ pounds of black powder or low explosives must be inclosed in a tight wooden box complying with specification No. 16. Each inside package containing less than 12½ pounds and more than 1½ pounds must be so placed in the outside box that the filling hole will be up. The boxes must be marked on top "THIS SIDE UP" and also as prescribed by paragraph 1536.

1534. Twelve and a half pounds or over of black powder or low explosives must be packed in packages that comply with specification No. 11 or 13. Kegs less than 8 inches long must be boxed as prescribed by paragraph 1533.

1535. **Weight.**—Packages must not weigh over 200 pounds gross.

1536. **Marking.**—Each outside package must be plainly marked, stamped, or stenciled "BLACK POWDER" or "LOW EXPLOSIVES," and may also show "BLASTING," "RIFLE," etc., as "BLACK BLASTING POWDER," "BLACK RIFLE POWDER," "LOW BLASTING EXPLOSIVE," etc. Additional marks, trade names, etc., may appear if desired by shipper, but such additional marking must not be more conspicuous nor must it obscure the marking prescribed herein.

High Explosives.

1551. **High explosives** consisting of a liquid mixed with an absorbent material must have the absorbent (wood pulp or similar material) in sufficient quantity and of satisfactory quality, properly dried at the time of mixing; nitrate of soda must be dried at the time of mixing to less than 1 per cent of moisture; and the ingredients must be uniformly mixed so that the liquid will remain thoroughly absorbed under the most unfavorable conditions incident to transportation.

1552. Explosives containing nitroglycerin must have uniformly mixed with the absorbent material a satisfactory antacid which must be in quantity sufficient to have the acid neutralizing power of an amount of magnesium carbonate equal to 1 per cent of the nitroglycerin.

1553. (*a*) **Packing.**—High explosives containing more than 10 per cent of nitroglycerin (except gelatin dynamite) must be made into cartridges not exceeding 4 inches in diameter or 8 inches in length, and must not be packed in bags or sacks; except that cartridges 5 inches in diameter and not exceeding 8 inches in length may be shipped, provided each cartridge of the explosive is completely enclosed in a shell made of strong paraffined paper, and thereafter enclosed in another such paper shell, the completed cartridge being dipped in melted paraffin.

(*b*) High explosives containing not more than 10 per cent of nitroglycerin may be shipped in bags or sacks. Each bag or sack must not contain more than 12½ pounds of explosive and must be placed in a box with filling end up. These bags or sacks and the coverings of all cartridges must be strong and so treated that they will not absorb the liquid constituent of the explosive.

1554. (*a*) All boxes in which high explosives in cartridges, bags, sacks, or in bulk are packed must be lined with strong paraffined paper or other suitable material. The lining must be without joints or other openings at the bottom or on the sides of the box, and must be impervious to water and to any liquid ingredient of the explosive.

(*b*) In packing cartridges of nitroglycerin explosives at least one-quarter of an inch thickness of dry, fine wood pulp or sawdust must be spread over the bottom of the lined box before inserting the cartridges, and all the vacant space in the top must be filled with this material.

(*c*) All cartridges exceeding 4 inches in length, except explosive gelatin or gelatin dynamite, must be placed horizontally in boxes.

1555. Inside packages containing not more than 1 pound each of dry uncompressed nitrocellulose, wrapped in strong paraffined paper or other suitable spark-proof material, will be accepted for shipment if securely packed in an outside package that complies with Specification No. 14 and is marked as prescribed in paragraph 1559. Outside packages must not contain more than 10 pounds of dry nitrocellulose.

1556. (*a*) High explosives containing no explosive liquid ingredient and not having, with their normal percentage of moisture, a sensitiveness to percussion greater than measured by the blow delivered by an 8-pound weight dropping from a height of 7 inches on a compressed pellet of the explosive three-hundredths of an inch in thickness and two-tenths of an inch in diameter confined rigidly between hard steel surfaces, as in the standard impact-testing apparatus of the Bureau of Explosives, may be shipped in bulk when securely packed in authorized containers. These explosives may also be packed in cartridges, and must be so packed when their sensitiveness is greater than the above limit. Wooden boxes or wooden kegs provided with suitable linings to prevent leakage must be used for chlorate powders.

(*b*) When the addition of water to any such explosive will make it non-explosive according to tests made by the Bureau of Explosives, the wet material may be shipped and handled in transit as prescribed by the Regulations for the Transportation of Dangerous Articles other than Explosives by Freight and by Express.

1557. **Containers** for high explosives must comply with Specifications Nos. 11, 14, 20, 21, and 22 as specified therein.

1558. (*a*) **Weights.**—Packages of high explosives containing an explosive liquid ingredient must not exceed 75 pounds gross weight.

(*b*) Packages of high explosives containing no liquid explosive ingredient whose shipment in bulk is authorized by paragraph 1556 must not exceed 125 pounds gross weight.

(*c*) The gross weight of an outside package containing not exceeding 10 pounds net of dry nitrocellulose, not compressed, packed as prescribed in paragraph 1555, must not exceed 35 pounds. Compressed sticks or blocks of dry nitrocellulose (guncotton) wrapped in strong paraffined paper may be shipped in outside packages complying with Specification No. 14, with a gross weight not exceeding 75 pounds.

1559. **Marking.**—Boxes must be plainly marked on top and on one side or end and kegs must be marked on one end " HIGH EXPLOSIVE—DANGEROUS," in letters not less than $\frac{7}{16}$ inch in height. The tops of boxes must be marked " THIS SIDE UP."

Smokeless Powder for Cannon.

1571. **Packing.**—Smokeless powder for cannon must be packed in tight wooden boxes free from loose knots and cracks, or in wooden barrels that comply with Specification No. 11, or in metal barrels or kegs that comply with Specification No. 13. Smokeless powder for cannon may be packed in water in metal barrels or in strong wooden barrels of the type used for alcohol.

1572. **Weight.**—Packages must not weigh over 180 pounds gross unless the powder is packed in water.

1573. **Marking.**—Each package must be plainly marked " SMOKELESS POWDER FOR CANNON," or " SMOKELESS POWDER FOR CANNON IN WATER."

Smokeless Powder for Small Arms.

1575. **Packing.**—Packages containing less than 9 pounds of smokeless powder for small arms must be inclosed in a tight wooden box complying with Specification No. 16. Each inside package containing less than 9 pounds and more than $1\frac{1}{2}$ pounds must be so placed in the outside box that the filling hole will be up. The box must be marked on top " THIS SIDE UP " and also as prescribed by paragraph 1578.

1576. Quantities of 9 pounds or over of smokeless powder for small arms must be packed in tight wooden boxes free from loose knots or cracks, or in wooden barrels that comply with Specification No. 11, or in metal barrels or kegs that comply with Specification No. 13. Kegs less than 8 inches long must be boxed as prescribed by paragraph 1575. Smokeless powder for small arms may be packed in water in metal barrels or in strong wooden barrels of the type used for alcohol.

1577. **Weight.**—Packages must not weigh over 150 pounds gross.

1578. **Marking.**—Each outside package must be plainly marked " SMOKELESS POWDER FOR SMALL ARMS," or " SMOKELESS POWDER FOR SMALL ARMS IN WATER."

Fulminates.

1591. **Packing.**—Fulminate of mercury in bulk must contain when packed not less than 25 per cent of water, and must in this wet condition be placed in a bag made of heavy cotton cloth of close mesh, equal in quality and weight to the cotton twill used for pockets in high-grade clothing. There must be placed inside the bag and over the fulminate a cap of the same cloth and of the diameter of the bag, and the bag must be tied securely and placed in a strong grain bag, which must, in turn, be tied securely and packed in the center of a cask or barrel in good condition and of the kind used for shipment of alcohol. The grain bag must not contain more than 150 pounds dry weight of fulminate and it must be surrounded on all sides by tightly packed sawdust not less than 6 inches thick. The cask or barrel must be lined with a heavy, close-fitting jute bag, closed by secure sewing to prevent escape of sawdust. After the barrel is properly coopered it must be filled with water and the bung sealed. The barrel must be inspected carefully and all leaks stopped.

1592. **Marking.**—Each cask or barrel must be plainly marked "WET FULMINATE OF MERCURY—DANGEROUS."

Small-Arms Ammunition.

1601. **Packing.**—Small-arms ammunition must be packed in pasteboard or other boxes, and these boxes must be packed in strong outside wooden or metal containers.

Small-arms ammunition in pasteboard or other boxes, and in quantity not exceeding a gross weight of 75 pounds, may be packed with non-explosive and non-inflammable articles and with small-arms primers or percussion caps (see par. 1619), provided the outside wooden or metal container is marked as prescribed in paragraph 1602.

1602. **Marking.**—Each outside package or case must be plainly marked "SMALL-ARMS AMMUNITION."

1603. No restrictions, other than proper packing and marking, are necessary for the shipment of small-arms ammunition.

Ammunition for Cannon.

1604. **Packing.**—Ammunition for cannon must be well packed and properly secured in strong wooden or metal containers.

1605. **Marking.**—Each outside package must be plainly marked "AMMUNITION FOR CANNON WITH EXPLOSIVE PROJECTILES," or "AMMUNITION FOR CANNON WITH EMPTY PROJECTILES," or "AMMUNITION FOR CANNON WITH SAND-LOADED PROJECTILES," or "AMMUNITION FOR CANNON WITH SOLID PROJECTILES," according as the projectiles do or do not contain a bursting charge, or "AMMUNITION FOR CANNON WITHOUT PROJECTILES."

Empty cartridge bags having attached black-powder igniters must be marked "EMPTY CARTRIDGE BAGS — BLACK POWDER IGNITERS."

1606. No restrictions other than proper marking are prescribed for shipments of material relating to ammunition for cannon, but containing no explosive or other dangerous article, such as cartridge cases, "dummy" or "drill" cartridges, etc.

Explosive Projectiles.

1607. **Packing.**—Explosive projectiles must be packed in strong wooden or metal boxes and each projectile must be properly secured. Projectiles exceeding 100 pounds in weight may be shipped without being boxed, if desired. When necessary, detonating fuses of an approved type may be assembled in explosive projectiles.

1608. **Weight.**—The gross weight of a box containing more than one explosive projectile must not exceed 250 pounds.

1609. **Marking.**—Each exterior package or projectile must be plainly marked "EXPLOSIVE PROJECTILE," "SAND-LOADED PROJECTILE," "SOLID PROJECTILE," "EMPTY PROJECTILE," "EXPLOSIVE TORPEDO," or "EMPTY TORPEDO."

No restrictions, other than proper marking, are prescribed for the shipment of sand-loaded projectiles, empty projectiles, solid projectiles, or empty torpedoes, and tracer fuses of an approved type may be assembled in projectiles.

Blasting Caps.

1611. **Packing.**—Blasting caps contain such a sensitive and dangerous explosive that very efficient packing is necessary, and the outside of all caps must be free from fulminate.

(*a*) Blasting caps must be packed in strong interior packages or containers, in which they must fit snugly, and the caps must be closed securely by teats projecting from a plate of suitable elastic material placed over the caps. Not more than 100 blasting caps may be packed in a single inside container. All inside containers must then be packed snugly in cartons or wrappings made of paper or pasteboard.

(*b*) For not more than 5000 caps the inside containers in cartons or wrappings must be packed in an outside box which must comply with Specification No. 15, and they must be separated from the outside box by at least 1 inch of tightly packed sawdust, excelsior, or equivalent cushioning material.

(*c*) For more than 5000 caps the inside containers, in cartons or wrappings, must be packed in an inside box made of sound lumber or in a hermetically sealed metal box made of not less than 30-gauge

United States standard, and this inside wooden or metal box must then be packed in an outside box made of sound lumber; both of these boxes must comply with Specification No. 15. Tightly packed sawdust, excelsior, or equivalent cushioning material at least 1 inch thick at all points must separate the inside box from the outside wooden box.

(*d*) More than 20,000 blasting caps must not be placed in one outside package.

(*e*) Not more than five tin boxes, containing not more than 100 blasting caps in each box, may be packed with safety fuse, each box to be placed in the center of a coil of fuse, and in this case the outside box must be made of sound lumber, complying with Specification No. 15.

(*f*) Electric blasting caps must be packed in pasteboard cartons containing not more than 50 caps each. These cartons must be packed in a wooden box, complying with Specification No. 15.

1612. **Weight**—The gross weight of an outside package containing blasting caps or electric blasting caps must not exceed 150 pounds.

1613. **Marking.**—Each outside package must be plainly marked "(number) BLASTING CAPS—HANDLE CAREFULLY," or "(number) ELECTRIC BLASTING CAPS—HANDLE CAREFULLY." In addition, each box must bear the marking "DO NOT STORE OR LOAD WITH ANY HIGH EXPLOSIVE."

Detonating Fuses.

1615. **Packing.**—Detonating fuses must be packed in strong, tight wooden boxes, and each fuse must be well secured.

1616. **Weight.**—The gross weight of one outside package must not exceed 165 pounds.

1617. **Marking.**—Each outside package must be plainly marked "DETONATING FUSES—HANDLE CAREFULLY."

Primers, Percussion and Time Fuses.

1619. (*a*) **Packing.**—Primers, percussion and time fuses must be packed in strong, tight wooden outside boxes, with special provision for securing individual packages of primers and fuses against movement in the box.

(*b*) Small-arms primers containing anvils must be packed in cellular inside packages, with partitions separating the layers and columns of primers, so that the explosion of a portion of the primers in the completed shipping package will not cause the explosion of all of the primers.

(*c*) Percussion caps may be packed in metal or other inside boxes containing not more than 500 caps, but the construction of the cap and the kind and quantity of explosives in each must be such that the explosion of a part of the caps in the completed shipping package will not cause the explosion of all of the caps.

(*d*) Small-arms primers and percussion caps may form a part of the gross weight of 75 pounds of small-arms ammunition that may be packed with other articles as authorized by paragraph 1601.

1620. **Weight.**—The gross weight of one outside package must not exceed 150 pounds.

1621. **Marking.**—Each outside box must be plainly marked "SMALL-ARMS PRIMERS—HANDLE CAREFULLY," or "PERCUSSION CAPS—HANDLE CAREFULLY," or "CANNON PRIMERS—HANDLE CAREFULLY," or "COMBINATION PRIMERS—HANDLE CAREFULLY," or "PERCUSSION FUSES—HANDLE CAREFULLY," or "COMBINATION FUSES—HANDLE CAREFULLY," or "TRACER FUSES—HANDLE CAREFULLY," etc.

1622. No restrictions other than proper packing and marking are necessary for the shipment of primers, percussion and time fuses. No restrictions other than proper marking are prescribed for the shipment of "Empty cartridge shells, primed."

Common Fireworks.

1631. **Packing.**—Common fireworks must be in a finished state, exclusive of mere ornamentation, as supplied to the retail trade, and must be securely packed in strong, tight, spark-proof wooden barrels that comply with Specification No. 11, or in boxes complying with Specification No. 12A.

1632. **Weight.**—The gross weight of one outside package containing common fireworks must not exceed 325 pounds, except that exhibition set pieces when specially packed may have weight not exceeding 400 pounds.

1633. **Marking.**—Each outside package must be plainly marked in letters not less than $\frac{7}{16}$ inch in height, "COMMON FIREWORKS—KEEP FIRE AWAY."

Special Fireworks.

1634. **Packing.**—Special fireworks must be in a finished state, exclusive of mere ornamentation, as supplied to the retail trade, and must not contain forbidden fireworks. (See par. 1501 (*f*) to (*l*), inclusive.)

Special fireworks must be securely packed in strong, spark-proof wooden barrels that comply with Specification No. 11, or in boxes complying with Specification No. 12.

1635. **Weight.**—The gross weight of one outside package containing special fireworks must not exceed 200 pounds, and the gross weight of a package containing toy torpedoes must not exceed 65 pounds.

1636. **Marking.**—Each outside package containing special fireworks or a mixture of common and special fireworks must be plainly marked in letters not less than $\frac{7}{16}$ inch in height, "SPECIAL FIREWORKS—HANDLE CAREFULLY—KEEP FIRE AWAY."

Safety Fuse, Cordeau Detonant, and Safety Squibs.

1638. **Safety fuse, cordeau detonant, and safety squibs** must be packed in strong wooden boxes or barrels properly marked, and may be loaded in any car with any other kind of an explosive or inflammable substance or with other freight. Cordeau detonant must not be packed in the same package with blasting caps, or with any high explosive.

SHIPPING ORDER.

1641. (*a*) **Shipper's certificate.**—The shipping order for any package containing an explosive named below must show each article under its proper name as specified in this paragraph, and must show the following certificate in the lower left-hand corner, over the written or stamped facsimile signature of the shipper or of his duly authorized agent:

This is to certify that the above articles are properly described by name and are packed and marked and are in proper condition for transportation according to the regulations prescribed by the Interstate Commerce Commission.

List of Shipping Names.

Low explosives.
Black powder.
High explosives.
Smokeless powder for cannon.
Smokeless powder for small arms.
Wet fulminate of mercury.
Ammunition for cannon with explosive projectiles.
Ammunition for cannon with empty projectiles.
Ammunition for cannon with sand-loaded projectiles.
Ammunition for cannon with solid projectiles.
Ammunition for cannon without projectiles.
Explosive projectiles.
Explosive torpedoes.
Detonating fuses.

(Number) blasting caps.
(Number) electric blasting caps.
Common fireworks.
Special fireworks.

(*b*) The following acceptable explosives must be described under their proper names as given below, but shipper's certificate is not required:

Small-arms ammunition.
Small-arms primers.
Cannon primers.
Empty cartridge shells, primed.
Empty cartridge bags—black powder igniters.
Combination primers.
Percussion caps.
Percussion fuses.
Time fuses.
Tracer fuses.
Combination fuses.
Safety fuse.
Cordeau detonant.
Safety squibs.
Empty projectiles.
Sand-loaded projectiles.
Solid projectiles.
Empty torpedoes.

(*c*) Whenever orders are placed in foreign countries for the importation of explosives, to be forwarded from port of entry by carriers subject to these regulations, the importer must furnish with the order to the foreign shipper, and also to the forwarding agent at the port of entry, full and complete information as to the necessary packing and marking required by these regulations. The forwarding agent must file with the originating carrier a properly prepared and certified shipping order as prescribed herein.

SECTION 3.

RULES FOR HANDLING.

1642. In handling packages of explosives at stations and in transferring them to and from cars the greatest care must be taken, and shocks or falls liable to injure the containing package must be avoided. Where an inclined chute is employed such chute shall be constructed of 1-inch planed boards, with side guards 4 inches high extending 3 inches above top face of bottom of chute and throughout its length fastened with brass screws. D-shaped strips or runners, not more than 6 inches apart and running lengthwise of chute, must be fastened to the upper surface of the bottom board by means of glue and wooden pegs extending through the bottom board and runners. Chutes must be occasionally wiped down with waste moistened with machine oil when dynamite packages are being handled.

A stuffed mattress, 4 feet wide by 6 feet long and not less than 4 inches thick, or a heavy jute or hemp mat of like dimensions, must be placed under the discharging end of the chute.

1643. (*a*) Careful men must be chosen to handle explosives, the platform and the feet of the men must be as free as possible from grit, and all possible precautions must be taken against fire. Suitable provision must be made, outside of the station when practicable, for the safe storage of explosives, and every effort possible must be made to reduce the time of this storage. If a shipment of explosives is not removed within forty-eight hours after notice of arrival at destination, it must be disposed of by return to the shipper, or by storage at the expense of the owner, or by sale, or when necessary to safety by destruction under supervision of a competent person.

(*b*) Unauthorized persons must not have access to explosives or other dangerous articles at any time while such articles are in the custody of the carrier.

Loading in Car.

1644. (*a*) Packages receive their greatest stresses in a direction parallel to the length of the car and must be loaded so as to offer their greatest resistance in this direction. Boxes of dangerous explosives named in paragraph 1661 when loaded in the car must rest on their bottoms and with their long dimension parallel to the length of the car.

(*b*) A car must not contain more than 70,000 pounds gross weight of explosives. This limit does not apply to shipments of small-arms ammunition or ammunition for cannon.

(*c*) When the lading of a car includes explosives and exceeds 5000 pounds the weight of the lading must be distributed in approximately equal parts in both ends of the car.

1645. Explosives packed in kegs, except when boxed, must be loaded on their sides with ends toward ends of the car; and packages of explosives must not be placed in the space opposite the doors unless the doorways are boarded on the inside as high as the lading.

Large casks, barrels, or drums may be loaded on their sides or ends as will best suit the conditions.

1646. (*a*) Packages containing any of the explosives for the transportation of which a certified and placarded car is prescribed (see par. 1661), must be stayed (blocked and braced) by the one who loads the car, by methods not less efficient than those described in Bureau of Explosives Pamphlet No. 6, to prevent change of position by the ordinary shocks incident to transportation. Special care must be used to prevent them from falling to the floor or from having anything fall on them or slide against them during transit.

(*b*) To prevent delays to local freight trains, when there are shipments of explosives for different destinations loaded in a " peddle car " or " way car," the shipments for each destination must be stayed separately.

(*c*) Forwarding and transfer stations for explosives must be provided with the necessary materials for staying.

(*d*) Shippers must furnish the material for staying packages loaded by them.

1647. Detonating fuses or blasting caps, or electric blasting caps, must not be loaded in a car or stored with high explosives of any kind, including explosive projectiles, nor with wet nitrocellulose.

1648. Wet fulminate of mercury must not be loaded with any explosive or inflammable article.

1649. Fireworks must not be loaded in the same car nor stored on carriers' property with any explosive named in paragraph 1661.

1650. Explosives named in paragraph 1661 that require a certified car, placarded " EXPLOSIVE," must not be transported in the same car with nor stored on railway property near any of the dangerous articles for which labels are prescribed by the Regulations for the Transportation of Dangerous Articles, other than Explosives by Freight.

1651. In a car containing explosives all packages of other freight must be loaded, and when necessary must be stayed to prevent injury to packages of explosives during transit. When practicable, explosives should be loaded so as to avoid transfer at stations.[1]

Waybilling.

1652. (*a*) The carrier must see that each shipment of explosives is properly described on the shipping order and on the revenue waybill under one of the names in the " List of shipping names," paragraph 1641, and that the correct gross weight is given.

(*b*) The revenue waybill or astray waybill for a shipment, and the card waybill and envelope containing revenue waybill when used as a card waybill for a car containing any of the following explosives:

[1] At stations where it is necessary to handle explosives at night it is recommended that incandescent electric lights be provided.

Low explosives,
Black powder,
High explosives,
Wet fulminate of mercury,
Ammunition for cannon with explosive projectiles,
Explosive projectiles,
Explosive torpedoes,
Detonating fuses,
Blasting caps, { In quantity exceeding 1000 blasting caps
Electric blasting caps, { or 1000 electric blasting caps,

must have plainly stamped or plainly written across the top, in letters not less than $\frac{3}{8}$ of an inch high, the word "EXPLOSIVES."

(*c*) The revenue waybill and astray waybill for a shipment, and the card waybill and envelope containing revenue waybill when used as a card waybill for a car containing:

Ammunition for cannon with empty projectiles,
Ammunition for cannon with sand-loaded projectiles,
Ammunition for cannon without projectiles,
Ammunition for cannon with solid projectiles,
Smokeless powder for small arms,
Smokeless powder for cannon,
Common fireworks, or
Special fireworks,

must have plainly stamped or plainly written across the top, in letters not less than $\frac{3}{8}$ of an inch high, the word "INFLAMMABLE."

(*d*) No indorsements are required on revenue waybills, card waybills, or other billing for shipments or for cars of the following explosives:

Small-arms ammunition.
Small-arms primers.
Cannon primers.
Empty cartridge shells, primed.
Empty cartridge bags—black powder igniters.
Combination primers.
Percussion caps.
Percussion fuses.
Time fuses.
Tracer fuses.
Combination fuses.
Safety fuse.
Cordeau detonant.
Safety squibs.

Empty projectiles.
Sand-loaded projectiles.
Solid projectiles.
Empty torpedoes.
Blasting caps. } In quantity not exceeding 1000 caps.
Electric blasting caps. }

Shipments from Connecting Lines.

1654. (*a*) Cars containing explosives named in paragraph 1661 which are offered by connecting lines must be carefully inspected by the receiving line on the outside, including the roof, and if practicable the lading must also be inspected. These cars must not be forwarded until all discovered violations have been corrected.

(*b*) If the car shows evidence of, or if there is any reason to suspect that it has received, rough treatment, the lading must be inspected and placed in proper condition before the car is permitted to proceed. When interchange occurs and inspection is necessary after daylight hours, electric flash lights or other suitable covered lights should be provided. Naked lights must not be used.

(*c*) Shipments of explosives offered by connecting lines must comply with these regulations, and the revenue waybill, freight bill, manifest of lading, card waybill, switching order, or other billing must bear the indorsement "EXPLOSIVES" or "INFLAMMABLES" prescribed by paragraph 1652 (*b*) and (*c*).

Disposition of Injured, Condemned, and Astray Packages.

1655. **Packages found injured or broken** in transit may be recoopered when this is evidently practicable and not dangerous. A broken box of high explosives that can not be recoopered should be reinforced by stout wrapping paper and twine, placed in another strong box, and surrounded by dry, fine sawdust, or dry and clean cotton waste, or elastic wads made from dry newspaper. A ruptured can or keg should be inclosed in a grain bag of good quality and boxed or crated. Injured packages thus protected and properly marked may be forwarded.

1656. **Condemned packages** of leaking explosives should (1) be returned immediately to shipper if at point of shipment; or (2) disposed of to a person who is competent and willing to remove them from railway property, if leakage is discovered while in transit; or (3) removed immediately by consignee if shipment is at destination.

When disposition cannot be made as above, the leaking boxes must be packed in other boxes large enough to permit, and the leaking box must be surrounded by at least 2 inches of dry, fine sawdust, or dry and clean cotton waste, and be stored in station magazine or other safe place until arrival of an inspector of the Bureau of Explosives or other authorized person to superintend the destruction or disposition of the condemned material.

1657. When name and address of consignee or shipper are known, an **astray shipment** must be forwarded to its destination or returned to the shipper by the most practicable route, provided a careful inspection shows the packages to be in proper condition for safe transportation.

When a package in an astray shipment is not in proper condition for safe transportation (see par. 1655), or when name and address of consignee and shipper are unknown, disposition will be made as prescribed by paragraph 1656.

Selection and Preparation of Cars.

DANGEROUS EXPLOSIVES.

1661. For the transportation of carloads or less-than-carload lots of—

Low explosives,
Black powder,
High explosives,
Wet fulminate of mercury,
Blasting caps, } In quantity exceeding 1000 blasting caps
Electric blasting caps, } or 1000 electric blasting caps.
Ammunition for cannon with explosive projectiles,
Explosive projectiles,
Explosive torpedoes, or
Detonating fuses,

only certified and placarded box cars may be used.

Certified Car Placarded "Explosives."

1662. Certified cars must be inspected inside and outside and must conform to the following specifications:

(*a*) Box car not less than 60,000 pounds capacity with friction draft gear. Steel underframe cars should be used when available. On narrow-gauge and other railroads, all of whose freight cars are of less than 60,000 pounds capacity, explosives may be transported in cars of less than that capacity, provided the available cars of greatest capacity and strength are used for this purpose.

(*b*) Must be equipped with air brakes and hand brakes in condition for service.

(*c*) Must have no loose boards or cracks in the roof, sides, or ends, through which sparks may enter.

(*d*) The doors must shut so closely that no sparks can get in at the joints, and, when necessary, they must be stripped. The stripping for doors should be on the inside and be fastened to the door frame where it will form a shoulder against which the closed door is pressed by means of wedges or cleats in door shoes or keepers. The openings under the doors should be similarly closed. The hasp fastenings must be examined with doors closed and fastened and must be cleated when necessary to prevent door shifting.

(*e*) The journal boxes and trucks must be carefully examined and put in such condition as to reduce to a minimum the danger of hot boxes or other failure necessitating the setting out of the car before reaching destination. The lids or covers of journal boxes must be in place.

(*f*) The car must be carefully swept out before it is loaded. For less-than-carload shipments the space in which the packages are to be loaded must be carefully swept.

Holes in the floor or lining must be repaired and special care taken to have no projecting nails or bolts or exposed pieces of metal which may work loose or produce holes in packages of explosives during transit.

(*g*) When packages of explosives are to be loaded over exposed draft bolts or kingbolts, these bolts must have short pieces of solid, sound wood with beveled ends (2-inch plank) spiked to the floor over them (or empty packages of the same character may be used for this purpose) to prevent possibility of their wearing into the packages of explosives.

(*h*) The roof of the car must be carefully inspected from the outside for decayed spots or broken boards, especially under or near the running board, and such spots must be covered or repaired to prevent their holding fire from sparks. A car with a roof generally decayed, even if tight, must not be used.

(*i*) The carrier must have the car examined by a competent employee to see that it is properly prepared, and must have a "Car certificate" signed in triplicate upon the prescribed form (see par. 1665) before permitting the car to be loaded.

(*j*) Cars not in proper condition, as above specified, must not be furnished to the shipper or used for the transportation of explosives.

Less Dangerous Explosives in Car Placarded "Inflammable."

1663. (*a*) Carloads or less-than-carload lots of—

Ammunition for cannon with empty projectiles,

Ammunition for cannon with sand-loaded projectiles,

Ammunition for cannon with solid projectiles,

Ammunition for cannon without projectiles,
Smokeless powder for small arms,
Smokeless powder for cannon, or
Fireworks,

may be loaded in any box car which is in good condition into which sparks cannot enter and whose roof is not in danger of taking fire through unprotected decayed wood. These cars do not require the car certificate but must have attached to both side doors and both ends the "INFLAMMABLE" placard prescribed by paragraph 1667, and the doors if not tight must be stripped to prevent the entrance of sparks.

Relatively Safe Explosives in Car Without Placards.

(*b*) Carloads or less-than-carload lots of—
Small-arms ammunition,
Small-arms primers,
Cannon primers,
Empty cartridge shells, primed,
Empty cartridge bags—black powder igniters,
Combination primers,
Percussion caps,
Percussion fuses,
Time fuses,
Tracer fuses,
Combination fuses,
Safety fuse,
Cordeau detonant,
Safety squibs,
Empty projectiles,
Sand-loaded projectiles,
Solid projectiles,
Empty torpedoes, and not exceeding 1000 blasting caps, or 1000 electric blasting caps,

may be loaded in any box car which is in good condition, without car certificate or placards.

(*c*) When specially authorized by the carrier, explosives in quantity not exceeding 150 pounds may be carried in construction or repair cars when the packages of explosives are placed in a "magazine" box made of sound lumber not less than 1 inch thick, covered on the exterior with metal, and provided with strong handles. This box must be plainly stenciled on the top, sides, and ends, in letters not less than 2 inches high, "EXPLOSIVES—DANGEROUS—HANDLE CAREFULLY." The box must be provided with strong hinges and with a lock for keeping it securely closed. Vacant space in the box must be filled with a cushioning material such as sawdust or excelsior, and the box must be properly stayed in the car to prevent movement. The car, when not occupied by a responsible employee, must be placarded "EXPLOSIVES."

Placarding of Cars and Certification of Contents.

1664. Uniform practice is important and the prescribed forms of car certificate and placards must be used.

1665. (*a*) **Car certificate.**—The following certificate (prescribed by par. 1662 (*i*), printed on strong tagboard measuring 7 by 7 inches, must be duly executed in triplicate by the carrier, and by the shipper if he loads the shipment. The original must be filed by the carrier at the forwarding station on a separate file, and the other two must be attached to the outside of the car doors, one on each side, the lower edge of the certificate not less than 4½ feet above the floor level.

(*b*) At stations where explosives are loaded into a properly certified and placarded car received with other shipments of explosives, or carload shipments are reconsigned, a record must be kept of the car, originating point, carrier's name and date of car certificate.

CAR CERTIFICATE.

No. 1. ——— Station, ———, 19—.

I hereby certify that I have this day personally examined ——— car No. ——— and that the roof and sides have no loose boards, holes, or cracks, or unprotected decayed spots liable to hold sparks and start a fire; that the kingbolts or draft bolts are properly protected and that there are no uncovered irons or nails projecting from the floor or sides of the car which might injure packages of explosives; also that the floor is in good condition and has this day been cleanly swept before the car was loaded; that I have examined all the axle boxes and that they are properly covered, packed, and oiled, and that the air brakes and hand brakes are in condition for service.

——— ———

Railway employee inspecting car.

———

No. 2 ——— Station, ———, 19—.

I hereby certify that I have this day personally examined the above car; that the floor is in good condition and has been cleanly swept and that the roof and sides have no loose boards, holes, cracks, or unprotected decayed spots liable to hold sparks and start a fire; that the kingbolts and draft bolts are protected, and that there are no uncovered irons or nails projecting from the floor or sides of the car which might injure packages of explosives; that the explosives in this car have been loaded and stayed and that the car has been placarded according to paragraphs 1644 to 1651, inclusive, 1661, and 1666, of the Regulations for the Transportation of Explosives prescribed by the Interstate Commerce Commission; that the doors fit or have been stripped so that sparks cannot get in at the joints or bottom.

——— ———,

Shipper.

——— ———,

Railway employee inspecting loading and staying.

NOTE.—Both certificates must be signed, Certificate No. 1 by the representative of the carrier. For all shipments loaded by the shipper he or his authorized agent must sign certificate No. 2, and the representative of the carrier must certify as to loading and staying and general condition. When the car is not loaded by shipper, certificate No. 2 must be signed only by the representative of the carrier. A shipper should decline to use a car not in proper condition.

"Explosive" Placard

1666. **Placard.**—A certified car for the dangerous explosives specified in paragraph 1661 must be protected by attaching to each outside end and side door, the lower edge not less than $4\frac{1}{2}$ feet above the car floor, a standard placard 12 by 14 inches, on which appears in conspicuous red and black printing on strong tag board the following notice:

14 inches

12 inches

_______________ R. _______ Company

EXPLOSIVES

(TO BE PRINTED IN RED)

HANDLE CAREFULLY

KEEP FIRE AWAY

(TO BE PRINTED IN RED)

_______________ STATION _______________ 19___

CONDENSED RULES FOR HANDLING THIS CAR

1. This car must not be placed in a passenger train, nor in a mixed train if avoidable.
2. Cars containing explosives must be near center of train and may be together if desired; must be at least 15 cars from engine and 10 cars from caboose when length of train will permit.
3. This car must not be placed next to cars bearing the inflammable or the acid placard or cars containing lighted heaters. Whenever it is possible to avoid so doing it must not be placed next to tank cars or flat cars or next to carloads of lumber, poles, iron, pipe, or other articles liable to break through end of car from rough handling.
4. The air and hand brakes on this car must be in service.
5. In shifting have a car between this car and engine whenever possible, and do not cut this car off while in motion.
6. Avoid all shocks to this car and couple carefully.
7. Avoid placing it near a possible source of fire.
8. Engines on parallel track must not be allowed to stand opposite or near this car when it can be avoided.
9. This placard must be removed from car when the explosives are unloaded.

NOTE.—Carriers will be permitted to use their present supply of placards, but all new supplies must conform to the amended form.

"Inflammable" Placard.

1667. A placard of diamond shape printed on strong tag board measuring $10\frac{3}{4}$ inches on each side and bearing in red and black letters the inscription prescribed by paragraph 1913 must be placed on each outside end and side door of a car containing any quantity of smokeless powder for cannon, or smokeless powder for small arms, or ammunition for cannon with empty projectiles, or ammunition for cannon with sand-loaded projectiles, or ammunition for cannon with solid projectiles, or ammunition for cannon without projectiles, or fireworks.

Handling Cars Containing Explosives.

1680. The phrase " cars containing explosives " as used in this and subsequent paragraphs, excepting paragraph 1697, refers to the dangerous explosives specified in paragraph 1661, which require a certified car placarded " EXPLOSIVES."

1681. Under no circumstances must a car known to require the " EXPLOSIVE " placard be taken from a station, including transfer stations or interchange points, a siding, or yard, unless it is properly placarded as per paragraphs 1665 and 1666, nor unless the car is in proper condition. The carrier must see that its representative in charge of a freight train makes a thorough check of the cars bearing " INFLAMMABLE," " ACID," and " EXPLOSIVE " placards with the billing to see that all placards required are attached and that those not required are removed.

1682. Every possible effort must be made to expedite the movement of cars containing explosives, and no unnecessary delay must occur at initial, interchange, or transfer stations.

1683. When a car containing explosives is in a train, the carrier must make proper provision for notifying its train and engine employees of the presence and location of such car in the train before leaving the initial station.

1684. Such cars must be frequently inspected to see that the placards and car certificates are intact. Whenever any of these become detached or lost in transit, they must be replaced on arrival at the next division terminal yard, if in through freight train, or at first station stop if in local freight train.

1685. On lines where regular trains are operated for freight service only, cars containing explosives must not be hauled in a train that carries passengers. Where only a mixed train service is operated or where passengers are carried in the caboose car of a freight train, a car containing a freight shipment of explosives may be hauled, but it must not be placed next to a car carrying passengers, and whenever it is practicable to do so, such a car or cars must be placed between freight cars not bearing " ACID " or " INFLAMMABLE " placards. Cars containing explosives must have air and hand brakes in service.

1686. (*a*) In through freight trains they must be placed near the middle of the train, and two or more such cars may be placed together if desired. They must be at least 15 cars from the engine and 10 cars from the caboose when length of the train will permit. They must not be placed next to cars bearing the " INFLAMMABLE " or the " ACID " placard, or cars containing lighted heaters. Whenever it is possible to avoid so doing they must not be placed next to tank cars or wooden-frame flat cars or next to carloads of lumber, poles, iron, pipe, or other articles to break through end of car from rough handling.

(*b*) These requirements apply also to local freight trains except that to avoid the danger of otherwise unnecessary switching at way stations, cars containing explosives may be placed not closer than the second car from caboose or the second car from engine. They must be placed between cars that do not bear the "INFLAMMABLE" or "ACID" placards, if such cars are in the train.

1687. (*a*) If shipments of explosives are accepted at nonagency stations, provision must be made for the proper certification and placarding of cars, examination of shipments, and loading and staying of packages in cars.

(*b*) Shipments of explosives must not be unloaded at nonagency stations unless the consignee is there to receive them, or unless proper storage facilities are provided at that point for their protection.

1688. When handling cars containing explosives in yards or on sidings, they must, if it is practically possible, be coupled to the engine, protected by a car between, and they must never be cut off while in motion.

Cars containing explosives must not be handled in switching or in trains, with doors open.

1689. They must be coupled carefully and all unnecessary shocks must be avoided. Other cars must not be cut off and allowed to strike a car containing explosives. They must be so placed in yards or on sidings that they will be subject to as little handling as possible, and be removed from all danger of fire. They must not be placed on tracks under bridges, and, when avoidable, engines on parallel tracks must not be allowed to stand opposite or near them.

1690. At points where trains stop and time permits, cars containing explosives and adjacent cars must be examined to see that they are in good condition and free from hot boxes or other defects liable to cause damage. If cars containing explosives are set out short of destination for any cause the carrier must arrange that proper notice be given to prevent accident.

1691. Whenever a car containing explosives is opened for any purpose and in every instance after such a car has received rough treatment, inspection must be made of the packages of explosives as soon as practicable without unnecessary disturbance of lading to see that they are properly loaded and stayed and in good condition. Upon the discovery of leaking or broken packages, they must be carefully removed to a safe place. Loose powder or other explosives must be swept up and carefully removed. If the floor is wet with nitroglycerin, the car is unsafe to use and a representative of the Bureau of Explosives should be immediately called to superintend the through mopping and washing of the floor with a warm, saturated solution of concentrated lye or sodium carbonate. If necessary, the car must be placed on an isolated siding and proper notice given.

1692. **Removal of Placards.**—The certificates and placards prescribed in paragraphs 1665, 1666, and 1667 must be removed from the car as soon as the explosives are unloaded.

In Case of a Wreck.

1697. In case of a wreck involving a car containing explosives, the first and most important precaution is to prevent fire. Before beginning to clear a wreck in which a car containing explosives is involved all unbroken packages should be removed to a place of safety and as much of the broken packages as possible gathered up and likewise removed, and the rest saturated with water. Many explosives are readily fired by a blow or by the spark produced when two pieces of metal or a piece of metal and a stone come violently together. In clearing a wreck, therefore, care must be taken not to strike fire with tools, and in using the crane or locomotive to tear the wreckage in pieces the possibility of producing sparks must be considered. With most explosives thorough wetting with water practically removes all danger of explosion by spark or blow; but with the dynamites wetting does not make them safe from blows. With all explosives mixing with wet earth renders them safer from either fire, spark, or blow. In case fulminate has been scattered by a wreck, after the wreck has been cleared the wet surface of the ground should be removed, and, after saturating the area with oil, be replaced by fresh earth. If this is not done, when the ground and fulminate become dry small explosions may occur when the mixed material is trodden on or struck.

REGULATIONS FOR THE TRANSPORTATION OF EXPLOSIVES AND OTHER DANGEROUS ARTICLES BY EXPRESS.

Prescribed under act of Mar. 4, 1909, and sec. 15 of the act to regulate commerce as amended June 18, 1910. Originally formulated and published Jan. 1, 1912, effective Mar. 31, 1912. Revision formulated and published July 2, 1914. Effective on October 1, 1914, and superseding the regulations published Jan. 1, 1912.

GENERAL NOTICE.

A. Special precautions are necessary in preparing for shipment by express packages of explosives and other dangerous articles. Any failure of the shipper or of a carrier to perform the duties imposed upon him in this respect may be the actual or contributory cause of a serious accident or fire.

B. Sections 235 and 236 of the act of March 4, 1909, require the shipper of explosives or other dangerous articles to describe and mark his package properly, and to inform the agent of the carrier of the true character of the contents. Heavy penalties are provided for the shipper who knowingly offers for transportation a dangerous article without complying with these requirements, as well as for the carrier that knowingly transports it.

C. To promote the uniform enforcement of the law and to provide for the safe transportation by express service in interstate commerce of such explosives and other dangerous articles as can be transported legally and safely on passenger trains, the following regulations are prescribed to define these articles for express transportation purposes, to state the precautions that must be observed by the shipper in preparing them for shipment and by the express carrier in handling them while in transit.

D. The Bureau for the Safe Transportation of Explosives and Other Dangerous Articles, hereinafter called Bureau of Explosives, organized by the railways under the auspices of the American Railway Association, is an efficient bureau in charge of an expert chief inspector. This bureau will make inspections and conduct investigations, and will confer with manufacturers and shippers and express carriers with a view of determining what specifications and regulations will, within reasonable limits, afford the highest degree of safety in packing and preparing dangerous articles for shipment and in transporting the same. The Commission will seek to avail itself of the expert knowledge thus developed and in formulating amendments to these regulations, while not bound thereby, will give due weight to such expert opinions.

E. These regulations apply to all shipments of explosives or other dangerous articles as defined herein, including carriers' material and supplies.

F. Shipments of dangerous articles which under these regulations may be transported by express, and which under individual carriers' regulations are transported as baggage on passenger trains, must be packed, marked, and labeled as prescribed herein for express shipments.

G. Express carriers that are subject to the act to regulate commerce and shippers must make these regulations effective, and must provide for the thorough instruction of their employees therein. Express carriers must not receive shipments of articles defined as dangerous by these regulations when the shipments are not packed, marked, labeled, described, and certified as prescribed herein.

H. Explosives and other dangerous articles the transportation of which by passenger trains is prohibited by law must not be offered for shipment. The method of manufacturing and packing articles defined as dangerous by these regulations, so far as it affects safe transportation, must be open to inspection by a duly authorized representative of the initial carrier, or of the Bureau of Explosives.

GENERAL RULES.

1. All shipments of articles subject to these regulations offered for transportation by express in interstate commerce must be properly described by the shipper, and the proper and definite name of the dangerous article must be marked on the outside of the package, in addition to the labels required herein.

2. (*a*) Articles for which detailed instructions for packing are not given herein must be securely packed in containers strong enough to stand without rupture or leakage of contents a drop of 4 feet to solid brick or concrete.

(*b*) Whenever orders are placed in foreign countries for the importation of dangerous articles to be forwarded from port of entry by express, the importer must furnish with the order to the foreign shipper and also to the forwarding agent at the port of entry, full and complete information as to the necessary packing, marking, and labeling required by these regulations. The forwarding agent must see that the packages are properly packed, marked, and labeled.

(*c*) Unless specifically authorized by these regulations, acceptable explosives must not be packed in the same outside package with each other or with other articles.

Information and Definitions.

Grouping.

3. For transportation by express, explosives and other dangerous articles are divided into the following groups:

1. **Forbidden articles.**
2. **Acceptable explosives.**
3. **Acceptable dangerous articles.**

Group 1.—Forbidden Articles.

4. Except when shipped by the War or Navy Department of the United States Government in time of war or of threatened war, the following articles must not be shipped by express, except as provided in section 232 of the act of March 4, 1909. (See par. 5 (*a*)):

(*a*) Low explosives or black powder.

(*b*) High explosives, including nitroglycerin explosives, chlorate powders, nitrate of ammonia powders, dry picric acid,[1] dry picrates, dry nitrocellulose (gun cotton, negative cotton), dry nitrostarch or dry trinitrotoluol.

(*c*) Smokeless powder for cannon.

(*d*) Smokeless powder for small arms, except in ammunition for small arms.

(*e*) Fulminate of mercury or of any other metal, except in manufactured products not forbidden.

(*f*) Blasting caps, including electric blasting caps.

(*g*) Ammunition for cannon, with or without projectiles.

(*h*) Detonating fuses.

(*i*) Explosive projectiles.

(*j*) Liquid nitroglycerin.

(*k*) Fireworks that combine an explosive and a detonator or blasting cap.

(*l*) Fireworks containing a match tip or head, or similar igniting point or surface, unless each such individual tip, head, igniting point, or surface is entirely covered and securely protected from accidental contact or friction with any other surface.

(*m*) Fireworks that ignite spontaneously or undergo marked decomposition when subjected for forty-eight consecutive hours to the temperature of 75° C. (167° F.).

(*n*) Firecrackers whose dimensions exceed 5 inches in length or

---

[1] Picric acid for medicinal purposes, and not exceeding 4 ounces in one outside package, may be shipped without other restrictions when in securely closed glass bottles, properly cushioned to prevent breakage.

$\frac{3}{4}$ inch in diameter, or whose explosive charges exceed 45 grains each in weight.

(*o*) Toy torpedoes exceeding $1\frac{1}{2}$ inches in diameter, or toy caps containing more than an average of thirty-five-one hundredths of a grain of explosive composition per cap.

(*p*) Fireworks that can be exploded en masse by a blasting cap placed in one of the units, or by impact of a rifle bullet, or otherwise.

(*q*) Explosives or other dangerous articles properly condemned by the Bureau of Explosives (except properly repacked samples for laboratory examination).

(*r*) Outside packages containing in the same compartment interior packages the mixture of whose contents would be liable to cause a dangerous evolution of heat, gas, or corrosive materials.

(*s*) Cylinders containing gases capable of combining chemically.

(*t*) Packages containing a dangerous article in a leaky condition, or in such an insecure condition as to make leakage probable during transit.

(*u*) Rags or cotton waste oily with more than 5 per cent of animal or vegetable oil, or wet rags.

(*v*) Boxes or kegs that have been previously used for high explosives must not be again used for shipments of any character.

(*w*) Carbon bisulphide, pyroxylin scrap (celluloid, fiberloid, pyralin, viscoloid, zylonite, etc., scrap), charcoal screenings, and white or yellow phosphorus except as provided in paragraph 42 (*b*).

Group 2.—Acceptable. Explosives.

5. The following explosives may be accepted for transportation by express, when offered in compliance with these regulations:

(*a*) **Samples of explosives for laboratory examination,** when properly packed and not exceeding a net weight of $\frac{1}{2}$ pound for each sample, and not exceeding twenty such samples at one time in a single vessel or vehicle. (See pars. 4, 6, and 14.)

(*b*) **Small-arms ammunition.**

(*c*) **Small-arms primers.**

(*d*) **Cannon primers.**

(*e*) **Percussion fuses, including tracer fuses.**

(*f*) **Time or combination fuses.**

(*g*) **Safety fuse.**

(*h*) **Cordeau detonant.**

(*i*) **Safety squibs.**

(*j*) **Common fireworks** and **special fireworks** except when forbidden. (See par. 4.)

DEFINITIONS OF ACCEPTABLE EXPLOSIVES.

6. The only **samples of explosives** that can lawfully be shipped by express are those intended for examination in a laboratory and not intended for use or demonstration.

7. **Small-arms ammunition** (such as is used in sporting or fowling pieces or in rifle or pistol practice, etc.) consists usually of a paper or metallic shell, the primer, and the powder charge—with or without shot or bullet—the materials necessary for one firing being all in one piece.

8. **Percussion and time fuses and primers** are devices used to ignite the black-powder bursting charges of projectiles or the powder charges of ammunition. For small-arms ammunition the primers are usually called "small-arms primers" or "percussion caps." **Tracer fuses** consist of a device which is attached to a projectile and contains a slow-burning composition, to show the flight of projectiles at night.

9. (*a*) **Safety fuse** consists ordinarily of a core of granular black powder, which is surrounded by yarn, tape, pitch, rubber, etc.

(*b*) **Cordeau detonant** is a fuse containing trinitrotoluol assembled in a drawn lead tube.

10. **Safety squibs** are small paper tubes containing a small quantity of black powder, one end of each tube being twisted and generally tipped with sulphur.

11. **Fireworks** include everything that is designed and manufactured primarily for the purpose of producing a visible or audible pyrotechnic effect by combustion or by explosion. They consist of common fireworks and special fireworks.

12. **Common fireworks** include all that depend principally upon nitrates to support combustion, and not upon chlorates; that contain no phosphorus and no high explosive sensitive to shock or friction; that produce their effect through color display rather than by loud noises. If noise is the principal object, the units must be small and of such nature and manufacture that they will explode separately and harmlessly, if at all, when one unit is ignited in a packing case. They must not be designed for ignition by shock or friction. Examples are Chinese firecrackers, Roman candles, pin wheels, colored fires, rockets, serpents, railway fusees, flash powders, etc.

13. **Special fireworks** include all that contain any quantity of phosphorus, a fulminate, or other high explosive sensitive to shock or friction; or that contain units of such size that the explosion of one while being handled would produce a serious injury; or that require a special appliance or tool, mortar, holder, etc., for their safe use; or that

are designed for ignition by shock or friction. Examples are giant firecrackers, bombs, and salutes not forbidden by paragraph 4, toy torpedoes and caps, ammunition pellets fired in a special holder, railway torpedoes, etc.

Packing and Marking of Acceptable Explosives.

14. **Packing.**—Samples of explosives for laboratory examination must be placed in well-secured metal cans or glass bottles, or in strong waterproof paper or cardboard packages, containing not more than ½ pound each, and the interior package must be placed in dry sawdust or similar cushioning material in a strong and tight wooden box, with ends not less than 1 inch thick, and top, bottom, and sides not less than ½ inch thick when a nailed box is used, or with ends, top, bottom, and sides not less than ½ inch thick when of lock-cornered construction.

15. **Weight.**—Not more than 20 half-pound samples of explosives for laboratory examination may be placed in one outside box or transported at one time.

16. **Marking.**—Each outside package containing samples of explosives for laboratory examination must have securely and conspicuously attached to it a square, red certificate label measuring 4 inches on each side and bearing in black letters the following:

Red label for samples of explosives. (Reduced size.)

PACKING AND MARKING OF SMALL-ARMS AMMUNITION, PRIMERS, FUSES, SAFETY FUSE, AND SAFETY SQUIBS.

17. **Small-arms ammunition** must be packed in pasteboard or other boxes, and these pasteboard or other boxes must be packed in strong outside wooden or metal containers. Small-arms ammunition, in pasteboard or other boxes and in quantity not exceeding a gross weight of 75 pounds, may be packed with nonexplosive and noninflammable articles, and with small-arms primers or percussion caps, provided the outside wooden or metal package is plainly marked "SMALL-ARMS AMMUNITION."

18. **Primers and percussion and time fuses** must be packed in strong, tight, wooden or metal containers, with special provision for securing individual packages of primers and fuses against movement in container.

19. **Small-arms primers** containing anvils must be packed in cellular packages with partitions separating the layers and columns of primers, so that the explosion of a portion of the primers in the completed shipping package will not cause the explosion of all of the primers.

20. **Percussion caps** may be packed in metal or other boxes containing not more than 500 caps, but the construction of the cap and the kind and quantity of explosives in each must be such that the explosion of a part of the caps in the completed shipping package will not cause the explosion of all of the caps.

21. **Small-arms primers and percussion caps** may form a part of the gross weight of 75 pounds of small-arms ammunition that may be packed with other articles as authorized by paragraph 17.

22. **Safety fuse, cordeau detonant, and safety squibs** must be packed in strong, tight wooden boxes or barrels.

23. **Weight.**—The gross weight of one outside package containing small-arms ammunition, primers, percussion caps, or percussion or time fuses must not exceed 150 pounds.

24. **Marking.**—Each outside box must be plainly marked "SMALL-ARMS AMMUNITION," or "SMALL-ARMS PRIMERS—HANDLE CAREFULLY," or "PERCUSSION CAPS—HANDLE CAREFULLY," or "CANNON PRIMERS—HANDLE CAREFULLY," or "COMBINATION PRIMERS—HANDLE CAREFULLY," or "PERCUSSION FUSES—HANDLE CAREFULLY," or "COMBINATION FUSES—HANDLE CAREFULLY,"

or "TRACER FUSES—HANDLE CAREFULLY," or "SAFETY FUSE," "CORDEAU DETONANT," or "SAFETY SQUIBS," etc.

Packing and Marking of Fireworks.

25. (*a*) **Common fireworks** must be in a finished state, exclusive of mere ornamentation, as supplied to the retail trade, and must be securely packed in strong, tight, spark-proof wooden barrels complying with specification No. 11, or in wooden boxes complying with specification No. 12A.

(*b*) **Special fireworks** must be in a finished state, exclusive of mere ornamentation, as supplied to the retail trade, and must not contain forbidden explosives. (See par. 4.) They must be packed in strong, spark-proof wooden barrels that comply with specification No. 11 or in wooden boxes that comply with specification No. 12.

26. **Weight.**—The gross weight of one outside package containing common fireworks must not exceed 200 pounds, and the gross weight of one outside package containing special fireworks must not exceed 100 pounds; the gross weight of an outside package containing only toy torpedoes must not exceed 65 pounds.

27. **Marking.**—All outside boxes or barrels containing common fireworks must be plainly marked "COMMON FIREWORKS—KEEP FIRE AWAY," and all outside boxes or barrels containing special fireworks or a mixture of common and special fireworks must be plainly marked "SPECIAL FIREWORKS—HANDLE CAREFULLY—KEEP FIRE AWAY," in letters not less than $\frac{7}{16}$ inch in height.

28. **Label.**—Each outside package containing common or special fireworks must have securely and conspicuously attached to it a square, red certificate label measuring 4 inches on each side and bearing in black letters the following:

Red label for fireworks. (Reduced size.)

Note—For information covering transportation of dangerous inflammable articles other than explosives, and for specifications for shipping containers, etc., consult latest complete Interstate Commerce Commission Regulations for Transportation of Explosives and Other Dangerous Articles by Freight and Express.

ACT OF MARCH 4, 1909, EFFECTIVE JANUARY 1, 1910.

By an act entitled "An act to codify, revise, and amend the penal laws of the United States," approved March 4, 1909, to take effect and be in force on and after the 1st day of January, 1910, the act entitled "An act to promote the safe transportation in interstate commerce of explosives and other dangerous articles, and to provide penalties for its violation," approved May 30, 1908, is repealed, and the following sections of said act to codify, revise, and amend the penal laws of the United States are substituted therefor:

SEC. 232. It shall be unlawful to transport, carry, or convey, any dynamite, gunpowder, or other explosive, between a place in a foreign country and a place within or subject to the jurisdiction of the United States, or between a place in any State, Territory, or District of the United States, or place noncontiguous to but subject to the jurisdiction thereof, and a place in any other State, Territory, or District of the United States, or place noncontiguous to but subject to the jurisdiction thereof, on any vessel or vehicle of any description operated by a common carrier, which vessel or vehicle is carrying passengers for hire: *Provided,* That it shall be lawful to transport on any such vessel or vehicle small arms ammunition in any quantity, and such fuses, torpedoes, rockets, or other signal devices, as may be essential to promote safety in operation, and properly packed and marked samples of explosives for laboratory examination, not exceeding a net weight of one-half pound each, and not exceeding twenty samples at one time in a single vessel or vehicle; but such samples shall not be carried in that part of a vessel or vehicle which is intended for the transportation of passengers for hire: *Provided further,* That nothing in this section shall be construed to prevent the transportation of military or naval forces with their accompanying munitions of war on passenger equipment vessels or vehicles.

SEC. 233. The Interstate Commerce Commission shall formulate regulations for the safe transportation of explosives, which shall be binding upon all common carriers engaged in interstate or foreign commerce which transport explosives by land. Said commission, of its own motion, or upon application made by any interested party, may make changes or modifications in such regulations, made desirable by new information or altered conditions. Such regulations shall be in accord with the best known practicable means for securing safety in transit, covering the packing, marking, loading, handling while in transit, and the precautions necessary to determine whether the material when offered is in proper condition to transport. Such regulations, as well as all changes or modifications thereof, shall take effect ninety days after their formulation and publication by said commission and shall be in effect until reversed, set aside, or modified.

Sec. 234. It shall be unlawful to transport, carry, or convey liquid nitroglycerin, fulminate in bulk in dry condition, or other like explosive, between a place in a foreign country and a place within or subject to the jurisdiction of the United States, or between place in one State, Territory, or District of the United States, or a place noncontiguous to but subject to the jurisdiction thereof, and a place in any other State, Territory, or District of the United States, or place noncontiguous to but subject to the jurisdiction thereof, on any vessel or vehicle of any description operated by a common carrier in the transportation of passengers or articles of commerce by land or water.

Sec. 235. Every package containing explosives or other dangerous articles when presented to a common carrier for shipment shall have plainly marked on the outside thereof the contents thereof; and it shall be unlawful for any person to deliver, or cause to be delivered, to any common carrier engaged in interstate or foreign commerce by land or water, for interstate or foreign transportation, or to carry upon any vessel or vehicle engaged in interstate or foreign transportation, any explosive, or other dangerous article, under any false or deceptive marking, description, invoice, shipping order, or other declaration, or without informing the agent of such carrier of the true character thereof, at or before the time such delivery or carriage is made. Whoever shall knowingly violate, or cause to be violated, any provision of this section, or of the three sections last preceding, or any regulation made by the Interstate Commerce Commission in pursuance thereof, shall be fined not more than two thousand dollars, or imprisoned not more than eighteen months, or both.

Sec. 236. When the death or bodily injury of any person is caused by the explosion of any article named in the four sections last preceding, while the same is being placed upon any vessel or vehicle to be transported in violation thereof, or while the same is being so transported, or while the same is being removed from such vessel or vehicle, the person knowingly placing, or aiding or permitting the placing, of such articles upon any such vessel or vehicle, to be so transported, shall be imprisoned not more than ten years.

APPENDIX III.

THE ROLE OF CHEMISTRY IN THE WAR.

THE ROLE OF CHEMISTRY IN THE WAR.[1]

The human race is living at the bottom of an ocean of atmosphere some 6 or 7 miles deep. Although it is not always realized by the unscientific mind, this aerial sea has weight and exerts a pressure upon all bodies of approximately 15 pounds to the square inch. Roughly speaking, and disregarding a small amount of rare gases and impurities, the air consists of about one-fifth oxygen and four-fifths of the inert gas nitrogen.

Every intelligent person knows that oxygen is the breath of life, and that nitrogen serves the purpose of just sufficiently diluting the oxygen so that the combustion of waste carbon conveyed by the blood to the body tissues goes on at the steady rate which conforms to the life processes of all animals. With this general knowledge in regard to the element nitrogen, the ordinary, well-informed, non-technical man rests content.

Educated people are, of course, aware that fixed nitrogen in combination with carbon, hydrogen, and some few other minor elements is built up by vegetable life and, in turn, assimilated into the bodies of animals, thus supplying our food of almost every variety. It is also fairly well understood that in the processes of digestion the complex nitrogenous bodies built up by plant life are broken down to simpler forms, in part supplying animal life energy and in part being voided by the animal, the manurial nitrogen products going back to the soil, thus completing what is known as the nitrogen cycle, caught in the wheel of which all material life, including the much-vaunted culture and progress of modern civilization, hangs suspended.

One thing that is not very generally apprehended by educated people, however, is that without fixed nitrogen in great abundance mankind could not wage war upon one another under modern conditions. Ever since gunpowder replaced the bow and arrow fixed nitrogen has been used by man to hurl destructive missiles at his adversaries. In fact, it should be stated that no explosive substance has ever been used in peace or war which did not depend for its activity on the extraordinary properties of the element nitrogen, which, as the major constituent of the air we breathe, could almost be said to content itself with the inert and pacific role of toning down the activities of its restless neighbor, oxygen.

[1] Communicated by the author. Reprinted from the Journal of the Franklin Institute, February, 1916.

I.

The Role of Nitrogen.

It becomes evident, from what has been said, that there must be some vital and important difference in character or quality between what may be termed fixed and unfixed nitrogen. In other words, it should be understood that all life and phenomenal existence, on this planet at least, depend upon the simple fact that the element nitrogen is able to assume two roles, in one of which it is unfixed, inert, sluggish, and slow to enter into combination with other elements, and in the other of which it is active, reactive, restless, ever ready to break down into new combinations, absorbing and giving out enormous energy as the restless changes take place. Whether the changes take place in a measured and orderly fashion, as in plant-cell growth and animal digestion, or with the most sudden and terrible violence, as in the case of high explosives, the energies either absorbed or released are equally potent and measurable. The celebrated chemist Berzelius once said of the element nitrogen as it occurs in the air, " It is difficult to recognize by any conspicuous property, but can only be recognized by means of properties which it does not possess."

Before pursuing our subject further it will be necessary to make quite clear what is meant by inert, unfixed nitrogen and active or fixed nitrogen. This explanation must be made in such a way that all apparent contradictions will immediately disappear. Gaseous nitrogen as it exists in the atmosphere has been proved by scientific methods to consist of a molecule made up of two atoms bound together by the equivalent of three bonds of affinity. What is meant is made clearer if we write a sort of alphabetical expression of the inert nitrogen molecule, as follows:

$$N \equiv N$$

It should not be supposed that the three bonds are actually arms or linkages holding the atoms together; they simply represent actually existent atomic forces, so that we may say that the element nitrogen is trivalent. In the same way we know that the element hydrogen is univalent, and we may express this by writing H—H, for the molecule of hydrogen is also known to be diatomic.

Now, suppose that by some means it is desired to combine or fix nitrogen to hydrogen; it is at once apparent that we should have to expend energy to tear apart the molecular bonds before we can fix the two elements together. In other words, the $N \equiv N$ would have to pass through the condition $N\equiv$ and $\equiv N$. Similarly, the H—H would have to split up into H— and —H. Subsequently the two elements might combine to form ammonia.

$$N\begin{array}{l}—H\\—H.\\—H\end{array}$$

For the purpose of this paper it is not necessary to go deeper into the combining valences of the different elements which it will be necessary to discuss. Only the simplest combination of nitrogen and hydrogen, viz., ammonia, has been mentioned in order to show the difference between fixed nitrogen and the inert or unfixed state of this gas as it exists in the air, with all its chemical affinities self-satisfied; in short, in the condition $N\equiv N$. If, however, this union is torn apart, $N\equiv$ is in an actively unsatisfied state and is prepared to fix itself into myriads of combinations with other elements. In other words, the molecule of nitrogen is quiet and well behaved, whereas the free atom of nitrogen is dynamically and even, in some combinations, very terribly reactive. It is this underlying chemical fact that has enabled men to slaughter and destroy each other on the gigantic scale now being demonstrated.

Those who have followed this explanation will readily see that it is not possible to maintain nitrogen in the condition of free unsatisfied atoms ($N\equiv$), for the simple reason that these atoms would return to the stable, quiescent molecule ($N\equiv N$), possibly with explosive energy. In order to take advantage of the reactive condition, it is necessary to lightly fix the nitrogen atom to some other atoms or groups of atoms in such a manner or in such a combination that the nitrogen at a blow can be suddenly released. Let us take the simplest example of what is meant. By an experiment so simple that the merest tyro in chemistry can perform it, ammonia can be made to react with the univalent element iodine to form the compound known as nitrogen iodide, in which iodine is made to replace the hydrogen, so that $N\begin{array}{l}—H\\—H\\—H\end{array}$ becomes $N\begin{array}{l}—I\\—I\\—I\end{array}$. Now, this nitrogen iodide is a brown powder which, when carefully dried, will remain innocently enough, resting quietly unchanged. If, however, we even so much as tickle this brown substance with a feather, or even if a door in the building in which it lies is rudely slammed, a terrible detonating explosion will occur, and the air will be filled with the stifling, violet-colored fumes of iodine. A quantity of this powder which could be heaped on the surface of a small silver coin would be sufficient to wreck everything in its neighborhood.

Whence this extraordinary energy? The thermodynamics of this and similar reactions are too complicated and mathematical to discuss here, but it is easy to see that the atomic forces at work in the sudden liberation of free nitrogen and iodine atoms, and their instantaneous rearrangement into inert molecules, involve enormous energy effects. Of course, nitrogen iodide is too treacherous a substance to be used as a high explosive, for in the dry condition the merest jar would cause it to detonate. It is obvious, therefore, that it has been the task of the chemist to find ways of locking nitrogen to other elements or groups of elements, with the result that it will be fixed tightly enough so that premature explosion will be avoided, but not so tightly but that it can be exploded by small quantities of more reactive nitrogen compounds made up in the form of percussion caps or detonators. All modern high explosives are just such chemical combinations of nitrogen as this, and we have, among others, nitroglycerin (dynamite), nitrocellulose (guncotton), trinitro-phenol (picric acid), nitrogelatine, trinitro-benzene, trinitro-toluene, etc. Masked under such trade names as lyddite, melinite, turpenite, cordite, etc., these nitrogen compounds are products of modern chemistry known and used by the armies and navies of the world.

In a later paragraph we shall have occasion to return to the constitutions of these more complicated nitro-substitution products. For the present, it will be necessary to inquire into the source and supply of the combined or fixed nitrogen on which modern warfare depends.

On the western coast of South America, in Chile and Peru, there occur vast natural deposits of nitrate of sodium, commonly known as Chile saltpeter. These deposits, with certain exceptions which will be noted later, constitute the world's supply of fixed or combined nitrogen, and in times of peace set the price for all combined nitrogen from whatsoever source derived. In sodium nitrate the nitrogen is linked to oxygen. Treated with sulphuric acid, which is cheap and abundant, sodium nitrate yields nitric acid. Various organic substances treated under certain conditions with nitric acid yield nitro-substitution bodies which are used as dyes, while other of these bodies are the high explosives referred to above. But sodium nitrate, for reasons that will now be apparent to the reader, is necessary as a fertilizer to keep up the fertility of the soil and thus make it possible for mankind to work out his destiny through the nitrogen cycle to which he is linked. It is indeed a curious thought that these natural deposits in a more or less remote corner of the world should exercise so great an influence on man and in so diverse a manner—on his life in the growth of the food he eats, and on his death in the production of destructive explosive agents, capable of killing thousands at a single blow.

In spite of the vastness of the Chilean nitrate beds, thoughtful scientific men have for many years given warning of the danger of their exhaustion. In 1898, Sir William Crookes, in his presidential address before the British Association for the Advancement of Science, dwelt in the most earnest manner upon the importance of this problem, and urged upon the attention of chemists and physicists the necessity for developing methods for fixing the inexhaustible supply of nitrogen in the air.

The present annual output of the Chilean fields amounts to about 2,500,000 tons. A recent scientific article [1] on this subject states:

> While there are a few scattered natural deposits other than those in Chile, there is none which has at the present time a chance of competing, most of them being of limited extent and situated in inaccessible regions. In Chile the deposits are easily worked, and even after years of careless mining, with no effort to effect economies, the present cost of producing nitrate is not excessive, varying from $10 to $20 per ton and selling in Liverpool for about $45 per ton. This leaves a profit of from $5 to $10 a ton on the operation, after paying the Government of Chile an export tax of about $12.25 per ton. In the past thirty years this export tax has netted the Chilean Government about $500,000,000. Of the total production of Chile, the United States imports about 600,000 to 700,000 tons per annum, the balance being practically all shipped to European countries. Chile saltpeter has sold for as high as $60 a ton, but since 1909, when the agreement among the producers expired, the price has approximated $45 per ton f. o. b. Liverpool, making a price of from $35 to $40 per ton f. o. b. Chile.

According to the same authority quoted above, 50 per cent of all the Chile saltpeter imported into the United States is used in the manufacture of explosives, while an additional 25 per cent is utilized in the arts requiring nitric acid. The balance, or 25 per cent, presumably finds its way to the soil as an intensive nitrogen fertilizer.

Let us now see what statistics will show in regard to the proportion of the two and one-half million tons of nitrate produced in Chile which is yearly imported into this country. The foreign commerce reports of the United States Department of Commerce have made the following figures available:

Nitrate of soda imported into the United States for the 12 months ending June—	
1913—	
Tons	589,136
Value	$20,718,968
1914—	
Tons	564,049
Value	$17,950,786
1915—	
Tons	577,122
Value	$16,355,701

---

[1] "Fixation of Atmospheric Nitrogen," Leland L. Summers, Trans. Amer. Electrochem. Soc., xxvii, 340–341, 1915.

These figures are interesting and can be interpreted in their relation to world conditions during the years included. They at once suggest a number of questions which will be interesting to note down.

1. Since the United States normally has taken about one-fifth of the world supply of nitrate, and since 75 per cent of this goes for the manufacture of explosives and nitric acid, what would happen to the United States if it were attacked by a strong naval power which would be able to blockade the western coast of South America and the entrance to the Panama Canal?

2. What reserve supply of fixed nitrogen have we in the United States in case we were called upon to wage a defensive war, and does anyone in authority in the United States show any indication of caring at all about this important matter? [1]

3. Since Germany and Austria have lost command of the sea, where are they getting the enormous supplies of fixed nitrogen necessary for their war against the world?

4. Assuming that England, France, Russia, Italy, and Japan are between them using 2,000,000 tons of Chilean nitrate for the manufacture of explosives, is the quantity adequate to the task they have undertaken?

The answers to these questions, though not treated seriatim, will be discussed in the following paragraphs.

In the great iron and steel industries of the world vast quantities of coal have to be partially burned to coke. In this process large quantities of valuable by-products are driven off and can be collected and used for many purposes, both in the peaceful arts and in war. To our everlasting shame be it said that we have for the most part, by the use of the open beehive-shaped coke ovens, allowed these valuable by-products to escape into the air and be lost. One at least of the good things that the present war has accomplished is in showing us the folly of such insane waste of valuable material. The substitution of the open type of coke oven by the inclosed by-product recovery oven is now at last going on apace. Our coking coals contain about 1 per cent of fixed nitrogen, and this can all be saved and converted into ammonia or the sulphate of ammonia. It will be remembered that once we have our nitrogen fixed we can use it as fertilizer or convert it by chemical processes into useful products, including high explosives.

[1] Since this article went to press, announcements have appeared in the daily press, stating that Brig. Gen. William M. Crozier, Chief of Ordnance of the United States Army, in his annual report has urged that the Nation take steps to be independent of the Chilean beds for the nitrates used in making gunpowder. In addition to this, it is reported that Mr. James B. Dukes announces that his company will turn out 4 tons daily at once of nitric acid made from nitrogen of the air. If these press reports are correct, answers to some of the questions considered in this paper are already in hand.

Among the coke recovery products we get coal tar, which in turn yields such valuable intermediates as benzene, toluene, and aniline, used in the manufacture of an infinite number of dyes, medicines, and explosives. The annual world's production of sulphate of ammonia from gas works and by-product coke ovens now amounts to about 1,250,000 tons, and the Liverpool price approximates that of sodium nitrate, varying from $45 to $60 per ton.

It will be seen from the above figures that while by-product recovery from coke-making offers an opportunity to eke out the needed supply of fixed nitrogen, taken by itself this source is insufficient to fill the demand in time of peace, and in time of war and embargo it would be quite inadequate, especially in the United States, where we are comparatively young in the application not only of by-product recovery but also in the chemical processes required to oxidize ammonia to the condition of nitric acid necessary for the manufacture of explosives.

Within the last fifteen years, or since Sir William Crookes sounded his note of warning to the world, chemists have paid especial attention to the problem of the fixation of the inexhaustible supply of nitrogen in the air. There are three lines of attacking this problem along which substantial success has been attained: First, the nitrogen can be made to combine directly with the oxygen of the air to form nitric acid; second, the nitrogen can be induced to combine with carbon to form cyanamide (C_2N_2), which can be used directly as a fertilizer or by subsequent treatment changed into ammonia and hence to nitric acid; and third, nitrogen can be directly linked tc waste or by-product hydrogen from other chemical industries to form ammonia. There are also other indirect processes which have been proposed for fixing atmospheric nitrogen, but these need not be discussed here.

It is apparent from what has been said in an earlier part of this article that enormous energy is called for in breaking up the linkage $N \equiv N$ and fixing the nitrogen atoms to other atoms, such as oxygen, carbon, and hydrogen. It will not be surprising, therefore, to learn that the success of such industries must hang, for the most part, on the successful harnessing of great water powers to this end. We have our Niagara and many other great potential water powers in North America; let us inquire, therefore, what we have done in this country toward the solution of this problem with which the future of the human race is so inevitably bound up.

Perhaps it might be permitted to begin this portion of the discussion with the statement that we in the United States have accomplished practically nothing at all along this line. It is a curious fact, often made a subject of comment, that in America have been made nearly all the inventions on which modern warfare depends, all of which, for lack of public interest and financial backing, have passed for their development to foreign countries. This is true of the aeroplane, the dirigible, the submarine, and it is equally true of the nitrogen-fixation processes. In 1902 two American pioneers, Lovejoy and Bradley, established at Niagara Falls their first industrial apparatus for fixing the nitrogen to the oxygen of the air by means of the electric arc and thereby directly producing nitric acid. These pioneers were on the right track, but nobody cared, least of all our Government, and so the infant industry died of inanition. I shall now dare to say that it is the development of this pioneer work in the hands of foreign scientists and engineers that made it possible for Germany to challenge a world at arms. Dr. L. H. Baekeland, in the Chandler lecture for 1914, has so excellently summed up the status of nitrogen fixation that no one could improve upon his brief capitulation, nor can Dr. Baekeland's summary be too often printed for the instruction of our countrymen. This eminent authority says:

The development of some problems of industrial chemistry has enlisted the brilliant collaboration of men of so many different nationalities that the final success could not, with any measure of justice, be ascribed exclusively to one single race or nation; this is best illustrated by the invention of the different methods for the fixation of nitrogen from the air.

This extraordinary achievement, although scarcely a few years old, seems already an ordinary link in the chain of common, current events of our busy life; and yet the facts connected with this recent conquest reveal a modern tale of great deeds of the race—an epos of applied science.

Its story began the day when chemistry taught us how indispensable are the nitrogenous substances for the growth of all living beings.

Generally speaking, the most expensive foodstuffs are precisely those which contain most nitrogen; for the simple reason that there is, and always has been, at some time or another, a shortage of nitrogenous foods in the world. Agriculture furnishes us those proteid- or nitrogen-containing bodies, whether we eat them directly as vegetable products, or indirectly as animals which have assimilated the proteids from plants. It so happens, however, that by our ill-balanced methods of agriculture we take nitrogen from the soil much faster than it is supplied to the soil through natural agencies. We have tried to remedy this discrepancy by enriching the soil with manure or other fertilizers, but this has been found totally insufficient, especially with our methods of intensive culture—our fields want more nitrogen. So agriculture has been looking anxiously around to find new sources of nitrogen fertilizer. For a short time an excellent supply was found in the guano deposit of Peru; but this material was used up so eagerly that the supply lasted only a very few years. In the meantime the ammonium salts recovered from the by-products of the gas works have come into steady use as nitrogen fertilizer. But here again the supply is entirely insufficient, and during the later period our main reliance has been placed on the natural beds of sodium nitrate, which are found in the desert regions of Chile. This has been of late our principal source of nitrogen for agriculture as well as for the many industries which require saltpeter or nitric acid.

In 1898 Sir William Crookes, in his memorable presidential address before the British Association for the Advancement of Science, called our attention to the threatening fact that at the increasing rate of consumption the nitrate beds of Chile would be exhausted before the middle of this century. Here was a warning—an alarm—raised to the human race by one of the deepest scientific thinkers of our generation. It meant no more nor less than that before long our race would be confronted with nitrogen starvation. In a given country, all other conditions being equal, the abundance or the lack of nitrogen available for nutrition is a paramount factor in the degree of general welfare or of physical decadence. The less nitrogen there is available as foodstuffs the nearer the population is to starvation. The great famines in such nitrogen-deficient countries as India and China and Russia are sad examples of nitrogen starvation.

And yet nitrogen as such is so abundant in nature that it constitutes four-fifths of the air we breathe. Every square mile of our atmosphere contains nitrogen enough to satisfy our total present consumption for over half a century. However, this nitrogen is unavailable so long as we do not find means to make it enter into some suitable chemical combination. Moreover, nitrogen was generally considered inactive and inert because it does not enter readily into chemical combination.

William Crookes's disquieting message of rapidly approaching nitrogen starvation did not cause much worry to politicians—they seldom look so far ahead into the future. But to the men of science it rang like a reproach to the human race. Here, then, we were in possession of an inexhaustible store of nitrogen in the air, and yet, unless we found some practical means for tying some of it into a suitable chemical combination, we would soon be in a position similar to that of a shipwrecked sailor, drifting around on an immense ocean of brine, and yet slowly dying for lack of drinking water.

As a guiding beacon there was, however, that simple experiment, carried out in a little glass tube as far back as 1785 by both Cavendish and Priestley, which showed that if electric sparks were passed through air the oxygen thereof was able to burn some of the nitrogen and to engender nitrous vapors.

This seemingly unimportant laboratory curiosity, so long dormant in the textbooks, was made a starting point by Charles S. Bradley and D. R. Lovejoy, in Niagara Falls, for creating the first industrial apparatus for converting the nitrogen of the air into nitric acid by means of the electric arc.

As early as 1902 they published their results, as well as the details of their apparatus. Although they operated only one full-sized unit, they demonstrated conclusively that nitric acid could thus be produced from the air in unlimited quantities. We shall examine later the reasons why this pioneer enterprise proved a commercial failure; but to these two American inventors belongs, undoubtedly, the credit of having furnished the first answer to the distress call of Sir William Crookes.

In the meantime many other investigators were at work at the same problem, and soon from Norway's abundant waterfalls came the news that Birkeland and Eyde had solved successfully, and on a commercial scale, the same problem by a differently constructed apparatus. The Germans, too, were working on the same subject, and we heard that Schoenherr, also Pauling, had evolved still other methods, all, however, based on the Cavendish-Priestley principle of oxidation of nitrogen. In Norway alone the artificial saltpeter factories use now, day and night, over 200,000 electric horse-power, which will soon be doubled; while a further addition is contemplated which will bring the volume of electric current consumed to about 500,000 horse-power. The capital invested at present in these works amounts to $27,000,000.

Frank and Caro, in Germany, succeeded in creating another profitable industrial process whereby nitrogen could be fixed by carbide of calcium, which converts it into calcium cyanamide, an excellent fertilizer by itself. By the action of steam on a cyanamide, ammonia is produced, or it can be made the starting point of the manufacture of cyanides, so profusely used for the treatment of gold and silver ores.

Although the synthetic nitrates have found a field of their own, their utilization for fertilizers is smaller than that of the cyanamide; and the latter industry represents to-day an investment of about $30,000,000, with three factories in Germany, two in Norway, two in Sweden, one in France, one in Switzerland, two in Italy, one in Austria, one in Japan, one in Canada, but not any in the United States. The total output of cyanamide is valued at $15,000,000 yearly and employs 200,000 horse-power, and preparations are made at almost every existing plant for further extensions. An English company is contemplating the application of 1,000,000 horse-power to the production of cyanamide and its derivatives, 600,000 of which have been secured in Norway and 400,000 in Iceland.

But still other processes are being developed, based on the fact that certain metals or metalloids can absorb nitrogen, and can thus be converted into nitrides; the latter can either be used directly as fertilizers or they can be made to produce ammonia under suitable treatment.

The most important of these nitride processes seems to be that of Serpek, who, in his experimental factory at Niedermorschweiler, succeeded in obtaining aluminum nitride in almost theoretical quantities with the use of an amount of electrical energy eight times less than that needed for the Birkeland-Eyde process and one-half less than for the cyanamide process, the results being calculated for equal weights of " fixed " nitrogen.

A French company has taken up the commercial application of this process which can furnish, besides ammonia, pure alumina for the manufacture of aluminum metal.

An exceptionally ingenious process for the direct synthesis of ammonia by the direct union of hydrogen with nitrogen has been developed by Haber in conjunction with the chemists and engineers of the Badische Aniline und Soda Fabrik.

The process has the advantage that it is not, like the other nitrogen-fixation processes, paramountly dependent upon cheap power; for this reason, if for no other, it seems to be destined to a more ready application. The fact that the group of the three German chemical companies, which control the process have sold out their former holdings in the Norwegian enterprises to a Norwegian-French group and are now devoting their energies to the commercial installation of the Haber process has considerable significance as to expectations for the future.

The question naturally arises: Will there be an overproduction and will these different rival processes not kill each other in slaughtering prices beyond remunerative production?

As to overproduction we should bear in mind that nitrogen fertilizers are already used at the rate of about $200,000,000 worth a year, and that any decrease in price, and, more particularly, better education in farming, will probably lead to an enormously increased consumption. It is worth mentioning here that in 1825 the first shipload of Chile saltpeter which was sent to Europe could find no buyer and was finally thrown into the sea as useless material.

Then, again, processes for nitric acid and processes for ammonia, instead of interfering, are supplementary to each other, because the world needs ammonia and ammonium as well as nitric acid or nitrates.

It should be pointed out also that ultimately the production of ammonium nitrate may prove the most desirable method, so as to minimize freight, for this salt contains much more nitrogen to the ton than is the case with the more bulky calcium salt, under which form synthetic nitrates are now put into the market.

Before leaving this subject let us examine why Bradley and Lovejoy's efforts came to a standstill where others succeeded.

First of all, the cost of power at Niagara Falls is three to five times higher than in Norway, and although at the time this was not strictly prohibitive for the manufacture of nitric acid it was entirely beyond hope for the production of fertilizers. The relatively high cost of power in our country is the reason why the cyanamide enterprise had to locate on the Canadian side of Niagara Falls, and why up till now, outside of an experimental plant in the South (a 4000-horse-power installation in North Carolina, using the Pauling process), the whole United States has not a single synthetic-nitrogen fertilizer works.

The yields of the Bradley-Lovejoy apparatus were rather good. They succeeded in converting as much as 2½ per cent of the air, which is somewhat better than their successors are able to accomplish.

But their units—12 kilowatts—were very much smaller than the 1000 to 3000 kilowatts now used in Norway; they were also more delicate to handle, all of which made installation and operation considerably more expensive.

However, this was the natural phase through which any pioneer industrial development has to go, and it is more than probable that in the natural order of events these imperfections would have been eliminated.

But the killing stroke came when financial support was suddenly withdrawn.

In the successful solution of similar industrial problems the originators in Europe were not only backed by scientifically well-advised bankers, but they were helped to the rapid solution of all the side problems by a group of specially selected scientific collaborators as well as by all the resourcefulness of well-established chemical enterprises.

That such conditions are possible in the United States has been demonstrated by the splendid teamwork which led to the development of the modern tungsten lamps in the research laboratories of the General Electric Co., and to the development of the Tesla polyphase motor by the group of engineers of the Westinghouse Co.

True, there are endless subjects of research and development which can be brought to success by efforts of single independent inventors, but there are some problems of applied science which are so vast, so much surrounded with ramifying difficulties, that no one man, nor two men, however exceptional, can furnish either the brains or the money necessary for leading to success within a reasonable time. For such special problems the rapid cooperation of numerous experts and the financial resources of large establishments are indispensable.

So much for the role of chemistry in the war in so far as it is affected by the nitrogen-fixation problem. Those who have read thus far will be able to formulate their own answers to the questions set down in an earlier paragraph and will understand how Germany, although cut off from the South America nitrate fields, has been able to assemble and use more high explosives in a shorter time than anyone would have believed possible previous to the year of grace (sic) 1914.

If we assume that the nations at war have provided themselves with adequate supplies of fixed nitrogen in the condition of nitric acid, let us now return to a more detailed discussion of the materials and methods which modern chemistry uses for the production of high explosives, without an abundant supply of which modern warfare must immediately come to an end. We have seen in what manner the nations are in fact fighting with fixed nitrogen. Indeed, it may be said that out of the atmosphere comes the power of making war, for there are geological reasons for believing that the nitrogen of the Chile nitrate beds was originally fixed by natural process from atmospheric nitrogen. We must now consider something of the chemistry of the element carbon and the wonderful role which it also plays in war.

II.

The Role of Carbon.

The element carbon, unlike nitrogen, does not appear in nature in the gaseous form. It is familiar to every one in an impure form as coal, as charcoal, and graphite, and in its pure crystallized form as the diamond. Considered as an atom in its chemical sense it is highly reactive and every ready to combine with other atoms and groups of atoms to form the endless variety of organic forms which make up the visible universe. The most characteristic attribute of the carbon atom is its power and tendency to link up with other carbon atoms, thus permitting an infinite variety of molecular architecture. It will be necessary to follow this statement a little further, on account of its bearing on the role of chemistry in the war.

We have seen that the free atom of nitrogen is called trivalent and is written N≡. Similarly, the free atom of carbon is known to be quadrivalent and might be written C≣. As a matter of fact, however, the quadrivalence of the carbon atom is expressed in the following form:

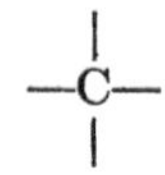

Really the carbon atom with its four bonds is thought of spacially as being at the center of a pyramid or tetrahedron. For our present purpose, however, we need not confuse ourselves with this conception, but think or it as written above. The point to be understood is that the free affinities of the carbon atom are easily saturated with other atoms or groups of atoms, as, for instance, in the following compounds:

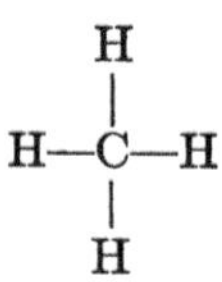

Marsh gas (methane)

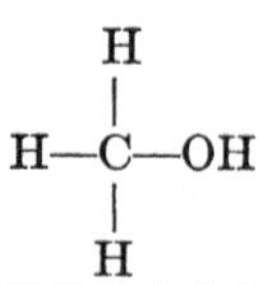

Methyl alcohol

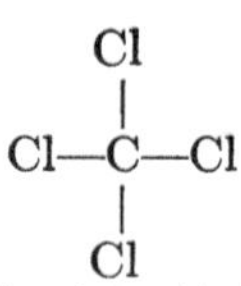

Carbon tetrachloride

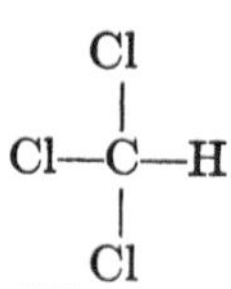

Chloroform

But the most interesting characteristic is the ability of carbon to link up as in 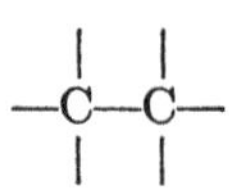 and 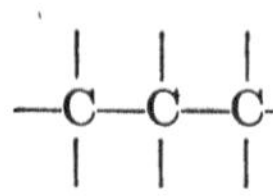 and so on until we reach a string or nucleus of six atoms, when in many cases the string acts as though it were unwieldy and, like a snake with its tail in its mouth, links up into form of a ring known as the benzene ring, and written:

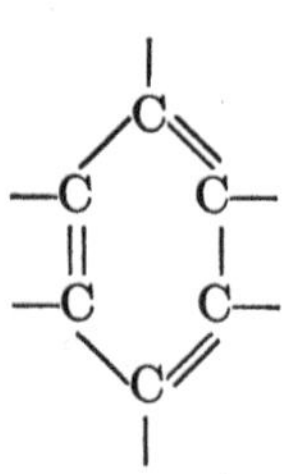

or further into double or even triple rings:

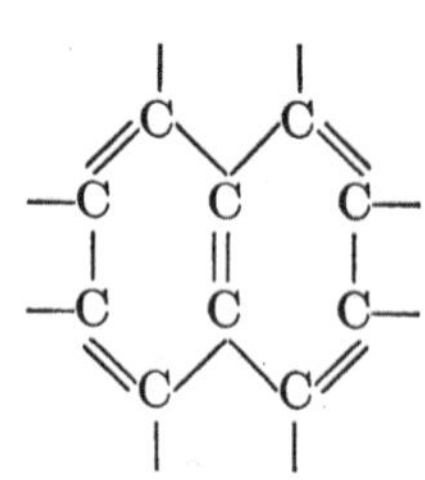

It may appear to the layman that we are involving ourselves pretty deeply in advanced chemistry, but we must be patient, because we are getting close to the secret of modern warfare as it is controlled by high explosives. We are also getting close to the secrets of the dye industry and modern medicinals, which subjects have been much discussed in this country since the outbreak of the war.

Benzene has already been referred to in an earlier paragraph as a by-product of the coke and gas industry. It is a limpid liquid substance which closely resembles gasoline in odor and properties. If cheap enough, it could be used in automobile engines, but its price before the war in this country was about 30 cents a gallon, which was prohibitive of its use for this purpose. It is an important raw material for the manufacture of high explosives, dyes, synthetic medicines, phonographic records, etc. Benzene has the chemical formula C_6H_6, and is to be considered as a ring of six carbon atoms attached as shown above, with one hydrogen atom fixed to each carbon. For the sake of brevity and simplicity, chemists no longer take the trouble to write in the carbon or hydrogen atoms into their ring formulæ, these being assumed, only the significant substituting atoms being placed and written in. Thus, for instance, the benzene, or, as the Germans call it, the benzol ring, is expressed by writing:

instead of the more cumbersome

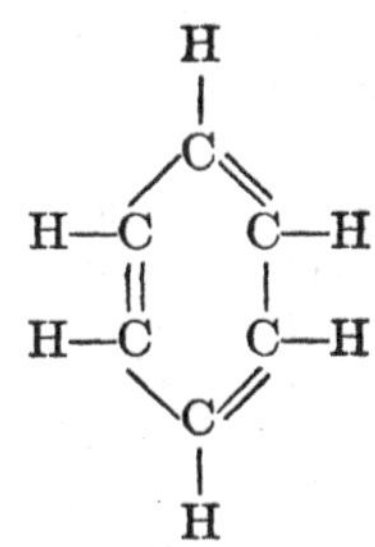

Now, suppose by treating benzene with certain chemicals, we replace one of the hydrogens by the group of atoms OH, we get

This body is carbolic acid, known to chemists as phenol. It was this substance that we have read in the newspapers Thomas A. Edison needed for making phonograph records after the German supplies ceased, and which he was able to make as soon as the recovered benzene began to be available from the American gas and coke plants. Now, if we start again with phenol and treat it with nitric acid in a special manner, we make trinitro-phenol, or picric acid, an intensely yellow substance which is used as a dye base and is also one of the most deadly of the high explosives. We write the formula of picric acid

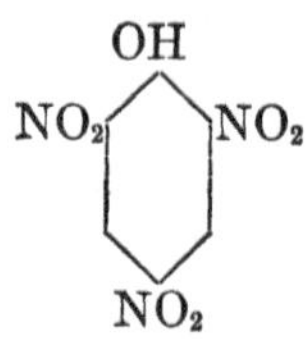

and designate it as a 2, 4, 6 substitution product, for the group or radical NO_2 must fix to just the right points in the carbon ring, or we should not get picric acid, but something else. Perhaps we have now succeeded in getting a glimpse into the wonderful molecular architecture that has been patiently worked out by chemists for the use of man in the arts of peace and war. Untold numbers of tons of picric-acid mixtures under the names of melinite and turpenite are being shot off on the European battlefields.

Looking at the graphic representation of the picric-acid molecule written above, it requires but a slight effort of the imagination to picture what takes place when this molecule is suddenly shattered into its elements. Large quantities of hot nitrogen, hydrogen, and oxygen atoms are instantly set free, seeking to expand and satisfy their various affinities. The chemical forces of disruption and rearrangement are titanic and when directed to that end scatter death and destruction round about.

Picric acid, when dry, melts down quietly at a little above the water-boiling temperature, with little danger of explosion unless it is detonated by something else. It is usually melted down with rosin or some other body which is used to dilute it. It is these other bodies which are partly responsible for the dense clouds of black smoke formed when shells loaded with picric-acid mixtures explode, and which on the European battlefields have earned for them the name of "Jack Johnsons."

Toluene [1] is a near relative of benzene; it is a liquid slightly less

[1] It is interesting to note that the price of benzene has risen from 30 cents to 90 cents a gallon since the war began. Toluene has risen in the same time from 40 cents to $5 per gallon.

volatile than the latter substance, and is also a by-product of the coke and gas industry. From it we can obtain trinitro-toluene—

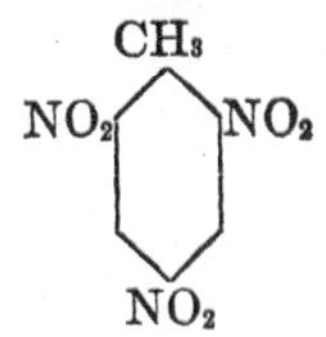

This product is also used as a modern high explosive, under the abbreviated name of T. N. T.

The German chemical industries are said to have accumulated vast stores of these and similar by-product nitro-substitution products from their great dye industries. The well-known blue dye, indigo, which used to be extracted from a plant grown in India, is now made synthetically in Germany, and the by-products from the synthesis furnish some of the raw material for the nitro bodies used in explosives. It is said to have taken the patient German chemists over twenty years to work out the synthesis of indigo, but when this was finally accomplished the natural indigo soon disappeared from the markets of the world.

Glycerin is a by-product from the manufacture of soap. By nitration we get nitroglycerin, which, when soaked up in an inert earthy powder, we call dynamite. Cotton consists mainly of cellulose. When cotton is nitrated, nitrocellulose or guncotton is obtained. When compressed into blocks or other forms, this guncotton is a detonating explosive of terribly high power, and is frequently used for loading torpedoes and mines. When made in a different way, however, nitrocellulose does not detonate, but burns rapidly, giving off large quantities of hot expanding gases. When in this form nitrocellulose is known as "smokeless powder." It is generally not used as a powder, however, but in the form of sticks (cordite) or short cylinders through which holes are bored to facilitate combustion. If cotton can not be obtained, as is probably now the case in Germany, wood-pulp cellulose may be substituted. It is said that gelatine made from slaughterhouse refuse or dead horses can in like manner be used in the production of nitrogelatine. The chemistry of these substances is more complicated than in the cases of nitrobenzene and toluene, so that we need not attempt to go more deeply into the subject in this place. Enough has been said to show the role of the chemistry of carbon and nitrogen as this applies to modern warfare.

III.

The Role of Hydrogen.

Hydrogen is the lightest gas known, being about seven times lighter than air, bulk for bulk. The manufacture of caustic alkali on an enormous scale is necessary to every civilized nation for the manufacture of paper, soap, and many other necessities of life. This alkali is in part manufactured by an electric water-power process which yields hydrogen as one of its by-products. At present about 30,000 electrical horsepower are employed in this branch of industry in the United States. In Germany all waste hydrogen is conserved for filling the balloons of Zeppelins and presumably for making synthetic ammonia by the Haber process, as well as for other industrial purposes. In this country practically all the hydrogen is allowed to escape into the air, to seek the outermost reaches of the atmospheric sea. It is probable, however, that the day is not far distant when we, too, will meet the necessity of conserving our hydrogen as well as our other useful industrial wastes.

The story of the development of the Haber process for fixing the nitrogen of the air to by-product hydrogen reads like a fairy tale of science wedded to patience. About twenty-five years ago American students of chemistry at a certain German university were surprised to find their professors carrying on patient researches on methods for oxidizing ammonia to nitric acid. Since in those days nitric acid cost less than ammonia, and as no methods for fixing the nitrogen of the air had even been proposed, so much patient work seemed at best to be premature. The fact is, however, that it was to come about that the great European war was to await the word "Go!" not alone from the statesmen and militarists, but also from the distinguished chemist-privy councillors (Geheimraths) of the German Empire. At the meeting of the Eighth International Congress of Applied Chemistry held in New York in September, 1912, the German professor, H. A. Bernthsen, in the course of his address,[1] said:

> I propose, however, to-day to deal, from my own direct experience with the development of the problem for the synthetical manufacture of ammonia from its elements. A few years ago the solution of this problem appeared to be absolutely impossible. It has recently been the object of very painstaking investigations by Prof. Haber and the chemists of the Badische Anilin und Soda-Fabrik, and numerous patents have been taken out with reference to the manufacture. Apart from what is already published in this way, however, we have refrained from any other announcements until we were in a position to report something final with reference to the solution of the technical question.
>
> This moment has now arrived, and I am in the agreeable position of being able to inform you that the said problem has now been solved fully on a manufacturing scale, and that the walls of our first factory for synthetic ammonia are already rising above the ground at Appau, near Ludwigshafen-on-Rhine.

So much for the accomplishment of Germany's independence of Chile saltpeter up to 1912. The fact that the contact process for changing synthetic ammonia to nitric acid had already been worked out tells us something of the vision that was in the minds of German scientists at least a quarter of a century ago.

But let us hear further from Prof. Bernthsen on some of the difficulties that had to be surmounted before this method of nitrogen fixation was an accomplished fact. He goes on to say:[2]

[1] "Synthetic Ammonia," H. A. Bernthsen, Trans. Eighth Internat. Cong. Applied Chem., vol. xxviii, pp. 185 and 186.

[2] Ibid., pp. 193–194, 199–200.

The problems to be solved were quite new and strange and demanded the mastery of very unusual difficulties. Although working with compressed gases under pressure at very low temperatures was already known in the industry, the problem here was the totally different one of constructing apparatus which should be large enough and at the same time able to withstand the high pressure with temperatures not far from a red heat. How well founded were the doubts as to the possibility of a solution of this task can be gathered from the instance of the wrought-iron autoclaves commonly used in the color industry. There, in spite of a very low range of temperature of at most 280° C., only pressures of from 50 to 100 atmospheres at the utmost come into consideration. But above 400° C. iron loses its solidity to a very extraordinary degree.

There is, further, the circumstance that the metals which come into consideration for the construction of the apparatus, and especially iron, are chemically attacked above certain temperatures by the gas mixture under pressure. Although the formation of iron nitride from iron and ammonia, which could have been expected according to the work of Fremy and others, can be avoided, yet it is found, for example, that steel containing carbon loses its carbon at the temperatures in question, owing to the action of the hydrogen, so that its capability of withstanding pressure is reduced to a minimum. It was further found, when using iron itself, that it is completely changed in its qualities, chiefly by taking up hydrogen. Again, at such high temperatures iron is pervious to quite a remarkable degree to hydrogen under high pressures. The question of materials for the apparatus therefore raised at once considerable difficulties, but at length these were more than overcome by suitable construction, details of which, I am sure, you will not expect from me to-day. The danger of serious explosions or of great, sudden flames of hydrogen, if the apparatus happens to become defective, can be guarded against by setting it up in bombproof chambers.

Great care must naturally be taken that oxygen or air does not get into the apparatus or the piping, for at the high pressure obtaining the explosion range is reached with merely a slight percentage of oxygen. Special devices are used to watch over this content of oxygen, and immediately a definite percentage is touched the alarm is automatically raised. Besides this, the proper constitution of the gas mixture in circulation is controlled by analysis from time to time.

The ammonia can be removed either by being drawn directly from the apparatus in liquid form, or an absorption agent can be suitably introduced into the apparatus. The simplest absorbent, water, has been found to be suited for this purpose; under the pressure used a concentrated solution of ammonia is secured. Any ammonia that may remain in the gas after the bulk has been removed by one or other of these methods can be further removed by special chemical means, if it is not preferred simply to leave it in the circulating gases.

* * *

The question has not yet been touched upon in the foregoing, how the elements nitrogen and hydrogen which are requisite for the new ammonia process can best be produced on a technical scale. Theoretically, the task would be unusually simple. If you remember that the terrestrial atmosphere, according to the studies of A. Wagener and others, consists of practically pure hydrogen at a height of about 120 kilometers—indeed, at a height of about 70 kilometers consists of almost exactly one volume of nitrogen and three volumes of hydrogen, besides a trace (about one-half per cent) of oxygen—you will understand that all the conditions were given for an ammonia factory according to Jules Verne; for it would then merely be necessary to suck down the gases from the higher strata of the atmosphere by a sufficiently long pipe line.

For us, poor mortals, matters are not so ideally simple, for, as the poet says:

Hart im Raume stossen sich die Sachen.

Fortunately, however, there is no great difficulty in separating nitrogen from the air, either by physical means, according to Linde's process or chemically, by removing the oxygen with glowing copper, burning hydrogen, or the like. And for the preparation of hydrogen in recent times a great deal of useful work has been done, too, owing to the extensive growth of its field of application. In certain works it is at disposal in large quantities as a by-product of the electrolysis of common salt. Besides this, it can be produced, for example, by passing steam over red-hot iron, or from water gas, for instance, by separating its constituents, hydrogen and carbon monoxide, by cooling to a very low tem-

> perature. All the methods of preparation which come into consideration we have, of course, minutely examined; owing to the comparatively trifling differences in the cost of production various methods can be employed. At all events, both elements, nitrogen and hydrogen, are at the disposal of the new industry to any extent and sufficiently cheap.
>
> As the production of these elements is not confined to the presence of cheap water power, all those countries where the manufacture of calcium nitrate, owing to the want of such power, is not practicable, as, for instance, in Germany, are now in a position to profit by the new industry.

The above-quoted matter makes interesting reading, in the light of the history of the past three years. The synthesis of ammonia by the Haber process takes place by heating a mixture of hydrogen and nitrogen gas under enormous pressures in great alloy-steel bombs buried in bombproof cellars and then rapidly withdrawing and cooling the

—H
synthetic ammonia (N—H) which is formed. But this is not all of
—H
the story.

Even when all these titanic arrangements have been made the synthesis of ammonia does not take place unless a catalyst is present.

Is this fairy tale of science getting too deep for ordinary comprehension, and is it time, as Lewis Carroll's walrus said, to talk of other things, such as ships and shoes and sealing wax and cabbages and kings?

But, as a matter of fact, could a world war be declared or waged without every one of those four things or the things they stand for that the walrus wanted to talk about?

And now we are also to learn that a world war cannot be waged without a catalyst. And what on earth is a catalyst?

Those who wish to follow this monster to its lair must be referred to the dictionaries and encyclopedias; but, after all, it is not so frightful as it sounds. In a few words, the meaning is this: A number of important and difficult chemical reactions will not take place, or only very slowly, unless some substance is present in such a manner that it can come into contact with the reacting bodies and thus act the role of a go-between promoter or introducer. Such a substance is not changed or necessarily wasted while the action is going on, but acts merely by contact.

Chemists have called this kind of action catalysis. Some one has said that the industrial development of a nation can be measured by the quantity of sulphuric acid that it produces, and it is interesting to note, in passing, that catalysts or contact agents have, in modern times, revolutionized the production of this most important substance. Catalysts consist usually of some metal or compound of a metal in finely powdered or subdivided form. As they are not used up or wasted in doing their work, and only comparatively small quantities are needed, the rarest and most expensive metals, such, for instance, as platinum, can be used even in very large-scale operations. Haber found that such rare and unusual metals as molybdenum and tungsten made excellent catalysts for promoting the synthesis of ammonia. But here we have to refer to one of the most curious facts that has been developed by modern chemical research. It has been found that these catalysts can be poisoned by certain things very much in the same way as though they were living cells. That is to say, there are substances which hinder or prevent or kill the activity of these catalysts, although the contact mass does not suffer a noticeable chemical change, envelopment, or destruction. For instance, traces of arsenic mixed with the sulphur from which sulphuric acid is made will poison the platinum catalyzer, whereas it has been found that, on the contrary, the presence of arsenic will act favorably (or medicinally, as it were) when iron oxide is used as a catalyst.

We are now prepared to hear Prof. Bernthsen[1] on this subject and to admire and marvel at the persistence and painstaking efforts made by modern science to overcome the obstacles with which Nature seems to surround her most profound secrets:

> It has now been ascertained that some of the poisons in the synthesis of ammonia are of quite a different nature from those of the sulphuric-acid process; they are, for instance, sulphur, selenium, tellurium, phosphorus, arsenic, boron, or the compounds of these elements, such as sulphuretted hydrogen, arsenic hydride, phosphorus hydride, as also many carbon compounds and certain metals of low melting-point which can readily be reduced by hydrogen from their compounds; for example, lead, bismuth, and tin, which do not act catalytically. Oxygen-sulphur compounds such as SO_2, which acts directly and smoothly in the sulphuric-acid catalysis, are very poisonous. Extremely minute quantities of these bodies, which are almost always present, even in the purest commercial products or in the so-called pure gases, suffice to render the catalysts absolutely inactive, or at least to diminish their action very seriously. Thus iron, for example, prepared from ordinary iron oxide with a content of one per thousand of sodium sulphate, is, as a rule, inactive. Iron containing one-tenth per cent of sulphur is generally quite useless, and even with one-hundredth per cent is of very little use, although in appearance and when examined with the ordinary physical and chemical methods no difference at all can be detected as compared with pure iron.
>
> The recognition of these facts gave rise to two problems:
>
> (*a*) The preparation of contact masses free from poison or the removal of such poisons from them; and
>
> (*b*) Freeing the gases to be acted on catalytically from all contact poisons.
>
> In order to free the contact bodies from these harmful substances the ordinary methods for removing them can, of course, be applied. The contact action can also be improved by heating contact metals which are inactive or of little use, owing to the presence of contact poisons, in the presence of oxygen or of bodies yielding oxygen. Or the metals can be heated, for instance, in the presence of oxygen, with the addition of suitable compounds, such as bases, and the resulting products reduced. These operations can be repeated if necessary. If more of such a body as mentioned is added than is necessary, it may act not merely by removing the poisons, but promote the yield, as I have already described to you.
>
> On the other hand, it is necessary, as I remarked, to take the greatest care that nitrogen and hydrogen are free or freed completely from all contact poisons.

[1] "Synthetic Ammonia," H. A. Bernthsen, Trans. Eighth Internat. Cong. Applied Chem., vol. xxviii, pp. 197–199.

Thus a trace of sulphur, one part per million, in the gas mixture can under certain conditions be injurious, so that even electrolytically prepared hydrogen must generally be further specially purified. The minute purification of the gases is even more important when hydrogen prepared, for example, from the water gas is used. The impurities, too, taken up from iron piping play sometimes an important part, and impurities which get into the gases during the compression, such as machine oil, often have a harmful effect.

The best method of removing impurities from the gas mixture depends, in turn, on the nature of these impurities and consists, for instance, according to the case, in filtering, washing, conducting over solid absorption agents, and so on. One good method is to bring the gases into contact with the material of which the contact mass is prepared at a raised temperature before passing them over the actual catalyst. The material takes up the impurities, and must, of course, be renewed from time to time. The negative results of earlier investigators in the formation of ammonia when using base contact metals (Wright, Ramsay, and Young, and more recently again, 1911, Neogi and Adhicary), according to which nitrogen and hydrogen do not combine in the presence of iron, are, in my opinion, probably due for the most part at least, to the use of metals or gases not free from contact poison. That previous inquirers had not the remotest idea that sulphur in the contact metal could be injurious is evident from the fact that they passed the gases without hesitation through concentrated sulphuric acid in order to dry them. The sulphuric acid thus taken up and the sulphur dioxide often contained in it can poison even the best catalyst very speedily and render it unfit for use. Or the contact metals were sometimes prepared directly from the sulphates, although a metal sufficiently free from sulphur can scarcely be obtained by this method.

A painstaking study, for which we are indebted principally to Dr. A. Mittasch and which involved literally many thousands of experiments, has afforded an insight into the importance of substances of the most varied nature as promoters and poisons, and thus a sure foundation has been prepared for a reliable continuous manufacture with a good yield of ammonia.

This is a wonderful story that has been here set forth, and should show very clearly that modern warfare depends not alone upon the work of a general military and naval staff or upon the drilling and marshalling of millions, but, first and foremost, upon the systematic and painstaking researches of scientists, who, to use a homely expression, might be said to be twisting reluctant Nature's tail in the effort to make her march in the desired path.

IV.

The Role of the Halogens.

No description of the role of chemistry in modern warfare would be quite complete without some reference to the poisonous gases and flames with which modern armies seek to ruin and devastate one another. A vast amount of terrifying descriptive matter has appeared in the press accounts of the great war, but very little accurate scientific information on this awful contribution of chemistry has as yet been made available.

We may safely discount as untrue the extraordinary stories of soldiers asphyxiated and left standing or sitting rigidly in the positions in which they were overcome. It is possible that bombs have been hurled that set free prussic acid or the deadly cyanogen gas (CN), another form of fixed nitrogen. It is more probable, however, that the most deadly effects have been produced by the use of a group of very active chemical elements known as the halogens. By the name of "halogens," meaning salt-formers, chemists distinguish the group of closely allied elements, chlorine, bromine, and iodine. At ordinary temperatures and pressure, chlorine, is a greenish-colored gas, bromine is a dark-red fuming liquid, and iodine is a beautiful violet-colored solid. At temperatures slightly above the ordinary bromine and iodine turn to heavy red and violet vapors. Chlorine is found in nature fixed to the element sodium in common salt (sodium chloride). Sodium and other bromides are found accompanying common salt, while the principal source of iodine is seaweed, as marine growths have the power of collecting and concentrating iodides from salt water. It is characteristic of the halogens that they are among the most active chemical agents known. When in a free state they seem anxious to combine with anything they can take hold of, and the mucous membrane of the human throat, lungs, and eyes is peculiarly sensitive to their corrosive action. Chlorine gas, the most active of the family, is obtained in enormous quantities in the same way that hydrogen is, as a by-product from the electric process of manufacturing caustic alkali from common salt.

In the arts of peace chlorine has an important use as a bleaching agent in the paper and other industries. When compressed into steel cylinders, it can be liquefied and thus shipped from one point to another, or it can be absorbed in lime to form the well-known chloride of lime or bleaching powder. In war liquefied chlorine contained in steel shells can be burst among the enemy, and when the conditions are favorable terrible effects can thus be produced.

Bromine has never found any extended use in the peaceful arts, in spite of the fact that it has been stated that the German Government has been for a number of years offering a prize for the discovery of a practical use for it. Bromine is also a by-product material which has been largely allowed to escape in this country, but which in Germany has been saved in increasing quantities for years.[1] In war bromine

[1] BROMINE.—Both technical and U. S. P. descriptions of this commodity continue in scanty supply and strongly held at a minimum of $5 by leading manufacturers, while maintained at $6 and even at $6.50 by second hands. Reports to the effect that the recent sharp uplift of prices for this article has been due to the operations of a syndicate of makers in Michigan, Ohio, Virginia, and West Virginia, are not credited by those in a position to know, as it is obvious that the growing shortage, due to recent heavy sales on contracts to European users, is alone responsible for this upward movement of prices. Seemingly this extensive export movement of bromine can not be forbidden by the United States Government authorities or virtually prohibited by a high export tax, without enactment to this effect by Congress. Efforts to prevent further heavy shipments of this commodity to foreign countries have recently been made in vain by manufacturers of bromide who complained of this export movement to the Secretary of Commerce, only to be referred to a special investigator in New York City, who informed him that the Government was powerless to interfere with this business.—Oil, Paint, and Drug Reporter, New York Dec. 20, 1915, p. 44.

is even more terrible than chlorine, for it possesses the property of especially attacking the eyes, even producing blindness in large doses. If it comes in contact with the skin it produces horrible burns which are slow to heal. When chlorine gas is allowed to come in contact with bromine under certain controlled conditions, they combine together to form a limpid fuming liquid, known as chlorine of bromine, which combines the properties of both elements. It is said that this awful substance is the material that the warring nations have been hurling at each other in shells, bombs, and hand grenades, although exact facts regarding these dreadful practices will probably not be available until after the war is over.

The role of iodine in the war is probably of a more kindly nature; it is not nearly so active an agent as the other two members of its family, and when dissolved in alcohol to make a tincture it has saved many lives on the battlefield when used as an antiseptic on open wounds.

Some of the newspaper accounts have described gas bombs which exploded with dense violet fumes. If true, this points to the use of iodine also for this destructive purpose, but it is not probable that such a use is common, owing to the higher cost, greater scarcity, and limited activity of iodine as compared with the other halogens.

Much has also been written in regard to the masks or breathers resorted to for protection from these dangerous gases. There are a number of chemical substances which are absorbents of chlorine and bromine and with which they can combine or fix themselves. Such chemicals spread between layers of fabric or liquids in which fabric masks can be soaked probably furnish such means as are available for protection.

CONCLUSION

Besides the role of chemistry in war, the allied art of metallurgy plays an important part. New steels and alloys have to be devised, tested, and finally manufactured, suitable for the varied needs of war. Aeroplanes, dirigibles, and submarines all require materials with special physical characteristics. It is said that the metallurgists of Germany had to devise a new steel alloy before the Haber process for fixing nitrogen could be successfully worked on the large scale of operation. Space here does not permit of a description of these special metals, but perhaps enough has been said to give the reader a general purview of the role of chemistry in war.

Although it has been shown that chemistry is the handmaiden of war, and that this last great struggle is indeed a chemists' war, as Dr. L. H. Baekeland has recently so happily phrased it: [1]

> Do not imagine that this is the first chemical war. The art of killing and robbing each other became chemical the day gunpowder was invented; at that time, however, the existing knowledge of chemistry was just of pinhead size. Napoleon knew very well how to use adroitly exact knowledge and chemistry for furthering his insatiable ambition to dominate the world; so he surrounded himself with the most able chemical advisers and scientists. Ever since then science, technology, and chemistry in particular have played a role of increasing importance in the armament of nations. . . .
>
> Do not reproach chemistry with the fact that nitrocellulose, of which the first application was to heal wounds and to advance the art of photography, was stolen away from these ultrapacific purposes for making smokeless powder and for loading torpedoes. Do not curse the chemist when phenol, which revolutionized surgery, turned from a blessing to humanity into a fearful explosive after it had been discovered that nitration changes it into picric acid.
>
> Let us hope, in the meantime, that war carried to its modern logical gruesomeness, shorn of all its false glamour, deceptive picturesqueness, and rhetorical bombast, exposed in all the nakedness of its nasty horrors, may hurry along the day when we shall be compelled to accept means for avoiding its repetition.

Americans, North or South, probably without exception will join eagerly in the hope thus eloquently expressed, but in the meantime and under present conditions a strong, rich nation can no more exist without adequate means of self-defense than a modern city could exist without a trained police force and fire department. This paper has shown that fixed nitrogen is the first and most important element of national defense. The paper will have well served the author's purpose in preparing it if it should succeed even to a slight extent in calling attention to the necessity for purchasing and storing at a central point an adequate supply of Chilean nitrate. As an alternative, arrangements should be made which would have the effect of inducing capital to exploit in this country the fixation of atmospheric nitrogen. The fact that fixed nitrogen will become an increasingly important factor in the production of food simply means that, come peace or war, foresighted preparation will not under any circumstances be unprofitable or in vain.

---

[1] "Chemical Industry," address by Dr. L. H. Baekeland before the American Chemical Society, Seattle, Wash., Aug. 31, 1915.

List of Atomic Weights Revised to 1917.

	Symbol.	Atomic weight.		Symbol.	Atomic weight.
Aluminum	Al	27.1	Molybdenum	Mo	96.0
Antimony	Sb	120.2	Neodymium	Nd	144.3
Argon	A	39.88	Neon	Ne	20.2
Arsenic	As	74.96	Nickel	Ni	58.68
Barium	Ba	137.37	Niton (radium emanation)	Nt	222.4
Bismuth	Bi	208.0	Nitrogen	N	14.01
Boron	B	11.0	Osmium	Os	190.9
Bromine	Br	79.92	Oxygen	O	16.00
Cadmium	Cd	112.40	Palladium	Pd	106.7
Cæsium	Cs	132.81	Phosphorus	P	31.04
Calcium	Ca	40.07	Platinum	Pt	195.2
Carbon	C	12.005	Potassium	K	39.10
Cerium	Ce	140.25	Praseodymium	Pr	140.9
Chlorine	Cl	35.46	Radium	Ra	226.0
Chromium	Cr	52.0	Rhodium	Rh	102.9
Cobalt	Co	58.97	Rubidium	Rb	85.45
Columbium	Cb	93.1	Ruthenium	Ru	101.7
Copper	Cu	63.57	Samarium	Sa	150.4
Dysprosium	Dy	162.5	Scandium	Sc	44.1
Erbium	Er	167.7	Selenium	Se	79.2
Europium	Eu	152.0	Silicon	Si	28.3
Fluorine	F	19.0	Silver	Ag	107.88
Gadolinium	Gd	157.3	Sodium	Na	23.00
Gallium	Ga	69.9	Strontium	Sr	87.63
Germanium	Ge	72.5	Sulphur	S	32.06
Glucinum	Gl	9.1	Tantalum	Ta	181.5
Gold	Au	197.2	Tellurium	Te	127.5
Helium	He	4.00	Terbium	Tb	159.2
Holmium	Ho	163.5	Thallium	Tl	204.0
Hydrogen	H	1.008	Thorium	Th	232.4
Indium	In	114.8	Thulium	Tm	168.5
Iodine	I	126.92	Tin	Sn	118.7
Iridium	Ir	193.1	Titanium	Ti	48.1
Iron	Fe	55.84	Tungsten	W	184.0
Krypton	Kr	82.92	Uranium	U	238.2
Lanthanum	La	139.0	Vanadium	V	51.0
Lead	Pb	207.20	Xenon	Xe	130.2
Lithium	Li	6.94	Ytterbium (Neoytterbium)	Yb	173.5
Lutecium	Lu	175.0	Yttrium	Yt	88.7
Magnesium	Mg	24.32	Zinc	Zn	65.37
Manganese	Mn	54.93	Zirconium	Zr	90.6
Mercury	Hg	200.6			

See Report of the International Committee on Atomic Weights, 1917.

INDEX.

www.ingramcontent.com/pod-product-compliance
Lightning Source LLC
LaVergne TN
LVHW102007210926
848483LV00009B/138

* 9 7 8 1 4 2 7 6 1 4 0 4 9 *